Trace Analysis

CHEMICAL ANALYSIS

A SERIES OF MONOGRAPHS ON
ANALYTICAL CHEMISTRY AND ITS APPLICATIONS

Editors

P. J. ELVING • J. D. WINEFORDNER
EDITOR EMERITUS: I. M. KOLTHOFF

VOLUME 46

A WILEY-INTERSCIENCE PUBLICATION

JOHN WILEY & SONS

New York / London / Sydney / Toronto

Trace Analysis

Spectroscopic Methods For Elements

Edited by

J. D. WINEFORDNER

Department of Chemistry
University of Florida
Gainesville

A WILEY-INTERSCIENCE PUBLICATION

JOHN WILEY & SONS

New York / London / Sydney / Toronto

Copyright © 1976 by John Wiley & Sons, Inc.

Library of Congress Cataloging in Publication Data:

Main entry under title:

Trace analysis.

  (Chemical analysis; v. 46)
  "A Wiley-Interscience publication."
  Includes bibliographical references and index.
  1. Trace elements—Analysis. 2. Spectrum analysis.
3. Trace elements—Spectra. I. Winefordner, James
Dudley, 1931–   II. Title. III. Series.
QD139.T7T7   545'.83   75-41460
ISBN 0-471-95401-2

Printed in the United States of America

10 9 8 7 6 5 4 3 2 1

AUTHORS

R. C. Elser
Department of Pathology
York Hospital
York, Pennsylvania

T. C. O'Haver
Department of Chemistry
University of Maryland
College Park, Maryland

M. C. Parsons
Department of Chemistry
Arizona State University
Tempe, Arizona

P. St. John
American Instrument
Company, Inc.
Silver Spring, Maryland

C. Veillon
Biophysics Research Laboratory
Harvard Medical School
Peter Bent Brigham Hospital
Boston, Massachusetts

J. D. Winefordner
Department of Chemistry
University of Florida
Gainesville, Florida

PREFACE

This book is not intended to be a textbook on atomic spectroscopic techniques nor a reference source for the fundamental aspects of the various methods. It is intended to be a reference source for information of importance for those involved with trace analysis of elements in real samples. Analytical figures of merit, including detection limits, sensitivities, selectivities, precisions, and accuracies are given where possible for each element by each atomic technique. Such data allow a critical comparison of techniques for a given application. For each atomic technique, the principles, instrumentation, methodology, and applications are briefly reviewed. Only those techniques that are truly trace analytical methods, that is, allow parts per million or lower detection limits, for elements are considered in this book. For example, Mössbauer spectroscopy, several types of electron spectroscopy for chemical analysis, including X-ray-induced electron spectroscopy and Auger spectroscopy, X-ray diffraction, NMR, microwave spectroscopy, and IR absorption spectroscopy are not considered separately here. These techniques as well as several nonspectroscopic ones are only briefly considered in the overall general comparisons given in Chapters 1 and 11, in which the methods are introduced and then summarized. Therefore this book is intended for the user of trace element techniques. It should be most useful to the student, the technician, or the researcher who needs to learn about or to perform a specific trace elemental determination. As a result, this book contains many tables, many data, and many figures to aid the reader in preparing the sample, in deciding on a method, and in interpreting the results in terms of quantitative analytical figures of merit.

We express our appreciation to Ms. Debbie Rooks for editorial assistance in reading the galley and page proof. Without her, we could not have completed this book.

J. D. WINEFORDNER

Gainesville, Florida
April 1976

CONTENTS

APPENDICES

J. D. Winefordner

INTRODUCTION

J. D. WINEFORDNER

Department of Chemistry
University of Florida
Gainesville, Florida

The importance of trace analysis of elements in chemical, biological, environmental, and metallurgical samples is well known. Often the requirements of the analysis are for the analysis of parts per million (ppm) or parts per billion (ppb), or even lower elemental concentrations of a variety of elements in an ultramicro sample, milligrams or below, consisting of a complex matrix. Such requirements dictate a thorough knowledge of the principles, methodologies, and uses of various analytical techniques in order to select the optimum one, if available, for the present analysis. However, if the technology is not present, then new research areas are indicated.

In order to understand the language used in trace and ultratrace analysis, several prefixes must be known (see Table 1.1). For example, 1 ng of an element X found in 1 Gg of sample represents a concentration of

$$\frac{1 \times 10^{-9}}{1 \times 10^{9}} \times 10^{6} = 10^{-12} \text{ ppm} \quad \text{or} \quad 10^{-9} \text{ ppb}, \quad \text{and so on}$$

Also, in this book, the SI system, of units is used where possible. The seven basic units are listed in Table 1.2. In Table 1.3, useful conversion factors and physical constants are given.

The criteria used (figures of merit) in the selection of a trace method of analysis are *limit of detection, sensitivity, precision,* and *accuracy.* Although these terms are discussed in several other chapters (particularly Chapter 2), perhaps a brief discussion of them here is necessary to obtain an overview of the important criteria in selecting a trace analysis method. The terminology used is that tentatively approved by IUPAC (1).

In all fields of analysis, the *measure x* of some physical parameter is related to the *concentration c* of the analyte in a certain sample (see Appendix A). The *sensitivity S* is the slope of the *analytical calibration curve,* which is the plot of the measure versus the concentration of the analyte in a series of *standards* having known analyte concentrations, that is, $H = (\partial x / \partial c)_{c_i}$ which is given by dx/dc if it is a straight line over the entire concentration range of interest. If c is plotted versus x, the plot is called an *analytical evaluation curve*. The term "sensitivity" must not be used to indicate either the limit of detection or the concentration required to give a certain signal, as in atomic absorption spectrometry. The *standard deviation* is:

$$\sigma = \sqrt{\sum_{i=1}^{n} \frac{(x_i - \bar{x})^2}{n-1}} \tag{1}$$

where x_i is the individual measure, and $\bar{x}$ is the mean of n measurements.

TABLE 1.1.   Prefixes for Units

Prefix	Symbol	Order
atto	a	10^{-18}
femto	f	10^{-15}
pico	p	10^{-12}
nano	n	10^{-9}
micro	μ	10^{-6}
milli	m	10^{3} 10^{-3}
centi	c	10^{-2}
deci	d	10^{-1}
deka	da	10^{1}
hecto	h	10^{2}
kilo	k	10^{3}
mega	M	10^{6}
giga	G	10^{9}
tera	T	10^{12}

TABLE 1.2.   SI Units

Length	Meter, m
Mass	Kilogram, kg
Time	Second, s
Amount	Mole, mol
Current	Ampere, A
Temperature	Kelvin, K
Luminous intensity	Candela, cd

If n is less than, say 10, σ is replaced by the term s. The relative standard deviation (RSD) is:

$$RSD = \frac{\sigma}{\bar{x}} \quad \text{or} \quad RSD = \frac{s}{\bar{x}} \tag{2}$$

and it should not be expressed as a percent by multiplying RSD by 100 to avoid confusion with percent concentration. *Variance* is the square of the standard deviation. The total normalized variance for an *analytical procedure* is:

$$\frac{s^2}{\bar{x}^2} = \frac{s_1^2}{\bar{x}_1^2} + \frac{s_2^2}{\bar{x}_2^2} + \frac{s_3^2}{\bar{x}_3^2} + \cdots + \frac{s_m^2}{\bar{x}_m^2} \tag{3}$$

where m is the number of independent factors contributing to the uncertainty, for example, background or blank corrections.

Precision is defined as the random uncertainty for the measure x or the corresponding uncertainty in the estimate of the concentration. It is often expressed as the RSD.

Accuracy is a measure of the agreement between the estimated concentration and the true value. There are three principal limitations imposed (1) on accuracy: (1) accuracy can never be guaranteed to be any better than the precision of the analytical procedure; (2) *bias* (systematic errors) in the calibration procedure always causes estimated uncertainties to disagree with the true value by an amount equal to the bias; and (3) in a multicomponent system of elements, the treatment of interelement effects always involves some approximations which can give good precision but still result in incorrect estimates (*systematic errors*) in the concentrations.

The *limit of detection* is the smallest concentration c_L that can be detected with reasonable certainty. This concentration is therefore:

$$c_L = \frac{x_L - \bar{x}_b}{S} \tag{4}$$

where x_L is the limiting detectable measure including the blank, and $\bar{x}_b$ is the average blank measure.

$$x_L = x_b + ks_b \tag{5}$$

where k is the protection factor (in Chapter 2 k is assumed to be Student's t value) and s_b is the standard deviation of the blank measures. Generally k

TABLE 1.3. Useful Conversion Factors and Physical Constants

A. *Conversion factors*

General units

Length	$1 \text{ Å} = 10^{-8} \text{ cm} = 10^{-10} \text{ m}$
Volume	$1 \text{ liter} = 10^{-3} \text{ m}^3$
Mass	$1 \text{ amu} = 1.66 \times 10^{-27} \text{ kg}$
Force	$1 \text{ dyn} = 10^{-5} \text{ N}$
	$1 \text{ N} = 1 \text{ kg m s}^{-2}$
Energy	$1 \text{ erg} = 1 \text{ dyn cm} = 10^{-7} \text{ J}$
	$1 \text{ J} = 1 \text{ kg m}^2 \text{ s}^{-2}$
	$1 \text{ eV} = 1.602 \times 10^{-19} \text{ J}$
	$1 \text{ eV} = 8068 \text{ cm}^{-1} \text{ or } 8068 \text{ K}$
	$1 \text{ cm}^{-1} = 1.985 \times 10^{-23} \text{ J}$
	$1 \text{ cal} = 4.184 \text{ J}$
Pressure	$1 \text{ torr} = 1 \text{ mm Hg} = 133.322 \text{ N m}^{-2}$
	$1 \text{ dyn cm}^{-2} = 10^{-1} \text{ N m}^{-2}$
	$1 \text{ Pa} = 1 \text{ N m}^{-2} = 1 \text{ kg m}^{-1} \text{ s}^{-2}$
Viscosity	$1 \text{ poise} = 1 \text{ g cm}^{-1} \text{ s}^{-1} = 10^{-1} \text{ kg m}^{-1} \text{ s}^{-1}$
	$1 \text{ stoke} = 1 \text{ cm}^2 \text{ s}^{-1} = 10^{-4} \text{ m}^2 \text{ s}^{-1}$
Radioactivity	$1 \text{ Ci} = 3.7 \times 10^{10} \text{ s}^{-1}$
Concentration	$1 \, M = 1 \text{ mole liter}^{-1} = 10^3 \text{ mole dm}^{-3}$

Magnetic units

Magnetic induction	$1 \text{ gauss} = 10^{-4} \text{ T} = 10^{-4} \text{ V s m}^{-2}$
Magnetic field strength	$1 \text{ oersted} = 79.58 \text{ A m}^{-1}$
Magnetic flux	$1 \text{ maxwell} = 10^{-8} \text{ Wb}$
	$1 \text{ gauss} \sim 1 \text{ Oe}$
Magnetic flux	$1 \text{ weber} = 1 \text{ V s} = \text{kg m}^2 \text{ s}^{-2} \text{ A}^{-1}$
Magnetic flux density	$1 \text{ T} = 1 \text{ V s m}^{-2} \; 10 = 1 \text{ kg J}^{-2} \text{ A}^{-1}$

Electrical units

Power	$1 \text{ W} = 1 \text{ J s}^{-1} = 1 \text{ kg m}^2 \text{ s}^{-3}$
Charge	$1 \text{ C} = 1 \text{ A s}$
Voltage (potential difference)	$1 \text{ V} = 1 \text{ J A}^{-1} \text{ s}^{-1} = 1 \text{ kg m}^2 \text{ s}^{-2} \text{ A}^{-1}$
Resistance (electric)	$1 \, \Omega = 1 \text{ V A}^{-1} = 1 \text{ kg m}^2 \text{ s}^{-3} \text{ A}^{-2}$
Conductance (electric)	$1 \text{ S} = 1 \text{ A V}^{-1} = 1 \, \Omega^{-1} = \text{kg}^{-1} \text{ m}^{-2} \text{ s}^3 \text{ A}^2$
Capacitance (electric)	$1 \text{ F} = 1 \text{ A s V}^{-1}$
Inductance	$1 \text{ H} = 1 \text{ V A}^{-1} \text{ s} = 1 \text{ kg m}^2 \text{ s}^{-2} \text{ A}^{-2}$

B. *Physical constants*

Avogadro's constant, N_A	$6.023 \times 10^{26} \text{ kg mole}^{-1}$
Bohr Magneton, μ_B	$9.274 \times 10^{-24} \text{ J T}^{-1}$
Boltzmann's constant, k	$1.38 \times 10^{-23} \text{ J K}^{-1}$
Electron charge, e	$1.60 \times 10^{-19} \text{ C}$
Electron mass (rest), m_e	$9.109 \times 10^{-31} \text{ kg}$

TABLE 1.3. (*Continued*)

Faraday constant, F	9.649×10^7 C kmole^{-1}
Gas constant (universal), $R(kN_A)$	8.314 J K^{-1} mole^{-1}
Neutron mass (rest), m_N	1.675×10^{-27} kg
Planck's constant, h	6.625×10^{-34} J s
Proton mass (rest), m_p	1.673×10^{-27} kg
Rydberg constant, R_∞	1.0974×10^7 m^{-1}
Velocity of light (*in vacuo*), c	2.9979×10^8 m s^{-1}
Vacuum permittivity, f_0	8.827×10^{-12} F m^{-1}
Vacuum permeability, μ_0	1.257×10^{-6} H m^{-1}

is taken as 3, and so:

$$c_L = \frac{3s_b}{S} \quad \text{or} \quad c_L = \frac{3N_b}{S} \tag{6}$$

which means that any concentration resulting in a signal three times the background standard deviation is considered just detectable. To evaluate $\bar{x}_b$ and s_b, at least 20 measurements should be used. If the major sources of variation are electrical noises, s_b can be replaced with N_b, the background noise level. If $3s_b$ is chosen, the confidence level (one-sided) is 99.86% for a purely gaussian distribution. At low concentrations, broader and asymmetric distributions are likely, and so in a practical sense $3s_b$ corresponds to a confidence level of about 90%. It should be noted that at the limit of detection the RSD is about 0.5.

In Table 1.4, a general comparison (2–4) of the limits of detection, the precision of measurement, and the uses of a large variety of analytical methods is given. The major portion of this comparison has been taken from the excellent survey of analytical methods for characterization of solids by Meinke (2). It should be noted that, for completeness, many nonspectroscopic methods are considered in Table 1.4. Also, it should be noted that not all techniques mentioned in Table 1.4 are discussed in detail in the chapters to follow; of course, some methods are nonspectroscopic, some are useful primarily for organic molecules, and some have too poor detection limits to be considered trace methods. The reader should refer to the excellent book by Morrison (3) for a thorough discussion of principles, technologies, and applications of several of the methods *not* discussed here. Also, the reader should refer to the fine treatises by Meinke (2) and by Meinke and Scribner (4) for a discussion of trace characterization of materials.

TABLE 1.4. Limits of detection, RSD, and Uses of Analytical Methods

Method	Limit of Detections		RSD	General Uses, Comments
	Absolute (g)	Concentration (ppm)	(fraction)	
Wet chemistry— gravimetry	10^{-1}–10^{0} 10^{-3}–10^{-2}	— —	0.00001 0.001	Weighing of reaction product of analyte; high accuracy
Wet chemistry— titrimetry	— — —	10^{3}–10^{4} 10^{0}–10^{3} 10^{-2}–10^{0}	0.0001 0.001 0.01	Visual indicator for major and minor concentrations; instrumental detection of end point for trace concentration
Organic micro-analysis	$> 10^{-5}$	—	0.005–0.01	Elemental analysis of mainly C, H, O, N, P, S, Cl, Br, I, and Si and functional group analysis; primarily for major constituents
Thermal analysis	10^{-5}–10^{-4}	—	0.01–0.2	Measurement of phase changes
Thin-layer chromatography	10^{-5}–10^{-3}	—	0.05–0.5	Semiquantitative method used mainly for qualitative analysis

Method				Comments
Gas chromatography	—	—	10^5–10^6	Solids, liquids, gases can be measured; special methods as derivatization pyrolysis, and indirect reaction may be needed
	—	—	10^4–10^5	
			10^3–10^4	
	—	—	10^2–10^3	Generally used for trace analysis of organics
	—	—	10–10^2	
			<10	
Liquid chromatography	—	—	10^{-3}–10^0	Generally used for trace analysis of organics
Controlled potential and controlled current coulometry	10^{-1}–10^0	—	—	Useful for any species convertible to another redox form
	10^{-3}–10^{-1}	—	—	
	10^{-9}–10^{-4}			
Ion-selective electrodes (direct measure)	—	—	10^2 to saturation	Measures activity, not concentration; precision can be improved via titrations
	—	—	10^0–10^2	
			10^{-2}–10^0	
Polarography, conventional	—	—	10^0–10^3	Useful for about 80 elements which can be oxidized or reduced
Polarography, special (cathode ray, pulse, anodic	—	—	10^{-3}–10^{-3}	Useful for about 80 elements which can be oxidized or reduced

(Precision column values, read left-to-right: 0.001, 0.002–0.005, 0.005–0.01; 0.01–0.05, 0.05–0.10, ≥ 0.10; 0.01–0.20; 0.00001, 0.0001, 0.01; 0.005–0.02, 0.01–0.05, 0.02–0.30; 0.02–0.2; 0.0002–0.2)

TABLE 1.4. (*Continued*)

Method	Limit of Detection		RSD	General Uses, Comments
	Absolute (g)	Concentration (ppm)	(fraction)	
stripping, differential, and so on)				
Molecular absorption spectrometry (UV-visible)[a]	—	$10^0 - 10^2$ $10^{-3} - 10^0$	0.01–0.05 0.05–0.10	Most used method for molecular analysis; one of the most used methods for elements
Molecular absorption spectrometry (IR)	—	$10^3 - 10^6$	0.05–0.20	Useful only for organic molecules
Molecular absorption spectrometry (microwave)	—	$10^0 - 10^3$	0.05–0.20	Useful only for small molecules in gas state
Molecular fluorescence spectrometry[a]	—	$10^{-3} - 10^1$ $10^0 - 10^4$ $10^{-3} - 10^1$	0.05–0.20 0.01–0.50 0.01–0.10	Organic analysis Rare-earth complexes Non-rare-earth complexes; in microspectro-fluorometry can detect $\sim 10^{-14}$ g

Method				
Molecular phosphorescence spectrometry	—	$10^{-3}-10^2$	0.01–0.20	Organic analysis only
Raman spectrometry (non resonance)	—	10^3-10^5	0.05–0.20	Useful for gases, liquids, solids
Raman spectrometry (resonance)	—	10^0-10^3	0.05–0.20	Useful for gases and liquids; air pollution studies
Nuclear magnetic resonance	—	10^1-10^5	0.01–0.10	Used mostly for solutions; mostly used for major constituents
Electron spin resonance	$10^{-9}-10^{-6}$	—	Semiquantitative	Used for trace analysis of paramagnetic species
Isotope dilution mass spectrometry	—	10^3-10^6 10^0-10^3 $10^{-5}-10^0$	0.001–0.002 0.002–0.005 0.005–0.5	Do ~40 elements by thermal ionization and ~10 more by electron imput; most accurate method for trace analysis

TABLE 1.4. (*Continued*)

| Method | Limit of Detection | | RSD | General Uses, Comments |
	Absolute (g)	Concentration (ppm)	(fraction)	
Spark source mass spectrometry[a]	—	10^{-3}–10^1	0.05–0.20	Best for conducting and semiconducting of solids; use photo graphic detector
Vacuum-fusion mass spectrometry	—	10^{-1}–10^2	0.05–0.20	Used mainly for inter stitial gases as H_2, O_2, and N_2 in materials
Spark source mass spectrometry with isotope dilution[a]	—	10^{-5}–10^{-1}	0.05–0.10	Used with polynuclidic elements; photographic detection
X-ray fluorescence spectrometry[a]	— —	10^1–10^2 10^{-1}–10^1 (precon- centration needed)	0.001–0.00 0.02–0.10	Used for major and minor elements in solids; improved detectors and energy dispersive X-ray fluorescence give better trace analysis results
Mössbauer spectroscopy	—	10^0–10^3	Semiquantitative	Used mainly for structure studies of solids; limited to a few elements

Method				
Neutron activation analysis[a]	—	10^{-1}–10^{-1} 10^{-3}–10^{-2}	0.02–0.05 0.02–0.10	Used for impurities with radiochemical separations; used for nondestructive multielement analysis; GeLi detectors; several separations needed and long irradiation and decay times; activation with fast-neutron, high-energy photon, charged particles is also possible
Electron probe microanalysis[a]	—	10^2–10^3 (1–5 μm scan diameter)	0.05	Used for microscopic studies; homogeneity of major and minor phases
Ion probe microanalysis[a]	—	10^{-1}–10^1 (1–3 μm scan diameter)	Semiquantitative	Used for microscopic studies; trace analysis of elements in surfaces
Laser probe microanalysis[a]	—	10^2–10^4 (10–20 μm scan diameter)	0.1–0.5	Used for microscopic studies; trace analysis of elements in surfaces

TABLE 1.4. (*Continued*)

Method	Limit of Detection		RSD	General Uses, Comments
	Absolute (g)	Concentration (ppm)	(fraction)	
Electron spectroscopy X-rays (ESCA)[a]	—	10^3–10^5	0.05–0.20	Used for surface analysis of major phase; used mainly for structural studies
Electron spectroscopy He photons (PES)[a]	—	10^0–10^3	0.05–0.20	Used for gas analysis
Auger spectroscopy[a]	—	10^3–10^5	0.05–0.20	Used for surface analysis of major phase; used mainly for structural studies
Atomic emission spectroscopy, AC spark[a]	—	10^1–10^3	0.05–0.10	Used for major, minor, and trace elements
Atomic emission spectroscopy, DC arc[a]	—	10^{-2}–10^2	0.10–0.20	Qualitative survey method

Atomic emission spectroscopy, **Rf plasma**[a]	—	10^{-4}–10^{2}	0.01–0.05	Used for minor and trace elements in solutions
Atomic emission spectroscopy, flame[a]	—	10^{-3}–10^{2}	0.005–0.05	Used for minor and trace elements in solution
Atomic absorption spectroscopy, flame[a]	—	10^{-3}–10^{1}	0.005–0.02	Used for minor and trace elements in solution
Atomic absorption spectroscopy, nonflame[a]	10^{-15}–10^{-9}	—	0.02–0.10	Used for trace elements in small samples
Atomic fluore-scence spectroscopy, flame[a]	—	10^{-3}–10^{2}	0.005–0.02	Used for minor and trace elements in solution
Atomic fluorescence spectroscopy, nonflame[a]	10^{-15}–10^{-9}	—	0.02–0.10	Used for trace elements in small samples

[a]These methods are discussed in some detail in the chapters to follow.

REFERENCES

1. International Union of Pure and Applied Chemistry, Commission V-4 Reprint, Part II, 1974.
2. W. W. Meinke, in *Treatise on Solid Chemistry*, N. B., Hannay, Ed., Vol. 1, Plenum Press, New York, 1973.
3. G. H. Morrison, Ed., *Trace Analysis Physical Methods*, Interscience, New York, 1965.
4. W. W. Meinke and B. F. Scribner, *Trace Characterization, Chemical and Physical*, NBS Monograph 100, U. S. Department of Commerce, Washington, D. C.

ANALYTICAL CONSIDERATIONS

T. C. O'HAVER

Department of Chemistry
University of Maryland
College Park, Maryland

A. SOURCES OF ERROR IN ANALYTICAL MEASUREMENT

Conceptually, the process of quantitative measurement by a spectroscopic technique may be broken down into three distinct steps: (1) measurement of the detector signal, (2) correction of the total signal for the portion *not* due to the analyte, and (3) determination of analyte concentration corresponding to the measured analyte signal. It is convenient to divide the sources of error in analytical measurements on this basis. Errors in measuring the detector signal, the most important of which are random (noise) errors, are discussed in Section A.1. The factors involved in the measurement of, reduction of, and correction for nonanalyte signals are considered in Section A.2. Methods for the accurate quantitative determination of concentration are treated in Section A.3.

Errors in analytical measurements that arise from the chemical or physical properties of the analytical sample are commonly referred to as *interferences.* Most commonly, interferences result in systematic errors. Basically two types of interferences can be discerned: additive and multiplicative. An *additive interference* occurs when a sample constituent generates a signal (i.e., emission, abosrption, or luminescence) that adds to the analyte signal. This additive signal is not normally a function of the analyte concentration, and thus an additive interference can be considered to change the intercept, but not the slope, of the plot of the analytical signal (measure) versus the analyte concentration (the analytical calibration curve). Additive interferences influence step 2 of the analytical process described above; they are discussed in Section A.2. In spectrometric analysis, most spectral interferences are additive in nature. Clearly, additive interferences are the most important at low analyte concentrations.

A *multiplicative interference* occurs when a sample constituent either increases (enhances) or decreases (depresses) the analyte signal by a certain factor, without generating a signal of its own, thus changing the slope but not the intercept of the analytical curve. Multiplicative interferences influence step 3 of the analytical process and are discussed in Section A.3. Common examples of multiplicative interferences include bulk, physical, chemical, and ionization interferences in atomic absorption, and quenching effects in molecular luminescence spectrometry.

The utility of making a distinction between additive and multiplicative interferences becomes particularly evident when considering techniques of correcting for interference, which may apply only to one type or to the other. For example, the *method of analyte additions* [see Section A.3.b(2)] is a very powerful tool for compensating for interferences, but it is *applicable only to multiplicative interferences*.

It should be pointed out that in some cases a sample constituent may cause both additive and multiplicative interferences. For example, in the molecular fluorimetric determination of pyrene, the presence of anthracene causes both an additive interference due to its own fluorescence, and a multiplicative interference due to the absorption by anthracene of a portion of the excitation and fluorescence radiation (*inner* or *prefilter effect*).

1.   ERRORS IN MEASURING SIGNALS

a.   Systematic Errors

Most analytical spectroscopic measurements require only that relative signals be measured; absolute signal or light intensity measurements are seldom needed. Thus systematic errors in measuring signals are seldom significant, as long as the errors remain constant for all measurements of samples, blanks, and standards. For example, a nanoammeter used to measure the photocurrent in a spectrofluorometer need not be absolutely calibrated for conventional analytical applications.

However, in absorption spectrophotometry, it is usually desirable for the absorbance or transmittance readout to be correct in the absolute sense. This is obviously necessary if molar (linear) absorption coefficients are to be measured.

Carefully calibrated photometric standards are available from the National Bureau of Standards. Alternatively, one may prepare solutions of certain stable inorganic salts for which the molar absorption coefficients at various wavelengths have been reasonably well ascertained by repeated measurements by many workers over the years; these salts include KNO_3, K_2CrO_4, and $K_2Cr_2O_7$. The absorption coefficients are listed in Table 2.1. Clearly, carefully prepared solutions made from the best available reagent-grade salts and distilled water are desirable.

TABLE 2.1   Solutions for Photometric Standards[a]

Material	Concentration (g liter^{-1})	Wavelength (nm)	Absorption coefficient at 21°C (liters g^{-1} cm^{-1})
KNO_3	10	302	0.0705 ± 0.0005
		262.5	0.014_8
K_2CrO_4	0.025 plus 2 g liter^{-1} KOH	373	24.85 ± 0.1
$K_2Cr_2O_7$	0.06 plus 0.2 ml liter^{-1} concentrated H_2SO_4	235	12.5
		257	14.5
		313	4.7
		350	10.7 ± 0.1

[a]J. R. Edisburg, *Practical Hints on Absorption Spectrometry*, Adam Hilger, London, 1966.

Wavelength calibration can be checked by means of the line emission of discharge lamps, for example, hydrogen, deuterium, neon, or mercury lamps. Some of the most useful lines are listed in Table 2.2. Hollow cathode discharge lamps and electrodeless discharge lamps also make excellent line sources for wavelength calibration.

Although emission and luminescence spectrochemical analyses are normally performed by means of relative intensity measurements, absolute measurements may be useful for determining the radiance of lamps, flames, plasmas, and other light sources. In this case, a standard light source of known spectral radiance can be used to calibrate a particular monochromator-photodetector system. Close attention must of course be paid to relevant optical parameters such as solid angles and transmittances of optics.

Measurements of luminescence decay times are subject to a positive systematic measurement error if the instrumental response time is not much less than the luminescence decay time. This is discussed further in Section A.1.b(2)(d), in connection with the effect of instrument response time on the trade-off of random and systematic errors.

TABLE 2.2   Wavelength Standards[a]

H_2	379.9, 486.1, 656.3
D_2	380.1, 486.3, 656.6
Ne	585.2 (brightest), 640.2, 540.0, 480.0
Hg	253.7, 313.1, 365.0, 546.1

[a]All wavelengths in nanometers.

b. Random Errors

(1) Reading Errors

Reading errors are caused (1) by mechanical limitations in the ability of the readout device to respond precisely to very small signals (or changes in signals) and (2) by the inability of the analyst to see these small signals because of limited visual acuity. For a conventional moving-coil (d'Arsonval) meter readout, reading errors are caused by bearing friction, parallax between the needle and the scale, and irregularities in the spacing of scale graduations. For a chart recorder readout, the dead band of the servomechanism and the width of the pen tracing contribute to reading errors. For both meters and recorders, visual errors in estimating the needle or pen deflection are important, especially for small deflections. For digital readout devices, visual error is (ideally) zero, but a reading error still remains in the form of a truncation or rounding error of ± 1 in the least significant digit. For example, a three-digit decimal readout of absorbance (say 0.000 to 0.999 absorbance units) has a large reading error (± 0.001) for a small absorbance, for example, ± 0.001 absorbance units for an absorbance of 0.001.

Reading errors should never be allowed to become significant, compared to other errors, because they are so easy to reduce to any desired level simply by increasing the gain (amplification) of the readout electronics. In emission measurements, this amounts to increasing the gain of the photodetector, nanoammeter, or lock-in amplifier. In absorption measurements with a logarithmic (direct absorbance) readout, the electronic gain *after* the log circuit is increased. This is usually called *scale expansion*. For transmittance readouts, increased gain and appropriate zero offset must be used.

(2) Noise

(A) TYPES. In general, we may define *noise* as any undesirable electrical fluctuation in the signal readout (meter reading, pen position, scope deflection, digital number, etc). Undesirable DC signals such as dark current, amplifier offset currents or voltages, and zero-adjust errors, are not normally considered noises; they are in any case easily eliminated or compensated for electrically. Under the term "noise," we include both *periodic* and *aperiodic* AC noises, as well as drift (see Figure 2.1).

(i) *Periodic Noise. Periodic noise* is characterized by a regular, repetitive waveform with a frequency spectrum consisting of one or more discrete frequency components. For example, power line pickup is a periodic noise consisting mainly of a 60 Hz component, usually with some

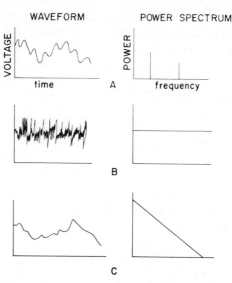

Fig. 2.1 Schematic representation of the waveforms and power spectra of (*A*) periodic noise, (*B*) white noise, and (*C*) 1/*f* noise.

weaker harmonic components at 120 Hz, 180 Hz, and so on. Stray pickup of radio and television signals, or of signals radiated by nearby instruments or other experimental equipment, may also occur. Periodic noises are, at least in principle, easy to eliminate by means of electronic filters placed at appropriate points in the readout system. If the noise consists of only one frequency component, a *notch filter* may be used to eliminate it specifically. Pretuned notch filters are available commercially for commonly encountered frequencies such as 60 and 120 Hz. Alternatively, a tuned amplifier or filter may be used to pass the desired signal frequency while attenuating all other frequencies. This approach is widely used, because it also provides some attenuation of aperiodic (random) noises.

(ii) Aperiodic Noise. Aperiodic or *random noises* are the largest and most important type of noise encountered in spectrometric systems. Unfortunately, this type of noise is also the most difficult to eliminate effectively. Aperiodic noise is characterized by a random, nonrepetitive waveform and by a broad-continuum frequency spectrum covering a wide range of frequencies. This is why aperiodic noises are so difficult to eliminate; even in a sharply tuned amplifier system, there is always some finite frequency component of aperiodic noise falling within the amplifier passband.

Different types of aperiodic noises are distinguished on the basis of their frequency spectra. *White noise* has a "flat" frequency spectrum; that is, noise power and voltage are the same at all frequencies. Common examples include *Johnson* (*thermal*) *noise* in a resistor and *shot noise* in the anode current of a photomultiplier. Another commonly encountered type of noise is *1/f noise*, so called because the noise power is inversely proportional to the frequency. Thus most of the noise power is concentrated at low frequencies. This type of noise is particularly troublesome in DC systems.

Other types of noise are imaginable. A *$1/f^2$ noise* would be produced by filtering white noise through a single-stage low-pass filter with a -6 db per octave attenuation rate. In a spectrometric system, this filtering action could be provided by a conventional electrical (e.g., *RC*) filter, by the long exponential decay characteristic of a luminescence emission, or by a thermal or mechanical lag (e.g., the thermal inertia of a tungsten lamp filament). For example, in the phosphorimetric measurement of a molecule with a long decay time (of the order of seconds), the white noise component of the excitation source fluctuation would be filtered into a $1/f^2$ noise and would contribute a $1/f^2$ component to the total noise at the instrument output.

Another type of low-frequency noise is *drift*. Drift errors tend to accumulate with time so that, for minimum errors, measurements should be made in the shortest possible time. However, short measurement times tend to increase the effect of white noises. Thus, in a real system subject to both white noise and drift, an optimum measurement time may exist that minimizes the total (drift plus noise) error.

(B) ELECTRICAL BANDWIDTH. The measurement and characterization of aperiodic noises requires knowledge of the *electrical bandwidth* of the measuring system. The electrical bandwidth is essentially a measure of the ability of the system to respond to rapid changes in signal at the input. The electrical bandwidth is manifested in these closely related parameters: *frequency response bandwidth, response time,* and *averaging time.*

(i) *Frequency Response Bandwidth.* A plot of the amplitude response of a measurement system to a sinusoidal variation in input signal amplitude versus the frequency of the input signal is called the *frequency response curve* of the system. The range of frequency Δf over which the amplitude (voltage or current) response is within 0.707 of the maximum ("midband") response is called the *frequency response bandwidth*, or simply *bandwidth*, of the system. In a DC system, the bandwidth extends from zero frequency (DC) up to the upper frequency limit f_u determined by the *time constant* τ of the system:

$$\Delta f = f_u = \frac{1}{2\pi\tau} \tag{1}$$

DC systems attenuate only noise frequency components above f_u. White noise may be reduced indefinitely by reducing f_u; but $1/f$ noises are not appreciably reduced by reducing f_u, because most of the noise power occurs at very low frequencies. The *effective noise bandwidth* Δf_n is not quite the same as the sinusoidal bandwidth discussed above:

$$\Delta f_n = \frac{1}{4\tau} \tag{2}$$

In a bandpass or tuned AC system, the bandwidth typically extends fro slightly below to slightly above the signal frequency. The smaller the the more selective the system is to the signal frequency and the greater extent to which noise is reduced.

The overall electrical bandwidth Δf of a system is given by the recip quadratic sum of the bandwidths Δf_i of the individual circuits or through which the signal passes:

$$\Delta f = \left(\sum_i \frac{1}{\Delta f_i^2} \right)^{-1/2}$$

(ii)   Response Time t_r. If the input signal (AC or DC) is c amplitude instantaneously at some time $t = 0$, the output sign change instantaneously, because of the finite electrical bandw system. Rather, the output signal will respond more slowly and value within 1% of the expected steady-state value in a tim *response time t_r* of the system. For a simple DC (low-pass response time may be shown to be 4.6 times the time consta

$$t_r = 4.6\tau$$

As a convenient rule of thumb, the response time is give by the reciprocal of Δf_n:

$$t_r \cong \frac{1}{\Delta f_n}$$

(iii)   Averaging Time t_a. Because of the finite band times of real systems, the output signal at any given not simply by the input signal at that instant, but ra the input signal from time $t - t_a$ to time t, where t_a *time* of the system. The averaging time is proportio and is given by:

$$t_a = \frac{1}{2\Delta f_n} = \frac{t_r}{2} = 2\tau$$

The importance of averaging time is apparent in comparing measured rms noises to the standard deviations of repetitively averaged (integrated) signal data. For example, some atomic absorption spectrometers are equipped with electronic integrators which determine the mean absorption signal over an (adjustable) averaging time t_a. The noise bandwidth of the integrator is given by the above equation, that is,

$$\Delta f_n = \frac{1}{2t_a} = \frac{1}{4\tau} \tag{7}$$

This relationship allows integrator times to be compared to damping time constants: A t second integration time is equivalent to (has the same noise bandwidth as) a $2t$ second time constant. Thus the standard deviation (rms noise) of the integrated and filtered output signals is equal under these conditions.

(C) EFFECT OF BANDWIDTH ON NOISE. The response of a DC readout system to a noisy signal at its input depends on how much of the noise power frequency spectrum falls within the noise bandwidth Δf_n. Noise frequencies outside Δf_n are attenuated. Thus, in general, reducing Δf_n would be expected to reduce the noise. But the extent to which noise can be reduced in this way depends on the type of noise frequency spectrum.

White noise has a constant noise power per unit frequency interval (i.e., WHz^{-1}). Thus the noise power passed by the system is simply the product of the input noise power density (in WHz^{-1}) and the electrical noise bandwidth Δf_n (in Hz). Because noise power is proportional to the square of voltage (or current), noise voltage (or current) is proportional to the square root of the noise bandwidth. This explains why we find a $\sqrt{\Delta f_n}$ term in the equations for white noises such as shot and Johnson noise.

However, $1/f$ noise is quite another story. The effect of Δf_n on $1/f$ noise is much less than that on white noise. In fact, it may be shown that, under typical conditions, the noise is approximately proportional to the 0.1 power of the bandwidth. This dependence on the bandwidth is so weak that it is usually convenient to think of $1/f$ noise as independent of bandwidth.

( ) UNDESIRABLE EFFECTS OF REDUCING BANDWIDTH. Attempting to reduce noise by reducing bandwidth excessively may introduce undesirable effects, because of the resulting long response time. In some cases, the reduction in random noise error may be more than offset by the increase systematic error produced by the slow system response. Some examples discussed here.

( ) *Analysis Time* t_{anal}. Every time the output signal is changed, such as changing the sample, the analyst must wait at least one response time

before taking a reading. Long response times (low bandwidths) are an obvious inconvenience. If an automatic sample changer or sequencer is employed, the system response time must be carefully tailored to the sample changing period.

(ii)   Sample Consumption. In conventional flame atomic emission, absorption, or fluorescence spectrometry the sample solution is consumed continuously at some rate during the measurement time; photodecomposition in molecular spectrometry has a similar effect [see Section A.1.b(2)(d)(iii)]. Longer response times therefore lead to greater total sample consumption. If it is desired to minimize the absolute (weight) detection limit, a bandwidth may exist that optimizes the trade-off between low noise at low bandwidths and low sample consumption at high bandwidths (short sampling time).

(iii)   Sample Stability. Occasionally the analytical sample is unstable and must be measured quickly. This obviously puts a lower limit on the bandwidth of the measurement system.

(iv)   Drift Errors. Drift in the instrument system generally results in greater analytical errors, the longer it takes to complete one set of measurements (e.g., sample, blank, standard). Thus reducing the bandwidth to reduce random noise error increases drift error. For example, the error produced by a linear, unidirectional drift is proportional to the measurement time. If it is assumed that the measurement time is proportional to the response time, the drift errors will be inversely proportional to the bandwidth. For white noise, the random noise errors will be proportional to the square root of the bandwidth. Thus *drift errors predominate at low bandwidths and noise errors at high bandwidths.* An optimum bandwidth, which minimizes the total error, may exist. If the noise is $1/f$ noise instead of white noise, the optimum bandwidth will be much higher.

(v)   Spectrum Scanning. When recording spectra, the scan rate r (in nm s^{-1}) must be compatible with the response time t_r. If r is too fast or t_r too long, or both, the recorded spectral peaks will be distorted and blurred out.

The maximum usable scan speed depends on Δf and on the structure of the spectrum. When recording line spectra, the scan speed r (in nm s^{-1}) should be:

$$r \leqslant \frac{\Delta\lambda_s}{t_r} \cong \Delta\lambda_s \Delta f \tag{8}$$

where $\Delta\lambda_s$ is the spectral bandwidth of the monochromator (in nm). For band spectra, r should be:

$$r \leqslant \frac{\delta\lambda_b}{t_r} \cong \delta\lambda_b \Delta f \tag{9}$$

where $\delta\lambda_b$ is the half-width (in nm) of the most narrow band or spectral feature to be recorded faithfully.

(vi)   Decay Time Measurements. In the measurement of luminescence decay times, there is a trade-off between increased random noise errors at large Δf_n and increased (positive) systematic errors at small Δf_n. In order to keep this systematic error less than about 3% relative, the response time t_r of the system should not exceed the decay time of the measured species τ_L. Under these conditions, the time constant of the electronics is about one-fifth of the decay time:

$$\tau \sim \frac{\tau_L}{5} \tag{10}$$

(vii)   Pulse Measurements. Both the width and height of measured pulses are affected by the bandwidth of the measurement system. For rectangular pulses, the error in both height and width will be small if the system response time t_r is equal to or less than the pulse width t_p. In nonflame atomic absorption and fluorescence techniques, narrow pulses with relative fast rise times (< 1 s) are often observed. For such pulses, the system response time t_r should be equal to or less than the rise time t_{rise}; this normally requires bandwidths somewhat higher than that permissible with conventional flame techniques:

$$t_r \leqslant t_{\text{rise}} \tag{11}$$

(viii)   Choice of Optimum Bandwidth. In principle, the system bandwidth should be chosen to optimize the trade-off between random noise errors and systematic response-time errors, that is, at the minimum of the plot of total error versus bandwidth. Fortunately, the minimum is generally quite broad and flat, so that the optimum bandwidth need only be estimated. Usually, the minimum bandwidth that does not introduce significant systematic errors is sufficiently close to the optimum.

(E)   SOURCES OF NOISE IN SPECTROMETRIC SYSTEMS

(i)   Photon Noise. Photon noise is due to the discontinuous (quantized) nature of light energy and electrical current. It is the most fundamental of all noises and remains when all other noises have been eliminated. When a beam of light falls on the cathode of a photoemissive detector, the arrival of the photons and the emission of photoelectrons occur randomly. The result is that the photoemission current of the phototube is not a pure steady DC current but also contains an AC noise component, called *shot noise*, as a result of the discontinuity of the photoemission process. It may be shown that the rms shot noise component $\overline{\Delta i_s}$ (in A) is given by:

$$\overline{\Delta i_s} = \sqrt{2ei_c\Delta f_n} \tag{12}$$

where e is the charge on the electron (in C), i_c is the average DC cathode current of the phototube (in A), and Δf_n is the noise bandwidth of the readout system (in Hz). For a photomultiplier tube, the noise is expressed in terms of the anodic current i_a:

$$\overline{\Delta i_s} = \sqrt{2eBM\Delta f_n i_a} \tag{13}$$

where M is the multiplication factor of the phototube:

$$M = g^z \tag{14}$$

and B is:

$$B = 1 + g^{-1} + g^{-2} + \cdots + g^{-z} \tag{15}$$

where g is the gain per dynode stage and z is the number of dynode stages.

Because shot noise increases with the square root of the photocurrent signal, the signal-to-shot-noise ratio *increases* with the square root of the photocurrent. Thus, all else being equal, larger photocurrents can be measured more precisely.

Photon counting techniques are routinely used in X-ray spectroscopy, and occasionally in Raman and other UV-visible methods. In these cases, the photon noise is expressed in terms of the counting statistics, as is common in radiochemical methods. The expected standard deviation of the total count is equal to the square root of the number of counts n:

$$\sigma_N = \sqrt{n} \tag{16}$$

In fact, the shot noise equation given above is derived directly from this expression by converting from total counts to electrical current.

(ii) Detector Noise. Detector noise is insensitive to the signal level of radiation reaching the detector. Detector noise and photon noise are inherent noise sources in spectroscopy and are always present. The detector noise for photoemission devices is given by equation 12 for a single-stage device, and by equation 13 for a multistage (photomultiplier) device, except that the anodic dark current i_d replaces the current i_a in both expressions. For a photon counting system, the detector noise count is simply $\sqrt{r_d t_c}$, where r_d is the dark count rate (in s^{-1}) and t_c is the counting (averaging) time (in s).

The radiant flux (in W), which must be incident on the detector surface of area S_d (in cm^2) per unit frequency response bandwidth in order to give a signal-to-noise ratio of unity is called the noise equivalent power of the detector NEP_λ (in $W\,Hz^{-1/2}$). The detectivity D_λ is related to NEP_λ:

$$D_\lambda = \frac{\sqrt{S_d}}{NEP_\lambda} \qquad (17)$$

where the subscript λ simply implies a wavelength dependence. All photon detectors (photoemission and semiconductive detecors) have a wavelength dependency, whereas all thermal detectors (thermocouples, bolometers, etc.) have a flat spectral response. The value of NEP_λ for UV-visible detectors is quite small, that is, 10^{-15} to 10^{-17} $W\,Hz^{1/2}$, whereas for the lower-sensitivity IR detectors $NEP_\lambda s$ may be in the range 10^{-10} to 10^{-15} $W\,Hz^{-1/2}$.

It is interesting to compare the photon flux r_p (photons s^{-1}), necessary to give a signal-to-noise ratio of unity in the UV (200 nm) for a photon detector with a NEP_λ of 10^{-16} $W\,Hz^{-1/2}$ and in the IR (20 μm) for a thermal detector of 10^{-12} $W\,Hz^{-1/2}$. If a counting time t_c of 1s is assumed, since

$$(r_p)_{S/N=1} = \left(\frac{NEP_\lambda}{h\nu}\right)^2 = \left[\frac{NEP_\lambda}{\frac{hc}{\lambda}}\right]^2 \qquad (18)$$

then

$$(r_p)_{\substack{200\,nm \\ S/N=1}} = \left[\frac{10^{-16}\,Js^{-1/2}}{\dfrac{6.6\times10^{-34}\,Js\times3\times10^8\,ms^{-1}}{2\times10^{-7}\,m}}\right]^2 = 10^4 s^{-1}$$

and

$$(r_p)_{\substack{20\,\mu m \\ S/N=1}} = \left[\frac{10^{-12}\,Js^{-1/2}}{\dfrac{6.6\times10^{-34}\,Js\times3\times10^8\,ms^{-1}}{20\times10^{-6}\,m}}\right]^2 = 10^{16}\,s^{-1}$$

Therefore, to obtain a signal-to-noise ratio of only unity in the IR, an unbelievably high count rate of 10^{16} s^{-1} is needed, whereas in the UV-visible, a much lower count rate of 10^4 s^{-1} is needed. Needless to say, photon noise is seldom important in the IR but is almost always of some consequence in the UV-visible.

(iii)   Fluctuation Noise. Fluctuation noise is a result of convective and diffusive processes resulting in a low-frequency ($1/f$ type) noise contribution. If the information signal can be modulated at a frequency considerably greater than the fluctuation noise frequencies, for example, fluctuation noises in flames used in atomic absorption spectrometry where the light source is chopped and a tuned detector is used, fluctuation noise is of little importance. However, if the fluctuation noise is carried on the information signal, it can be of great importance.

Fluctuation noise is proportional, and the fluctuation noise current (rms) is given by:

$$\overline{\Delta i_f} \cong \xi_f\, i \tag{19}$$

when ξ_f is the fluctuation factor, that is, the fractional extent of the signal that is fluctuation noise and i is the signal level (in A); ξ_f is a function of the measurement frequency f. It should be noted that fluctuation noise is essentially independent of the frequency response bandwidth Δf.

(F)   ADDITION OF NOISES. Each of the various sources of noise in a spectrometric system contributes to the total noise in the output signal. Drifts and DC offsets add algebraically. Independent random noises add quadratically; thus

$$\overline{\Delta i}_T = \sqrt{\sum_j \left(\overline{\Delta i_j}\right)^2} \tag{20}$$

where $\overline{\Delta i}_T$ is the total random noise, and $\overline{\Delta i_j}$ is the contribution of the jth source of independent random noise.

In many cases, it is possible to learn something about the relative contribution of the various noise sources by measuring the output noise under appropriately chosen experimental conditions. For example, *detector dark current noise and electronics noise can be evaluated by simply blocking the light to the detector.* (An electrically null double-beam spectrometer which uses dynode voltage feedback, such as the Beckman DB, must be operated in the single-beam mode at a fixed dynode voltage, if this is to work properly.) In an atomic absorption spectrometer, flame emission noise can be evaluated by turning the flame on and the source off. The noise observed (again in the single-beam mode) is due to the sum of the flame emission noise, the photon shot noise due to the flame emission signal, the phototube dark current noise, and the electronics noise. Of course, only $\%\,T$ (or percent absorption readout) can be used if the source is off; the direct absorbance readout would be off-scale. Source noise can be evaluated if the flame is turned off and the source is turned on. But in this case, the output noise is the sum of the source fluctuation noise, source photon shot noise, phototube dark current noise, and electronics noise. These last two noises can be evaluated independently as explained above, and thus can be subtracted out (quadratically) from the total observed noise. Source fluctuation noise and source photon (shot) noise are more difficult to separate. Fortunately, however, the dependence of these noises on the total photosignal is different; source fluctuation noise, being a certain fraction of the total source intensity, increases linearly with the total photosignal, whereas source photon (shot) noise varies as the square root of the photosignal. Thus it is easy to recognize whether either of these noises is dominant or whether a mixture of the two exists. For this purpose, the photosignal is conveniently varied by changing the monochromator slit width or by inserting neutral density filters into the light beam. The same technique can be used in condensed-phase molecular absorption and fluorescence spectrometry. In flame atomic fluorescence and in low-temperature phosphorescence measurements, the flame or sample cell should be replaced by a diffuse reflector or a solid-state phosphor, respectively, to eliminate the contribution due to the sample cell or flame noise. (The decay time of the phosphor must be included in calculation of the total system bandwidth.)

(G)   MEASUREMENT OF NOISE. The most generally useful measure of noise is rms noise, which is equivalent to the standard deviation. However, rms noise may not be easy to measure directly. Conventional AC voltmeters are actually average reading devices which are calibrated to read the rms values of sine waves only; other waveforms, including random noise, give erroneous results. True rms AC voltmeters are available at somewhat higher cost, but these are useful only for measuring noise frequencies above about 5 Hz. If appreciable noise power exists below 5 Hz, even these meters are unsuitable. We are generally most interested in measuring the total noise at the instrument output. The output signal is invariably DC, even if an AC or lock-in detection system is used, and the signal is usually filtered by a low-pass filter. Thus we are really concerned only with low-frequency noises between the direct current and the high-frequency cutoff of the output filter (or of the chart recorder, whichever is lower).

This low-frequency limit problem can be overcome by measuring the noise directly from the chart recording of the signal. This is the most convenient and most widely used technique. It has the advantage of responding to very low-frequency noises, including drift, which would be missed by an AC meter. However, rms noise cannot be measured directly this way; usually the peak-to-peak noise is estimated and converted to rms by dividing by a factor of 5. In this approach, some estimation on the part of the operator is involved, and it is assumed that the noise is sufficiently random and gaussian to allow use of the factor of 5. Furthermore, the dead band of the recorder may cause a systematic negative error in the measured noise at low noise amplitudes, because a portion of the noise deflection is absorbed into the dead band.

Probably the most satisfactory method of measuring noise is to use an electronic integrator to determine the averages of the signal over a series of equal time intervals Δt and then to compute the standard deviation of the averages. This is equal to the rms noise measured at a noise bandwidth of $(2\Delta t)^{-1}$ Hz. At least 10 averages, and preferably more than 30, should be taken to ensure adequate precision of the noise measurement.

(3) Signal-to-Noise Ratio*

(A) OPTIMIZATION OF SIGNAL-TO-NOISE RATIO. Neither the maximization of signal nor the minimization of noise is, by itself, sufficient to *ensure optimum instrumental performance*. Rather, the signal-to-noise ratio should be maximized. If it is assumed that the various chemical manipulations are performed properly, *maximization of the signal-to-noise ratio will result in the best precision and lowest detection limits*. If mathematical expressions of the signal and the noise can be obtained, the signal-to-noise ratio can be optimized with respect to any independent variable by conventional derivative maximization. A graphical solution is often most practical.

The experimental optimization of instrument parameters should ideally be based on the signal-to-noise ratio, but it is admittedly a temptation to maximize only the signal, because it can be done so quickly and easily. The use of the chart recorder, however, allows the signal and noise data to be recorded together and then converted to a signal-to-noise ratio with a minimum of work.

(B) EFFECT OF RANDOM NOISE ON ANALYTICAL PRECISION. The precision of an analytical method is determined by a large number of factors, not the least of which are the skill and care with which the sampling, sample preparation, and other checmical manipulations are carried out. However, once these procedures have been optimized to the greatest possible or practical extent, the analytical precision may begin to depend on the instrumental signal-to-noise ratio. This is particularly true of measurements made near the detection limit.

For a measurement whose precision is limited by the instrumental signal-to-noise ratio, the relative standard deviation (RSD) is given by:

$$RSD = \frac{\sqrt{(\overline{\Delta i_s^2} + \overline{\Delta i_b^2})/n}}{i_s - i_b} \qquad (21)$$

*Refer to Appendices B and C for a discussion of the signal-to-noise ratio optical measurement systems.

where i_s, $\overline{\Delta i_s}$, i_b, and $\overline{\Delta i_b}$ are the output readings and rms noises of the sample (including blank) and blank solutions, respectively, and n is the number of pairs of sample and blank readings considered to constitute one measurement of the analytical signal. (In this case, the analytical signal is $i_s - i_b$.) The above equation can also be used to calculate the smallest analytical signal that can be determined with a given RSD.

(C) EFFECT OF THE SIGNAL-TO-NOISE RATIO ON LIMITING DETECTABLE SAMPLE CONCENTRATION. In many cases, it is desirable to know the smallest analytical concentration that can be just barely detected with a given level of confidence. This is called the *limiting detectable sample concentration*, or simply the *detection limit*. It can be shown that this is the sample concentration that gives a signal-to-noise ratio of $t\sqrt{2}/\sqrt{n}$, where n is the number of sample-blank measurement pairs on which the detection is to be based, the $\sqrt{2}$ comes about according to the law of propagation of errors in multiplication, and t is the value of the Student's t statistic for the specified confidence limit and number of measurement pairs. The noise is assumed to be rms noise. Simpler definitions are commonly used, in which the signal-to-noise ratio at the detection limit is taken as some fixed number, for example, if only one measure of the sample is made but, say, 15 measures of the blank, the signal-to-noise ratio at the detection limit is just t which is often taken as 3—sometimes 2. For peak-to-peak noise measurements, the signal-to-noise ratio at the detection limit is usually defined as 1, which is *roughly* equivalent to an rms signal-to-noise ratio of 2.

It should be pointed out that detection limits are inherently imprecise numbers, that is, they have a RSD of about 0.5. For one thing, they are usually not measured directly, but rather are extrapolated from measurements made at higher concentrations, where the higher signal-to-noise ratio permits easier measurement of the signal. Furthermore, noise measurements are difficult to make precisely. *As a result, differences between detection limits less than about a factor of 2 should not normally be assumed significant.* It should also be kept in mind that a concentration at the detection limit can only be *detected*, as the term "detection limit" implies, and not measured quantitatively. It should be stressed that the detection limit applies to a *complete analytical procedure* and not to a given instrument or instrumental method.

Detection limits reported by manufacturers for their instruments, and by researchers for proposed new analytical techniques, are usually obtained under ideal conditions with highly optimized parameters. Simple solutions in low-background solvents are invariably used, and the longest time constant or integration time is often used. Under practical laboratory conditions, it may not be possible to approach these quoted detection limits by closer than a factor of 5 or 10.

2. REDUCTION OF AND CORRECTION FOR NONANALYTE SIGNALS

Once the total instrumental signal has been measured precisely, the next step in the analytical process is to determine what portion of this total signal is due to the analyte itself. It is clearly desirable to maximize this portion by the proper selection of analytical and/or instrumental conditions and, if necessary, to correct for the remaining nonanalyte signals. Of course, all spectroscopic methods are to some extent inherently selective by virtue of wavelength (or energy) selection. Thus the discussion of resolving power in Section 2.B is particularly pertinent to the problem of minimizing nonanalyte signals. In atomic spectroscopy, the inherent widths of the atomic lines are generally narrow enough so that the required analytical selectivity can usually be obtained by using a wavelength (energy)-selective system of sufficiently high resolution. However, in most types of condensed-phase molecular spectroscopy, the spectral bands are inherently much broader, so that instrumental resolution alone may not be sufficient to prevent overlap of the bands of the components of a mixture. Even in atomic spectroscopy, overlap of atomic lines with broad molecular bands or background continua cannot be eliminated by means of instrumental resolution alone. In these cases it is necessary to reduce, measure, and correct for the remaining nonanalyte signal. This may be done by (a) appropriate chemical methodology; (b) the use of a suitable blank solution, which provides a direct measure of the nonanalyte signal; and (c) the use of data treatment techniques which reduce or eliminate the effect of the nonanalyte signal.

a. Chemical Methodology

Analytical spectroscopic methods may be divided into *direct methods*, which measure an inherent spectroscopic property of the analyte itself; *chemical* or *reaction methods*, in which the analyte is converted by the addition of suitable reagents into a form with more suitable spectroscopic characteristics; and *indirect methods*, in which the analyte is involved in a reaction or other physiochemical interaction (e.g., energy transfer or quenching) which causes a change in the spectroscopic properties of another chemical species which is measured. Most atomic absorption techniques are direct, although indirect atomic absorption techniques exist for SO_4^{2-} (via Ba^{2+}), Cl^- (via Ag^+), and F^- (via interference on Mg^{2+}).

As for molecular absorption spectrometry, IR methods are usually direct, whereas UV-visible luminescence techniques may be direct, indirect, or reaction methods. For example, bilirubin in infant blood serum can be measured by direct colorimetry on an untreated sample, because essentially nothing else absorbs at the same wavelength. However, serum cholesterol is conventionally determined by a reaction method using a color reagent, because by itself cholesterol absorbs only near 210 nm, a region where many other serum constituents also absorb. Colorimetric and fluorimetric methods for trace elemental analysis are examples of chemical reaction methods; ordinarily, inorganic ions in solution do not exhibit sufficiently specific and intense light absorption or fluorescence for direct analysis. A highly specific analytical reagent is obviously desirable to eliminate nonanalyte signals in chemical reaction methods. Truly specific reagents are rarely available. One example is the Griess reagent (a mixture of sulfanilic acid and β-naphthol) used for the colorimetric determination of NO_2^-; only NO_2^- forms a pink color with this reagent. In most cases, additional chemical reagents are added, such as *masking agents* which tie up interfering species, or the solution conditions (especially pH) are adjusted to prevent the formation of interfering species. For example, in the colorimetric determination of molybdenum by SCN^-, the interference of Fe^{3+}, which like molybdenum forms a red complex with SCN^-, is eliminated by reducing Fe^{3+} to Fe^{2+}. In some cases, an analysis may be based on a specific change in analyte signal on changing the solution conditions. The requirement is that only the analyte signal be changed, while all other signals remain unchanged. The ΔpH *technique* in absorption spectrophotometry is an example. In this case, the absorbance of the sample solution is measured before and after an adjustment in solution pH. The difference between these two absorbances is then a measure of the concentration of material whose spectra are pH-sensitive. This technique is particularly valuable for the measurement of acidic and basic drugs in blood serum extracts, because changes in the pH of normal serum have little effect on the spectrum of its extracts. Phenobarbital, morphine, chlorpromazine, and salicylic acid are examples of drugs whose spectra change markedly with pH.

For many complex samples, specific reagents and conventional chemical treatment are not enough, and *chemical separations* are required. The separation methods most commonly used in conjunction with spectroscopic methods include *solvent extraction; ion exchange*; and *thin-layer, gas-liquid*, and *liquid-liquid chromatography*. In fact, thin-layer plate scanners and liquid chromatographic detectors based on UV absorption and fluorescence are now commercially available.

b. The Blank

In the strictest terms, the blank is a solution that contains everything in the sample solution except the analyte. The blank signal is simply subtracted from the total analytical signal. If it is assumed that there are no analyte–matrix interactions, such a procedure completely corrects for nonanalyte signals. However, it is seldom practical or possible to prepare an exact blank in real analyses of complicated samples. Thus some approximation must be made in practice. A *solvent blank*, the simplest, consists of the solvent used to dilute the samples. This type of blank corrects for signals produced by the solvent itself or by impurities in the solvent. A *reagent blank* contains in addition to the solvent any reagents used in the sample preparation, color or fluorescence development, or other pretreatment steps, but does not contain any of the analyte or sample matrix components. This type of blank additionally corrects for signals produced by the added reagents or their impurities. The solvent and reagent blanks also correct for contamination of the solvent and reagent with the analyte species. In analytical procedures in which the analytical signal is chemically produced by reaction of the analyte with an analytical (color or fluorescence) reagent, *a sample blank*, consisting of an aliquot of the sample diluted as usual but without the addition of the analytical reagent, may be useful. This blank corrects for signal-producing matrix constituents that do not react with the analytical reagent. Occasionally, both a reagent blank and a sample blank may be used, both blank signals being subtracted from the total analytical signal. This procedure is used, for example, in some colorimetric methods for serum constituents in clinical chemistry, in the event that both the untreated sample and the color reagents themselves absorb to some extent.

In some cases, it may be possible to approximate a rigorous blank by preparing a *matched-matrix blank*, which is made up to resemble the matrix constitution of the sample (without the analyte, of course). This approach is useful only if the sample matrix is well known and predictable. Blood serum and sea water are examples of samples for which suitable matched-matrix blanks may be useful. An even closer approximation to a rigorous blank is an *internal blank*, prepared from an aliquot of the sample solution by selective destruction of the signal producing form of the analyte. This technique is occasionally possible in condensed-phase molecular absorption and luminescence spectrometry. For example, in the fluorimetric analysis of urinary estrogen (a common clinical analysis), an internal blank is prepared by the addition of a nitrite salt, which specifically quenches the fluorescence of the estrogens.

In analytical procedures involving *sample preparation, separation,* or *preconcentration steps*, it is almost always essential that a blank be carried through the entire procedure. It is also desirable that the blank signals be measured and recorded individually, rather than simply setting the instrument response to zero on the blank. In this way, excessively high blanks values can be recognized. This is particularly important in trace work, where high blanks from contamination impurities in reagents, and so on, are fairly common.

c.   Data Treatment Techniques

The choice of an appropriate data treatment technique to reduce or eliminate the effect of nonanalyte interfering signals depends on the relative spectral distribution of the analyte line or band and the interfering band (background). Several cases can be distinguished, and these are illustrated in Figure 2.2.

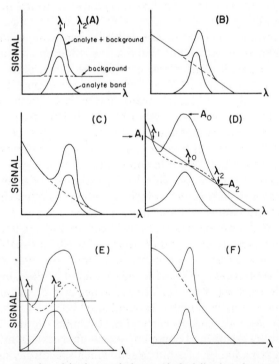

Fig. 2.2 Some examples of background shapes (dashed lines) and appropriate correction methods (see text for explanation). (*A*) Flat linear background; single-point correction. (*B*) Sloped linear background; baseline method. (*C*) Simple curved background; coordinate transformation. (*D*) Curved inflected background; the allen correction. (*E*) Peaked background; dual-wavelength techniques. (*F*) General broad background; derivative techniques.

(1)   Flat Linear Background; Single-Point Correction

In this case, the interfering (background) signal has a flat linear spectral distribution in the vicinity of the analyte maximum. This is a common situation in atomic spectroscopy where scattering, molecular band absorption, and molecular emissions are essentially flat and linear in the very narrow spectral interval around the atomic line(s). In these cases, a single reading of the background signal at a wavelength just offset from the analyte line is made and is subtracted from the total signal measured on the line (see Figure 2.2A).

In atomic absorption spectrometry, where line sources are most often used, absorption measurements at a wavelength off the analytical line must of course be made at a wavelength that is emitted by the source but which is not absorbed by the analyte. This puts some restrictions on the choice of background correction wavelength, particularly because the background correction line must be close to the analyte line (within a few 0.1 nm, ideally) so that the background will be the same. Nonresonance lines of the analyte and rare-gas lines (from the fill gas) are commonly used for background correction. Occasionally a line from another element lamp is useful, but only if that element is not in the sample or is not atomized under the existing analytical conditions. For example, a rhenium lamp emits a large number of lines, some of which fall near the resonance lines of other elements. Besides being a rather rare element, rhenium is not atomized in the most commonly used air–acetylene flame. The availability of continuum-source background correctors should also be pointed out; in this case, light from a continuum source, such as a deuterium lamp, and from the line source are alternately (via a chopper) allowed to pass through the flame. If both beams are of equal "intensity," broad-band absorption or scatter will attenuate both beams equally, whereas atomic vapor attenuates significantly only the analytical radiation from the line source (the ratio of the two beams is measured).

In molecular absorption and luminescence spectrometry, the analyte spectral band half-widths are usually not narrow enough, nor the interfering bands wide enough, to allow method 1 to be used. However, in condensed-phase molecular absorption spectrometry, this approach can be used to reduce the effect of random variations in sample cell position and cell wall reflectivity from sample to sample. These variations, which are often larger than the inherent instrumental noise level, result in changes in the position of the zero-absorbance baseline and can therefore be corrected for by measuring the absorbance at some wavelength off the analyte maximum and subtracting this from the absorbance measured on the analyte maximum. Note that it is not necessary that the reading off the maximum be at a wavelength where the analyte absorbance is zero, although it is desirable that it be as low as possible to avoid loss of sensitivity. A nearby minimum is ideal. Needless to say, all standards and blanks have to be measured in exactly the same way.

(2) Sloped Linear Background; The Baseline Method

In this case, the slope of the background is sufficiently great in the vicinity of the analytical line or band that a single-point background correction is not accurate. Two approaches are possible. If the spectrum can be scanned, a tangent baseline may be constructed on the recorded spectrum at the base of the analyte peak. The perpendicular distance from the analyte maximum and the constructed baseline is taken as the analytical measure. This is illustrated in Figure 2.2B. If spectra cannot be scanned, readings taken at two wavelengths equally above and below the analyte maximum are averaged and subtracted from the reading at the analyte maximum. If the analyte signal is not zero at these wavelengths, the standards and blanks must be measured the same way. Clearly, these methods work for the case of a flat linear background as well.

(3) Simple Curved Background; Coordinate Transformation

A simple curved background without maxima, minima, or inflections is illustrated in Figure 2.2C. In this case, the baseline method (method 2) is only approximate. However, the baseline method can be made to give accurate results by transforming the coordinates (wavelength and/or intensity) in such a way that the background is linearized. For example, if the background is exponential in shape, the spectrum can be replotted on semilog paper. Or if the background is shaped like the skirt of a gaussian band, replotting on "probability" paper is possible. One practical situation in which coordinate transformation has been used sucessfully is in the correction for scattering of turbid samples in absorption spectrophotometry. Here the background has a λ^n distribution and is linearized by replotting on log-log paper. Coordinate transformations can be done on-line by a hardware analog electronic device or by a small minicomputer.

(4) Inflected Curved, Background; The Allen Correction

A technique that can easily handle an inflected curved background is the Allen correction. The background must have an inflection point, as illustrated in Figure 2.2D, and its exact shape must be known. A straight line is drawn so that it intersects the background at the wavelength of the analyte maximum λ_0 and at two other wavelengths λ_1 and λ_2. The analytical signals (intensity or absorbance) are measured at these three wavelengths and are designated S_0, S_1, and S_2, respectively. The corrected analyte signal S is given by

$$S = S_0 - S_1 - \left(\frac{\lambda_0 - \lambda_1}{\lambda_2 - \lambda_1}\right)(S_2 - S_1) \tag{22}$$

The correction compensates for changes in the intensity of the background, but its shape must remain the same. The Allen correction can be used for linear or peaked backgrounds but not for simple noninflected curved backgrounds.

(5)   Peaked Background with Maximum and/or Minimum; The Dual-Wavelength Technique

A peaked background, illustrated in Figure 2.2E, is a very common type of interference. It may be handled by means of the Allen correction or, somewhat more simply, by the use of the *dual-wavelength* technique. Two wavelengths are selected (λ_1 and λ_2 in Figure 2.2E) for which the background signals are identical, but for which the analyte signal is not the same. The difference between the reading at these two wavelengths is thus a measure of the analyte alone and is not a function of the background intensity, as long as the background *shape* remains the same. Many choices of λ_1 and λ_2 are possible—the best choice generally being the one giving the largest analyte difference signal. The dual-wavelength technique is most conveniently used with commercially available *dual-wavelength spectrophotometers* which automatically switch back and forth between two adjustable wavelengths and continuously compute the difference signal. On a conventional spectrophotometer, wavelength resetability limits the precision of manual dual-wavelength measurements. The dual-wavelength technique also works in flat linear background cases, but not for sloped backgrounds without maxima or minima.

(6)   General Broad Backgrounds; Derivative Spectrometry

A powerful technique, which works especially well for any shape of background that is much broader than the analyte band, is derivative spectroscopy. In this technique, the rate of change of spectral intensity or absorbance with respect to wavelength is recorded, that is, $dS/d\lambda$. Narrow bands, which have a high rate of change with wavelength, are emphasized compared to broader bands and background. The particular advantage of derivative techniques, compared to the dual-wavelength, Allen correction, and linearization techniques, is that the exact shape of the background is much less important, as long as the background is much broader than the analyte band. In fact, the background shape can even change from sample to sample with little effect.

Clearly, the derivative technique applies rigorously to the *flat linear background case*, for the *derivative* of such a background is zero. For the *sloped linear background, only a second-derivative technique is rigorously correct*, that is, $d^2S/d\lambda^2$. In fact, second-derivative techniques have been successfully used for the correction of continuum background interferences in atomic flame emission spectrometry. For *curved, inflected,* or *peaked backgrounds* the *derivative technique is not exact*, because the background generated a finite derivative signal. However, if the analyte band width is substantially less than that of the background, the derivative approach may well be more accurate than other correction techniques. Furthermore, if the background has a maximum, minimum, or inflection, its derivative will have zero crossing of the signal at some wavelength. Derivative measurements made at this wavelength are therefore free of background interference.

Derivative spectrometry can be performed with a conventional spectrometer by scanning the spectrum and generating the derivative electronically. Alternatively, wavelength modulation (via vibrating entrance slit, output mirror, or quartz plate before the exit slit) and synchronous detection can be used (see Section D.1). Wavelength modulation instruments may be able to achieve a higher signal-to-noise ratio in measurements of noisy signals. They also have the further advantage that they can obtain derivatives at fixed wavelengths in real time, which makes it much easier to use the zero crossing discussed in the previous paragraph.

(7) Computer-Based Techniques

With the increasing use of on-line minicomputers in chemical instrumentation, sophisticated computer-based "curve resolution" and deconvolution techniques may become more widely used for routine analytical purposes. These extremely powerful techniques are capable of resolving a highly overlapped mixture spectrum into its spectral components. However, they suffer from the disadvantage that the shapes of each component must usually be known, and fairly good initial guesses for the positions and intensities of the component peaks may be required to allow reliable convergence in a reasonable time.

3.   QUANTITATIVE DETERMINATION OF CONCENTRATION

a.   Calibration Methods

In most spectrometric methods, analyte concentrations are determined by comparison to one or more *standards* of known analyte concentration. A series of standards of different concentrations serves to establish the *analytical calibration curve*, the relationship between concentration and analytical signal (calibration curve versus inverse relationship). Ideally, the analytical curve should be linear, should have zero intercept (pass through the origin), and should have a reproducible slope independent of the presence of other constituents in the sample (see Chapter 1). The linearity depends mainly on the inherent characteristics of the spectroscopic technique and on various instrumental factors. The intercept is a function of nonanalyte signals, which have been dealt with above. The slope is a function of both fundamental and instrumental factors and may also be influenced by nonanalyte matrix constituents in the sample. This last effect is what constitutes a *multiplicative interference*. Such interferences can often be controlled by the appropriate preparation and use of standards.

In the simplest cases, that is, in the absence of significant multiplicative interferences, standards may be prepared simply by dilution of purified analyte in a suitable diluent. *Solution standards* are the easiest to handle and are the most widely used in spectroscopic methods, but *solid standards* are more often used in arc or spark emission spectrographic and in X-ray methods. In such cases, the diluent is an appropriate solid matrix material such as graphite powder.

Once appropriate standards are prepared, they may be used in one of several ways.

(1) Calibration Curve Method

A series of standard solutions encompassing the expected concentration range of the analyte in the sample solutions is prepared and measured. The plot of analytical signal versus solution concentration, called the *analytical calibration curve* (working curve, standard curve, and analytical curve are incorrect terms) is prepared. In the calibration curve method, the concentrations of the sample solutions are interpolated from the calibration curve. This method is time-consuming, particularly since the calibration curve is likely to have to be rerun day to day or even from hour to hour (because of uncontrolled variations in instrumental parameters). However, this is the only suitable method for materials whose calibration curves are highly nonlinear. A least-squares fit to the calibration data, most easily done by computer, is helpful in this case. However, care should be taken to use the lowest-degree polynomial that gives a satisfactory fit to the data, so as not to force an artificial fit to an unrealistically high-order polynomial. For example, it is possible to obtain a "perfect" fit to any five-point calibration plot with a fifth- or higher-order polynomial, but this is clearly meaningless. *Random scatter in the calibration points must be distinguished from systematic nonlinearities.*

(2) The Factor Method

If the calibration curve is known to be linear, the calibration curve method may be replaced by the much simpler *factor* or *single-standard method*. In this method, a single standard solution close to the estimated sample concentration is used. The sample concentration c_x is calculated:

$$c_x = \left(\frac{S_x - S_b}{S_s - S_b} \right) c_s \tag{23}$$

where S_x, S_s, and S_b are the measured analytical signals of the sample, standard, and blank solutions, respectively, and c_s is the concentration of the standard. Because only one standard solution is involved, it is convenient to run the standard repeatedly, before and/or after each sample, and thereby avoid errors due to changing experimental conditions.

(3) Two-Standard or Bracketing Method

This method is something of a compromise between the factor and calibration curve methods; it partially compensates for nonlinearity in the analytical curve and yet requires only two standard solutions. A preliminary calibration curve is measured (but need not be repeated regularly). A sample of solution is then measured, and its approximate concentration is estimated from the preliminary calibration curve. Next, the two standards closest to the estimated sample concentration are measured. The sample concentration may be determined mathematically or graphically. This method is especially useful for high-precision work; in this case, the standards are made up to be very close to the estimated sample concentration, and scale expansion is used to allow more precise readings.

b. Compensation for Multiplicative Interferences

(1) Matched-Matrix Standards

Multiplicative interferences can sometimes be reduced or eliminated by adjusting the matrix of the standards to match that of the unknown samples. At the very least, the solvent, pH, and temperature of the standards and samples should be the same. Additionally, if the other constituents of the sample are known, these may be added to the standards. For example, in the atomic absorption determination of trace metals in distilled liquors, the standards are made up to the proof of the samples by the addition of 95% ethanol. This avoids a multiplicative interference due to the different surface tension, viscosity, and density of alcoholic solutions. Matched-matrix standards are also common in sea water and blood serum analysis.

(2)   Analyte Addition Method

The most direct way to make an ideal matched-matrix standard is to dilute the standard material in some of the unknown sample solution instead of in pure solvent. In this way, the matrix of the sample and standards would be nearly identical. This is in fact the essence of the *analyte addition* method, in which standards are prepared by adding small volumes of concentrated standard solution to one or more aliquots of the sample solution. Of course, the concentrations of the resulting standard solutions cannot then be calculated a priori, because some analyte is already in the sample solution aliquots, but this is taken into consideration in the calculation step.

The simplest version of the analyte addition method is the *single-addition method*. The procedure is as follows. Two identical aliquots of the sample solution, each of volume V_x, are taken. To the first (labeled A) is added a small volume V_s of a standard analyte solution of concentration c_s. To the second (labeled B) is added the same volume V_s of the solvent. The analytical signals of A and B are measured and corrected for nonanalyte signals. The unknown sample concentration c_x is calculated:

$$c_x = \frac{S_B V_s c_s}{(S_A - S_B) V_x} \tag{24}$$

where S_A and S_B are the analytical signals (corrected for the blank) of solutions A and B, respectively. The analytical curve must be linear for this equation to be valid. If c_x can be estimated in advance, V_s and c_s are chosen so that S_A is roughly twice S_B on the average. It is best if V_s is made much less than V_x, and thus c_s is much greater than c_x, to avoid excess dilution of the sample matrix. If a separation or concentration step is used, the additions are best made first and carried through the entire procedure.

A more elaborate version is the *multiple-addition method*, in which a series of several sample aliquots are spiked with increasing amounts of analyte. The analytical signals of the resulting solutions are measured and plotted versus the amount of analyte added (analytical curve plot). The unknown concentration is determined by extrapolation of this plot, either graphically (see Figure 2.3) or via a least-squares fit. This procedure has the advantage of revealing nonlinearity in the analytical curve, which would not be obvious in the single-addition method. However, if the curve is not linear, a reliable extrapolation is difficult.

The analyte addition method is an extremely valuable tool, because it is the only standardization technique that compensates for multiplicative interferences when the sample matrix is unknown or varies from sample to sample. However, several conditions must be met for analyte addition to be valid:

1. The *analytical curve must be linear*, especially for the *single-addition method*.

2. The *added standard must be in the same chemical form* (as to oxidation state, complexation, etc.) as the analyte.

3. The *multiplicative interferences must be constant over the concentration range of the samples and addition standards*.

4. *Additive interferences must be absent or corrected for*, that is, the analytical signals used in the analyte addition calculation must be due only to the anayte. This is an important point; analyte addition corrects only for multiplicative interferences, not additive interferences.

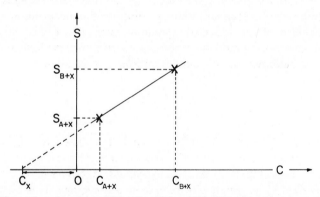

Fig. 2.3 Illustration of analyte addition (multiple additions) to determine concentration c_x of unknown.

(3) Reference Standards

If the stability of the instrumental sensitivity is the largest source of analytical error, a reference standard may be helpful. This is an element or compound that gives an analytical signal which is precisely measureable and is distinct from the signal from the analyte, for example, appears at a different wavelength. The reference standard is added in the same concentration to all samples, standards, and blanks, and the *ratios* of the analyte signals S_A to the reference standard signals S_R are measured, that is, S_A/S_R. Calibration curves are prepared by plotting the measured ratio S_A/S_R versus the concentration c_A of analyte in the standards. The advantage of this technique is that any instrumental variations that have the same relative effect on the analyte and on the reference standard will cancel out of the ratio. Reference standardization is used mainly in arc or spark emission spectrography, although it has found some limited use in flame atomic absorption and molecular luminescence methods.

(4)   Swamping Method

Occasionally a multiplicative interference can be overcome by adding to all sample standards and blanks an excess of the substance causing the interference. In this way, the relatively smaller and possibly variable amounts of interfering substance already in the samples, but not in the standards, have a proportionally smaller effect. This technique is obviously most suitable if the interference is an enhancement rather than a suppression. An example is ionization interference in atomic flame methods, in which the presence of an easily ionizable element suppresses the ionization of the analyte and thereby increases the analytical signal. This interference is conventionally overcome by adding an excess of any easily ionizable element to suppress the analyte ionization completely.

Dilution can also often be used, if the detection limit is sufficiently low, to dilute out certain types of interferences. For example, in flame spectrometric methods, various types of chemical or condensed-phase interferences can be reduced by dilution.

B.   RESOLVING POWER OF OPTICAL SPECTROSCOPIC METHODS

1.   SPECTRAL BANDWIDTH (also see Chapter 5)

The *spectral bandwidth* $\Delta\lambda_s$ of a monochromator is a measure of its ability to produce a monochromatic beam of radiation from a polychromatic source. For large slit widths, $\Delta\lambda_s$ is given by:

$$\Delta\lambda_s = \frac{d\lambda}{dx}w \qquad \text{(in nm)} \qquad (25)$$

where $d\lambda/dx$ is the monochromator's reciprocal linear dispersion (in nm mm^{-1}), and w is the slit width (in mm). If a monochromator is set to a wavelength of λ, radiation within $\lambda \pm \Delta\lambda_s/2$ will be passed. If a monochromatic line is scanned, a triangular "slit function" of base width $2\Delta\lambda_s$ will be recorded, neglecting diffraction.

2. RESOLUTION AND RESOLVING POWER

The *resolution* $\Delta\lambda$ of a monochromator is the minimum separation between two equally intense monochromatic lines which can just barely be resolved. For complete (baseline) separation, the bases of the slit functions of the two lines may touch but not overlap; thus in this case the resolution $\Delta\lambda$ is equal to $2\Delta\lambda_s$. Slightly less than complete resolution may be tolerated in some cases, but the separation of the lines must in any case be greater than one spectral bandwidth.

The *resolving power R* is defined as:

$$R = \frac{\bar{\lambda}}{\Delta\lambda} \tag{26}$$

where $\Delta\lambda$ is the resolution measured at a mean wavelength $\bar{\lambda}$. Resolving powers of the order of 10^3 to 10^4 are typical of bench-top UV-visible monochromators. If $\Delta\lambda$ is replaced by $\Delta\lambda_0$, the limiting resolution as determined (see Section B.3) by fundamentally limiting processes, as diffraction, the resolving power is the theoretical value R_0 and is:

$$R_0 = \frac{\bar{\lambda}}{\Delta\lambda_0} \tag{27}$$

If it is assumed that two spectral components with a separation of $\Delta\lambda_s$ can just be resolved, the practical resolving power R_p is:

$$R_p = \frac{\bar{\lambda}}{\Delta\lambda_s} \tag{28}$$

3. OPTICAL LIMITATION

It must be recognized that the spectral bandwidth of a monochromator can not be reduced indefinitely by reducing the slit width w. In fact, the equation given above for $\Delta\lambda_s$ is not valid at very small slit widths because of certain optical considerations which limit the actual spectral bandwidth to some nonzero minimum value $\Delta\lambda_l$ as the slit width w approaches zero. These optical limitations include diffraction, optical aberration, coma, astigmatism, and instrument maladjustment, including poor focus adjustment and nonparallelism of slit jaws. Only the first of these, diffraction, is a fundamental and predictable limitation; the others depend on the design, quality, and adjustment of the monochromator. The diffraction term itself limits $\Delta\lambda_l$ to:

$$\Delta\lambda_l \gtrsim \frac{d\lambda}{dx}\left(\frac{\lambda f_{co}}{a}\right) \tag{29}$$

where f_{co} is the collimeter focal length, and a is the effective aperture of the monochromator. In practice, $\Delta\lambda_l$ is larger than this (perhaps twice), because of the optical imperfections mentioned.

A useful test of a monochromator's characteristics is a plot of the experimentally measured spectral bandwidth $\Delta\lambda_s$ as a function of slit width w. (The spectral bandwidth is equal to one-half the base width of the slit function obtained by scanning an isolated monochromatic line, for example, from a mercury pen lamp assuming a medium- or low-resolution bench-top monochromator is used.) The diffraction term may be calculated as indicated above and subtracted (quadratically) from $d\lambda/dx$, leaving the optical imperfection term.

4. CHOICE OF OPTICAL MONOCHROMATOR SLIT WIDTH

The choice of slit width w (assuming entrance and exit slit widths are the same) is basically a trade-off between intensity and resolution. For scanning molecular spectra, the slit width is normally adjusted so that the spectral bandpass is less than the half-width of the narrowest band or spectral feature to be recorded. For atomic line spectra, the lines recorded (with bench-top monochromators) are actually slit functions with half-width equal to one spectral bandwidth. Thus the choice of slit width depends on the separation of the spectral lines (isolation of line or choice) to be resolved.

The effect* of slit width on intensity (signal level) must also be considered. The measured intensity (signal level) of a monochromatic line is directly proportional to the slit width w; while the measured intensity (signal level) of a continum source is proportional to the *square* of the slit width. Thus, in measuring a line emission superimposed on a continuum background, a narrower slit width results in a higher line-to-continuum-intensity ratio, which is especially desirable if the background is noisy. However, an excessively narrow slit width reduces the overall level of the measured signals to the point where detector noise becomes excessive. The optimum slit width often maximizes the signal-to-total-noise ratio, assuming the absence of spectral interferences or sufficiently large variation of the slit widths is possible.

C.   ELECTRONIC MEASUREMENT METHODS

1.   MODULATION OF SIGNALS

Spectroscopic methods normally involve the measurement of the output signal of the detector as a function of wavelength, time, analyte concentra-

---

*This entire discussion is premised on the assumption of a bench-top medium- or low-resolution monochromator. If the resolution is sufficient to allow partial or complete determination of the spectral line profile (atomic line), it is necessary to convolute the spectrometer slit function and the line profile function to obtain the exact influence of the slit width.

tion, and so on. The most widely used detectors are electrical transducers which convert photon flux into an analog (current, voltage, resistance, etc.) or digital (count rate) signal. We deal with analog signals first and comment on digital signals later.

In their simplest form, many spectroscopic experiments produce DC signals which are amplified by DC amplifiers and displayed on meters, recorders, or digital voltmeters (which normally require DC signals). Simple DC systems of this type are fairly common in some areas of spectroscopy (e.g., molecular fluorimetry). However, high-gain DC amplifiers are subject to significant drift and offset errors. Furthermore, the presence of $1/f$ noise in the signal seriously restricts the extent to which the signal-to-noise ratio can be improved by simple low-pass filtering. For these reasons, it is often desirable to *modulate* (chop) the signal in order to transform the signal information from direct current up to some AC frequency high enough to avoid the drift and $1/f$ noise problems. The resulting AC signal is then amplified by an AC amplifier and converted back into direct current by means of a *demodulator* or *rectifier*. (The conversion to direct current is necessary because all commonly used readout devices require DC signals.)

Modulation can be performed in several different ways. The simplest way is to chop, that is, periodically interrupt, the DC electrical output signal from the detector by means of an electromagnetic relay or switch. This results in a square wave signal whose peak-to-peak amplitude is equal to the DC signal voltage. The square wave signal may then be amplified by an AC amplifier. However, drift and DC errors generated in the detector (e.g., dark current) would also be chopped, hence would not be rejected by the AC amplifier. For this reason, it is much better to *chop the light beam striking the detector* by means of a rotating sectored disk or similar device. The position of the chopper in the optical path of the instrument may be important. In atomic absorption and fluorescence spectrometry, the light is preferably chopped between the source and the flame, so that the steady thermal emission from the flame is not modulated and is rejected by the AC electronics. Other types of modulation occasionally used in spectroscopic instrumentation include *wavelength modulation* and *sample modulation*. In wavelength modulation, the wavelength of the monochromator is caused to vary rapidly back and forth over a small spectral interval. This modulation technique is used when the information of interest is the *difference* between the spectral intensities at two different wavelengths. For example, wavelength modulation has been used to correct automatically for continuum background emission in flame emission spectrometry. In this case, mechanical chopping is not suitable because both the line and continuum emission would be chopped and therefore could not be separated by the detection system. Wavelength modulation has also been used extensively in derivative spectroscopy [see Section A.2.c(6)].

Sample modulation includes several related techniques in which modulation of the signal is produced by changing the measured sample in some way. For example, in a flame spectrometer, the flow rate of the sample solution can be modulated or the flow of the solution can be switched off and on and alternated with the flow of another (e.g., blank) solution. In these cases, the desired information is the *difference in the optical signals produced by the two solutions. In condensed-phase spectrometry, a type of sample modulation may be obtained by switching the optical light path between two samples, one usually the blank. This is the basic idea behind a double (dual)-beam absorption spectrophotometer.*

The best modulation method modulates only the information of interest, leaving unmodulated as many of the undesirable offsets, drifts, and noises as possible. In some cases, two (double) modulation methods may be used simultaneously. This is useful if there are two or more different interfering signals which cannot all be eliminated by a single modulation.

In spite of the obvious advantages of a properly applied modulation system, there is one fundamental disadvantage (besides the cost and extra complexity) which should be recognized. This is the *modulation signal power loss.* By its nature, the modulation process must periodically attenuate or cut off the physical quantity being measured (i.e., light intensity). Thus the average signal power at the detector is less than that for an unmodulated DC system. This effect can become very significant in a signal-limited experiment in which shot noise is the predominant noise. In such a case, the use of any modulation system actually reduces the signal-to-noise ratio. The choice between modulation and nonmodulation must be made on the basis of the overall quality of the analytical signal. The benefits of eliminating the unmodulated interfering signals must outweigh the disadvantage of decreased signal-to-noise ratio due to shot noise; otherwise an unmodulated DC system is better overall.

In determining the expected noise-reduction capability of a modulation system, it is necessary to distinguish between additive and multiplicative noises and drifts. An *additive* noise or drift adds to the DC level of the signal but does not change its AC amplitude (see Figure 2.4C). Examples include changes in the offset of an AC amplifier, changes in phototube dark current, variations in the background continuum intensity in a wavelength-modulated atomic fluorescence or emission measurement, and variations in the thermal emission of the atomizer in an atomic fluorescence measurement with a chopped excitation source. In each of these cases, only the DC component of the total signal is affected; the AC component is unchanged. Thus a modulation system with AC detection is

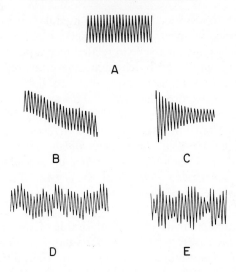

Fig. 2.4 Representation of a sinusoidal AC signal waveform with (*A*) no noise or drift, (*B*) additive drift, (*C*) multiplicative drift, (*D*) additive noise, (*E*) multiplicative noise. The signal-to-noise ratios of noisy signals *D* and *E* are roughly the same, yet the signal in *D* can be measured more precisely by an AC electronics system, because the noise is additive.

capable of rejecting additive noises (including $1/f$ noises) and drift. *Multiplicative noises*, however, affect the *amplitude* of the AC signal (see Figure 2.4*C*). Examples include gain variation noise or drift, variations in photodetector sensitivity (e.g., due to an unstable high-voltage power supply), source intensity fluctuations in any type of light intensity or wavelength-modulated absorption or luminescence spectrometric measurements, and variations in solution flow rate or nebulization in flame atomic emission measurements. Any *noises, drifts, or other variations that are added to the signal before the modulation and are therefore modulated along with the signal, or any factors that affect the overall system gain or sensitivity, are multiplicative noises. An AC system cannot reject multiplicative noises any more effectively than a DC system with equivalent noise bandwidth; both systems must rely solely on their low-pass filters to reject such noises.* Multiplicative drift and $1/f$ noise therefore cannot be effectively filtered out by a low-pass filter and thus are still a problem in AC detection systems.

2. ELECTRONIC SIGNAL PROCESSING METHODS

a. Basis of Comparison

(1) Bandwidth

The most important characteristic of an electronic signal processing system with respect to signal-to-noise ratio enhancement is its equivalent noise (frequency response) bandwidth Δf_n. If a white noise signal of amplitude N (in rms V Hz^{-1}) is applied to the input of a unit-gain signal processing instrument, the noise amplitude at the output will be $N\Delta f_n$ (in rms V) for a continuous analog readout; or, for a discrete, sequential (digital) readout, the standard deviation of the output readings will be $N\Delta f_n$ (V). Noise bandwidth can generally be reduced at the expense of increased measurement time and, for this reason, it is best to compare different signal processing systems at the same measurement time. For example, if a DC integrator is being compared to a low-pass filter, the integration time should equal the filter's response time t_r; this puts them on an equal basis as far as measurement time is concerned and makes the comparison more realistic. Of course, the advantage of a particular system may be that it *allows long measurement times to be used without significant drift errors*. Thus both the *maximum practical measurement time* and the noise bandwidth at a given measurement time are important characteristics of a signal processing instrument.

(2) Input and Output Impedance

The input impedance requirements of a signal processing system depend on whether the detector output voltage or current is to be measured. For a voltage measurement, the input impedance of a given stage of the signal processing should be *much larger* than the output impedance of the previous stage. If not, a *loading error* will be introduced—the measured voltage being lower than the actual open circuit voltage. However, for a current measurement, the input impedance of the signal processing system should be *much less* than the output impedance of the current source. Otherwise, a similar loading error will result. The output of a vacuum phototube or photomultiplier tube is basically a current signal and should ideally be measured with a *low input impedance current measuring device* (a nanoammeter or picoammeter). In this particular case, however, the output impedance is so high that measuring devices of relatively high input impedance are satisfactory. Thus it is possible to measure phototube output currents by connecting the anode to ground through a high-value resistor and measuring the resulting voltage with a very high input imped-ance (an electrometer-type) voltmeter. However, a low input impedance current-measuring device has some advantages, particularly if *fast response times* are important.

(3) Range of Signals

It is obviously desirable that the chosen signal processing system be able to handle the expected range of signal levels to be encountered. Most commercial systems have calibrated range switching to facilitate measurement of signals of varying amplitude. If measurements of a widely varying signal must be taken unattended, then an autoranging system is desirable. Autoranging digital voltmeters, with digital outputs for printers or other recording devices, are available commercially. Logrithmic amplifiers may also be used to "compress" data of very wide dynamic range.

b. Types of Electronic Measurement Systems

(1) DC Amplifier and Low-Pass Filter

A DC amplifier, followed by or including a low-pass filter, is the simplest signal processing system. The noise bandwidth of a single-section RC low-pass filter is given by:

$$\Delta f_n = \frac{1}{4\tau} \tag{30}$$

where τ is the *time constant* and is equal to RC. Thus, in principle, the noise bandwidth may be reduced without limit by increasing the time constant τ. With modern electronic circuitry, time constants of 100 s or even longer are obtainable. However, there are two reasons why extremely long time constants may not be profitable. First, a low-pass filter attenuates only noise frequencies *above* the noise bandwidth. Offsets, drift, $1/f$ noise, and other low-frequency noises are therefore not attenuated significantly by low-pass filtering. Second, the 1% response time of a low-pass filter is equal to 4.6 τ and will therefore increase if τ is increased in an effort to reduce Δf_n. The significance of response time is discussed in Section A.2.d.

In spite of these shortcomings, the DC amplifier low-pass filter system may be the best choice, particularly in purely white noise limited situations. Furthermore, advances in electronics have made available highly stable DC amplifiers, for example, operational amplifiers with insignificant amounts of offset, drift, and noise.

(2) DC Integrator

An alternative to the use of a low-pass filter for reducing noise at the output of a DC amplifier is a DC integrator. This is an electronic circuit which determines the time integral of a DC electrical signal. If an amplified photosignal is gated into an integrator for a fixed integration time t_i, the integrator output at the end of time t_i will be proportional to the *average* value of the input signal over the interval t_i. Integration times from microseconds to hundreds of seconds are obtainable with presently available instrumentation. The longer the integration time, the larger the signal-

to-noise ratio, because random noise tends to average out to zero over a long period of time.

The effective noise bandwidth of a DC integrator is $(2t_i)^{-1}$. The *rms noise output of an integrator is the standard deviation of a series of sequential averages obtained from a given signal input*. Thus, a 1.0 s integration is equivalent (in noise-reduction capability) to a low-pass filter with a 0.5 s time constant; both have equivalent noise bandwidths of 0.5 Hz. However, the difference is that the effective response time of the integrator is less than half that of the filter. The response time of a 0.5 s time constant filter is 2.3 s, while the effective response time of an integrator is just its integration time, 1.0 s in this case. Thus we can say that the integrator allows a faster data rate than a low-pass filter. A disadvantage of the integrator is that its inherently discontinuous output signal is less suitable for recording spectra and other continuous variables.

(3)   AC Amplifiers

An AC amplifier system is required when a modulation technique is utilized. The important sections of an AC amplifier instrumentation system are (1) the AC amplifier proper, (2) the detector, which converts the AC amplifier output into direct current, and (3) the postdetector amplifier and low-pass filter stages. The detector and output stages are necessary because most conventional readout devices (e.g., recorders, digital readouts) require DC signals.

(A)   THE AMPLIFIER PROPER.   Most of the gain of an AC system is in the predetector amplifier stages. Either wide band or tuned AC amplification is possible, and each has its particular advantages and limitations. A wide band system, with a frequency response extending from well below to well above the modulation frequency, is the simplest to construct and operate. However, it has the disadvantage that the noise amplitude at its output may be quite large, because of the wide noise bandwidth, and the detector circuitry may be over loaded. This difficulty is avoided in a tuned (narrow band) amplifier, whose frequency response bandpass encompases only a narrow frequency interval centered on the signal frequency. All noise frequency components falling outside the passband are rejected, and consequently the noise input to the detector is much less than for a wide band system. A disadvantage of the tuned amplifier is that the frequency of tuning has to be readjusted if the modulation frequency drifts. Nevertheless, tuned amplifiers are generally considered more useful.

(B)   THE DETECTOR CIRCUIT.   The purpose of the detector circuit is to convert the amplified AC signal into direct current. There are basically two kinds, asynchronous and synchronous. An *asynchronous detector*, the simpler and less expensive, is essentially a type of full-wave recifier; it passes the AC signal unchanged whenever the instantaneous polarity of the signal is positive and inverts (reverses the polarity of) the signal whenever it is negative. Thus the output of the detector is always positive in polarity, and the average DC value of this signal is proportional to the amplitude of the AC input to the detector. The DC component is extracted by filtering the detector output through a low-pass filter. The asynchronous detector is called *asynchronous* because it is not synchronized in any way with the frequency and phase of the AC signal. Rather, it simply inverts any portions of the AC signal that are negative, and therein lies the fundamental disadvantage of this kind of detector. For example, if the signal is noisy, the noise will *also* be rectified into direct current. In fact, a pure noise input, even with no real signal component at the modulation frequency, still produces a nonzero detector output because of the rectification of the noise (this is called the *noise offset* and is illustrated in Figure 2.5). In principle, the noise offset could be zeroed out electrically, but the real problem is that the noise, and therefore the noise offset, is liable to drift and change unpredictably. A tuned AC amplifier, as shown in Figure 2.6C, greatly reduces the noise offset by reducing the noise bandwidth, but there is still a small offset due to the small amount of noise falling within the amplifier passband. Even this becomes significant, however, when very small signals buried in relatively large amounts of random noise are being measured.

The noise offset problem is completely eliminated by the *synchronous detector* (see Figure 2.7). Like the asynchronous detector, the synchronous detector exists in one of two states, either passing the signal unchanged (+) or inverting its polarity (−). However, its state at any instant is determined not by the polarity of the detector input signal, but rather by the *polarity of* a *reference waveform*, which has the same frequency as, and a fixed-phase relationship to, the actual signal waveform. The reference waveform is obtained directly from the modulating device and therefore has a very high signal-to-noise ratio at all times, no matter how low the signal-to-noise ratio of the measured signal is. Thus the synchronous detector is not "confused" by the presence of large amounts of noise at its input. The phase relationship between the signal and the reference waveforms is adjusted electronically (in an operation called *phasing* the detector) so that the detector will be in the noninverting state (i.e., the detector input signal will be passed unchanged) whenever the modulation phase is such as to cause positive excursions in the measured AC signal, and in the inverting state when the modulation causes negative signal excursions. In this way, only the desired AC signal component at the modulation

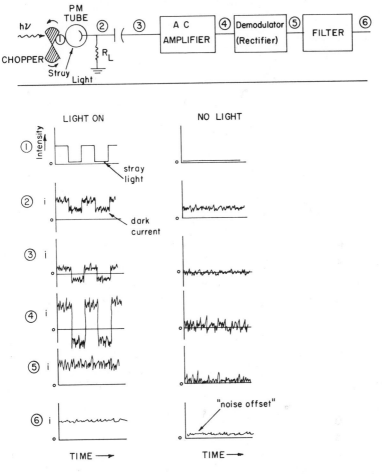

Fig. 2.5 Schematic diagram of a wide band asynchronous system. Signals at numbered points in circuits: (1) Light signal at photomultiplier tube, (2) photomultiplier tube current, (3) amplifier input (AC component of photomultiplier tube current), (4) amplifier output (amplified AC component of photomultiplier tube current), (5) demodulator output (rectified), (6) output to readout meter or recorder.

frequency is effectively full-wave-rectified into a DC signal. Random noise or any other extraneous signals not synchronized with the reference waveform do not produce a net DC output signal but produce only an AC (noise) component which is filtered out by the low-pass filter following the detector (see Figure 2.7). Thus the detector is synchronized or "locked in" to the frequency and phase of the signal. AC amplifier systems with synchronous detectors are popularly referred to as *lock-in* amplifiers. Since

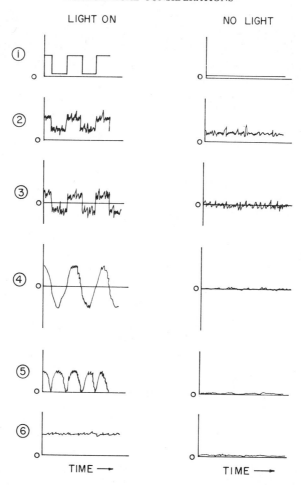

Fig. 2.6 Schematic diagram of a tuned AC asynchronous system (see top of Figure 2.5). Signals at numbered points in circuits: (1) Same as in Figure 2.5, (2) Same as in Figure 2.5, (3) Same as in Figure 2.5, (4) same as Figure 2.5 (amplifier output is approximately a sine wave), (5) same as Figure 2.5 (full-wave rectified sine wave), (6) same as Figure 2.5 (noise offset is much lower than in Figure 2.5).

a lock-in system has no noise offset, a wide-band "front-end" amplifier can be used. A wide-band amplifier is easier to use than a tuned amplifier because it requires no tuning, and also because it avoids the phase shift introduced by a tuned amplifier when the signal frequency changes slightly. Thus a wide band lock-in system can more easily "track" changes in the modulation frequency. However, a disadvantage of a wide band system is that it responds to some extent to the odd harmonics of the signal

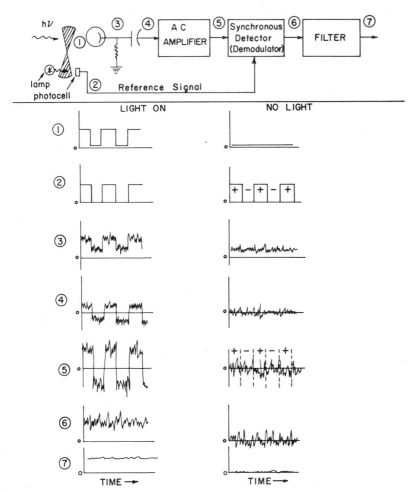

Fig. 2.7 Schematic diagram of synchronous (lock-in) system. Signals at numbered points in circuits: (1) Same as (1) in Figure 2.5, (2) reference signal ($-$, invert; $+$, do not invert), (3) same as (2) in Figure 2.5, (4) same as (3) in Figure 2.5, (5) same as (4) in Figure 2.5, (6) same as (5) in Figure 2.5 (periodically inverting random noise does not affect average value), (7) same as (6) in Figure 2.5 (no noise offset here)

frequency. This can occasionally be a problem with complex signal waveforms containing significant higher harmonics.

(C) THE POSTDETECTOR LOW-PASS FILTER AND AMPLIFIER. The output signal from the detector circuit is further amplified and low-pass-filtered by the postdetector stages. Any DC errors introduced in these stages would be detrimental but are usually insignificant because of the relatively high signal level at this point in the circuit. Commercial instruments have several switch-selected low-pass time constants and often allow a choice of attenuation rates of -6 or -12 db per octave. The -12 db per octave attenuation rate gives greater rejection of high-frequency noise at the expense of longer response time. The effective overall bandwidth of a lock-in amplifier system is equal to that of the low-pass filter alone, even if the front end is wide band. That is, a lock-in system behaves very much like a sharply tuned filter (except for the harmonic response problem mentioned above). For example, a low-pass filter time constant (τ) of 10 s is equivalent to a low-pass bandwidth ($1/2\pi\tau$) of 0.015 Hz. A simple tuned AC amplifier with a bandwidth this small would be very difficult to construct and to keep tuned. Thus, the *lock-in system is capable of much greater overall frequency selectivity than any practical tuned AC amplifier system.*

(4) The Boxcar Integrator

A boxcar integrator is basically an electronically gated DC amplifier followed by a low-pass filter or integrator. The amplifier is gated "on" for an adjustable interval of time (the *gate width*) each time a trigger pulse is received. A *delay time* between the trigger pulse and the gate may also be selected. The boxcar system is used when it is desired to measure a small "time slice" of a repetitive (repeatable but not necessarily periodic) transient signal. The gate width and delay time are adjusted so that the gate is open only during the interval of interest. Alternatively, the trigger can be obtained by an internal pulse generator which also provides a pulse to initiate each repetition of the signal transient.

A boxcar system can be used in either the fixed delay mode or the sweep delay mode. In the fixed delay mode, the delay time is constant. This mode is used when *only one* particular time slice of the signal is of interest. For example, when making flame atomic fluorescence measurements with a pulsed hollow cathode lamp excitation source, the fluorescence is observed only in short pulses when the lamp is pulsed "on." Between the lamp pulses, only flame background emission and phototube dark current are observed. Thus it is convenient to use a boxcar system to pick out and measure the fluorescence pulses and reject the remainder of the signal. In this case, the trigger can be obtained from the hollow cathode lamp pulser electronics.

In the sweep delay mode, the delay time is set initially at a low value and is slowly increased during the experiment, thereby scanning the gate interval through the period of the repetitive signal. The output of the gated amplifier is low-pass-filtered and recorded as a function of the delay time, resulting in a plot of the signal waveform. This technique allows fast, repetitive waveforms to be extracted from a great deal of random noise without undue distortion of the signal waveform. The reason is that the *effective time resolution of the system is determined by the gate width*, whereas the *effective noise bandwidth is determined by the time constant of the low filter*. Thus it is possible to record a fast, noisy waveform by using a short gate width (to obtain the required time resolution) and a long, slow, delay time sweep with a large output time constant (to obtain the required noise reduction capability). The only requirement is that the signal be repetitive and that there is a way to synchronize the signal repetitions with the boxcar gate.

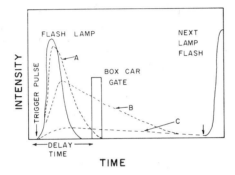

Fig. 2.8 Schematic diagram of the time-dependent processes with a boxcar integrator in time-resolved spectroscopy. With the delay time and gate width shown, the system mostly responds to species B with little interference from species A or C. Dashed lines represent decaying spectroscopic signals after the source is terminated.

A good example of the application of a boxcar system is in time-resolved spectrometry, for example, pulsed source, time-resolved phosphorimetry. The waveforms are illustrated in Figure 2.8. A high-speed flash lamp produces short, repetitive pulses of light which excite the phosphorescence of a sample containing a mixture of three phosphors (A, B, and C) of progressively longer decay times. The boxcar gate width and delay time are adjusted to optimize the measurement for any one of the three components. For example, the conditions shown in Figure 2.8 are optimized for B. Alternatively, the sweep delay mode can be used to record the overall decay characteristics of the mixture. A real advantage of the boxcar system for measuring phosphorescence decays, compared to simply observing the decay in real time on an oscilloscope, is that fast decays of low intensity can be measured with much greater precision (lower noise) and much greater accuracy (lower systematic errors resulting from insufficiently fast system response time).

A general disadvantage of a boxcar system when it is used to recover a waveform in the sweep delay mode is that all the waveform information falling outside the gate interval is wasted. Thus the overall efficiency of the system becomes very low when short gate widths and long sweep times are used, as is required when very fast, noisy waveforms are measured. In such cases, it is very helpful if the repetition rate of the waveform is as high as possible so as to avoid excessively long measurement times.

The Fourier of Time Resolution is Frequency (phase) resolution. In frequency-resolved spectroscopy the sample (for example, a mixture of phosphors) is excited with a modulated source of excitation to produce radiation signals. As the frequency of modulation f increases, the slower-decaying (longer-lived) species exhibits a phase shift and a decreased amplitude with respect to the exciting (modulated) source. As a result, by means of either frequency or phase variation of the measurement system (lock-in or boxcar), the slower-decaying species can be effectively "phased out" (the signal but not its noise), and only the faster-decaying species can be measured. A mixture with more than two decaying components causes complications.

A schematic comparison of the time-resolved and frequency-resolved methods for two decaying species (A and B) in a mixture is given in Figure 2.9.

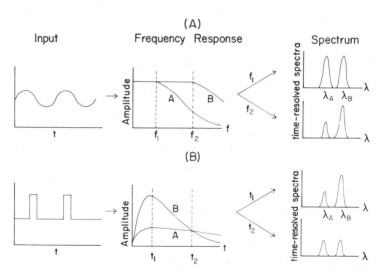

Fig. 2.9 Comparison of frequency-resolved (A) and time-resolved (B) spectroscopy. Input = input or exciting waveform; Frequency response = detector output as function of frequency (A) or time (B); spectrum = intensity versus wavelength plots for frequency (A) and time (B) resolved measurements

(5) Multichannel Averagers

The poor efficiency of boxcar systems in the measurement of fast, noisy waveforms is overcome in a *multichannel signal averager*. This system may be viewed as an array of sequentially gated boxcar integrators, each one of which is assigned to a unique time segment of the signal waveform and is gated "on" only during that time segment. Each time segment is averaged over many repetitions of the signal waveform, and thus none of the signal information is wasted. Random noises are averaged toward zero as the number of repetitions is increased. Multichannel averager systems are very useful in luminescence decay studies (using a repetitively pulsed flash lamp). They can also be used for measuring spectra although, in this case, the spectra must be repetitively scanned and there must be a way to synchronize the spectrometer scan to the multichannel averager, for example, in an NMR spectrometer. The specific advantage of a multichannel averager for measurements of noisy spectra, as compared to simply scanning the spectrum very slowly with a long time constant, is that drift and $1/f$ noise have much less effect. Thus *it is possible to obtain a good quality spectrum from an unstable light source*, whereas the conventional slow scanning method produces a spectrum full of peaks and dips due to intensity fluctuations rather than to geniune spectral features.

Both digital and analog multichannel signal averagers are available commercially. Analog systems, using capacitors as storage elements, are produced with 50 and 100 channels. Digital systems are more complex and costly, and of course require conversion of analog signals to digital, but are better when a large number of channels (up to 4000) or long storage times are necessary. For photon counting systems, digital signal averages are ideal, because the data already in digital form can be fed directly from the counter to the multichannel memory. General-purpose minicomputers can also be programmed for multichannel signal averaging. These offer outstanding versatility but are not as fast as the hardware signal averagers (called *waveform educators* by one manufacturer).

An overall feature of all signal processing systems for enhancing the signal-to-noise ratio is that *they all trade measurement time for a reduction in noise*. Some may be more efficient than others or less susceptible to drift or $1/f$ noise, but they all ultimately depend on averaging out random noise over a more-or-less long time period. The basic assumption is that the *noise is not correlated with the signal*.

(6) Pulse Systems

(i) Photon Counting.   Photon counting is a very sensitive method for low-level light intensity measurements, which is being increasingly used in analytical spectrometry, particularly in Raman and luminescence spec-

trometry. Photon counting is possible because the multiplication factor M of modern multiplier phototubes is sufficiently high (usually 10^5 to 10^8) so that each photoelection emitted by the photocathode produces a· discernible pulse in the anode current. These pulses can be amplified and counted by conventional electronic counters or count rate meters. The average count rate $\bar{r}$ (in Hz) is proportional to the light flux Φ (in W) on the photocathode:

$$\bar{r} = \frac{i_a}{Me} = \frac{\gamma_a \Phi}{Me} \cong \frac{\gamma_c \Phi}{e} \tag{31}$$

where γ_a and γ_c are the anode and cathode sensitivity, respectively (in A W^{-1}), and e is the charge on the electron (in C). In a typical analytical spectroscopic measurement, average count rates of 10^2 to 10^6 can be expected. The mean amplitude (height) of the pulses is given approximately by:

$$\bar{V}_p \cong \frac{Me}{C_s} \tag{32}$$

where $\bar{V}_p$ is the mean pulse height (in V), and C_s is the total stray capacitance (in F) from the anode to the signal ground (the stray capacitance is due mainly to the capacitance of the shielded cable connecting the phototube to the input of the preamplifier or counter). The temporal half-width of the pulses (in seconds) is given approximately by:

$$t_p \cong 0.7 R_L C_s \tag{33}$$

where R_L is the phototube load resistance (in Ω). In equations 29 and 30, it is assumed that $R_L C_s$ is much larger than the transit time t_t spread of the phototube (about 2 ns for a typical 1P28). This assumption is almost always valid. The maximum average count rate r_{max} which can be measured before pulse overlap becomes significant is:

$$r_{max} \cong \frac{1}{20 R_L C_s} \tag{34}$$

As an example of the above equations, consider a typical photometric system using a photomultiplier with a multiplication factor M of 10^6, a load resistance R_L of 1 kΩ in a circuit with a stray capacitance C_s of 100 pF. For this system, the mean pulse height V_p is approximately 1.6 mV, the pulse width t_p is about 70 ns (much larger than the typical transit time spread), and the maximum count rate r_{max} is roughly 500 kHz. As viewed on an oscilloscope, the pulses have very fast rise times followed by a relatively longer exponential "tail" defined by the time constant $R_L C_s$.

A block diagram of a typical photon counting system is shown in Figure 2.10. The amplifier must have a wide bandwidth Δf sufficient to provide output pulses compatible with the discriminator. The discriminator is a pulse height sensing circuit which passes only those pulses whose pulse height falls within a certain range (the "window"). All pulses with heights below the *lower discriminator level*, or above the *upper discriminator level*, are rejected. The discriminator levels are adjusted to pass photoelectron pulses and reject dynode emission pulses, cosmic radiation scintillation pulses, and any other pulses that do not have the correct voltage height. Unfortunately, not all photoelectron pulses are exactly the same height, because of the random fluctuations in M from pulse to pulse. Thus the pulse height window must be opened up enough to let through the majority of photoelectron pulses without accepting too many noise pulses. The optimum conditions must usually be found experimentally.

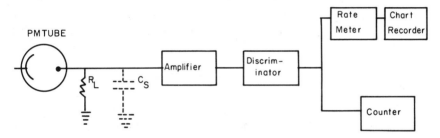

Fig. 2.10 Diagram of a photon counting system.

Any type of electronic counter capable of count rates of 10 MHz or higher can be used. A rate meter (frequency-to-voltage converter) can also be used for recorder readout. Commercial photon counting systems are now available, and it is also possible to construct a system from standard nuclear instrumentation modules.

The advantages of photon counting, compared to a simple (analog) DC amplifier system, include: lower electronic drift, therefore allowing longer counting (integration) times to be used; DC leakage and nonphotocathodic pulses are rejected; and direct digital readout is obtained without need for an analog-to-digital converter. This last advantage makes the design of a digital multichannel signal averager for photon counting particularly simple.

It is instructive to compare digital (photon counting) measurement systems to the analog (current measuring) systems discussed above. A photon counting system with an analog ratemeter output is essentially the *digital equivalent* of a simple DC system with low-pass filter. The overall system noise bandwidth, in either case, is determined by the output time constant τ. The standard deviation of a series of successive counts is equivalent to, and ultimately the source of, the rms noise in an analog measurement. A photon counting system with a simple totalizing counter readout is *equivalent* to a DC integrator system; the counting time is the integrating time. All the comments made in Section C.2(b)(1) and (2) concerning DC low-pass filters and integrators also apply to the equivalent photon counting system. The use of modulation methods and AC detection is less common in photon counting systems. The photon counting equivalent of a lock-in amplifier is an *up-down (bidirectional) counter*, whose counting direction is determined by a reference signal. That is, the photon pulses are counted *up* (forward) for one half-cycle and down (backward) during the next half-cycle of the reference waveform. This is quite analogous to the synchronous detector [Section C.2.b(3)]. A boxcar detection system is especially easy in photon counting; the gate circuit can be a simple "and" gate, an elementary digital integrated circuit module. As already mentioned, a multichannel signal averager would be especially straightforward to design for a photon counting experiment; parallel binary output words from the counter could be accumulated directly in a digital memory.

(ii)   Other Pulse Measurements.   Pulse measurements are encountered in other spectroscopic experiments using conventional analog photocurrent measurements. These include: pulsed source, time-resolved spectroscopy [Section C.2.b(4)]; atomic fluorescence or absorption using pulsed hollow cathode lamps; and nonflame atomization and transient gas evolution techniques in atomic absorption and fluorescence. In each case, it is required that the electrical bandwidth of the measurement system (including the recorder, for nonflame and transient gas evolution techniques) be sufficiently high to avoid excessive distortion of the pulse height and shape. However, excessively high electrical bandwidth increases random noises and may reduce measurement precision.

CHEMICAL ASPECTS
OF ELEMENTAL ANALYSIS

T. C. O'HAVER

Department of Chemistry
University of Maryland
College Park, Maryland

A. SAMPLE PREPARATION

Many spectroscopic analytical methods require samples in the form of solutions and, whether required or not, solutions have several general advantages over solid samples. In particular, it is much easier to prepare, mix, aliquot, and dilute samples and standards if they are in the form of solutions. It is the purpose of this section to describe the ways in which various types of samples may be put into solution and to discuss the errors that may be introduced by the solution procedure.

1. METALS (1, 2)

Most metals are easy to dissolve in mineral acids. HNO_3 and HCl, either singly or in combination, are commonly used. If large quantities of tungsten are present, the use of H_3PO_4 helps keep the tungsten in solution. HF may be added to destroy any residue of silica. Zirconium and its alloys can be dissolved in HF and HNO_3. Some alloys may contain carbide, nitride, oxide, or intermetallic inclusions which are resistant to simple acid treatment. The use of fusion or Teflon bomb techniques (see Section 2) may be required in such cases.

2. MINERALS AND RELATED GEOCHEMICAL MATERIALS (1–4)

Fusion, often in Na_2O_2 or Na_2CO_3, is the classic way of solubilizing geochemical materials, especially if the analysis of silicon is desired. A more recent development is the lithium metaborate ($LiBO_2$) fusion introduced by Suhr and Ingamells (1). In this procedure, 0.2 g samples are fused with 1 g $LiBO_2$ and poured while still molten into 100 ml 3% HNO_3. Slavin (2) used this procedure to obtain clear solutions of cement, silica brick, and alumina refractories.

Fusion techniques retain silicon in solution and are thus useful if silicon is to be determined. The major disadvantages are the time and special equipment required (platinum ware, muffle furnace), and the large excess of flux added to the sample. Obviously, the major cation in the flux cannot be determined and, furthermore, trace impurities in the flux material may lead to excessively high blanks.

HF, in combination with other mineral acids, is often used to dissolve silicaceous material. The use of HF drives off silicon as the tetrafluoride and thus is suitable only if silicon is not to be measured. Teflon containers are recommended. If large amounts of calcium are present, CaF_2 may precipitate. Also, HF does not dissolve zircon or carbon residue.

Bernas (4) has developed a dissolution technique based on a pressure decomposition vessel, commonly called a Teflon bomb. The sample is placed in a specially designed Teflon container, treated with aqua regia and HF, tightly sealed in a stainless steel bomb, and heated at 110°C for 30–40 min. After cooling, H_3BO_3 is added to tie up excess F^- and keep the silicon in solution. This technique has been found useful with glasses, nitrides, and other refractory material that is difficult to dissolve by other methods. The advantages are that excess alkali salts are not added and that silicon is retained.

Cement samples can be treated with 4 N HCl. Silica is not dissolved. Glass can be dissolved in HF in a Teflon beaker, evaporated to dryness, and the residue taken up in HCl. H_3BO_3 may be added to aid in the elimination of F^-. Coal ash is basically a silicate material; dissolution in HF, sometimes with $HClO_4$, has been found useful if silica is not determined.

3.   ORGANIC MATTER (5, 6)

Samples containing large amounts of organic matter represent a large and important class of materials for which the sample preparation steps are often long and involved and sometimes a major source of analytical error. This includes solid biological samples such as plant and animal tissue, man-made organic materials such as plastics, and liquid samples such as blood, urine, oils, and liquid fuels, when analyzed for elements at concentration levels too low to allow the use of dilution methods.

The two most widely used methods for the destruction of organic matter are dry ashing and wet digestion.

a. Dry-Ashing

In dry-ashing, the organic matter is decomposed at high temperature in the presence of atmospheric oxygen. In a typical procedure, the sample is weighed into a clean silica or platinum dish, covered with a crystallizing dish, and dried under a heat lamp until the water has evaporated and the sample has a brittle, charred appearance. The sample dish is then placed in a muffle furnace, usually at 500°C, until ashing is complete. After cooling the ash is taken up in dilute mineral acid.

Dry-ashing is relatively simple, can accommodate large samples, and does not require the addition of large amounts of potentially contaminating reagents. However, its principal disadvantage is the serious losses of trace elements that can occur because of (1) volatilization, (2) retention on the walls of the ashing dish, and (3) retention in the acid-insoluble fraction of the ash. Volatility losses are especially serious for mercury and selenium, for which dry-ashing procedures are not recommended. Under certain conditions, arsenic, boron, cadmium, chromium, iron, lead, phosphorus, vanadium, and zinc have also been reported to be lost. Elements that occur as volatile organic complexes, such as copper, iron, nickel, and vanadium porphyrins in petroleum, can be lost through volatilization even at comparatively low temperatures. Nonmetals form many volatile compounds which are easily lost.

Wall retention has been found to be a problem, especially with cobalt, copper, iron, silver, aluminum, and manganese, when using silica dishes. Some elements, for example, silicon, aluminum, calcium, copper, tin, beryllium, iron, niobium, and tantalum, may react to form acid-insoluble compounds in the ash, especially at ashing temperatures above 500°C.

To avoid losses through volatilization and wall retention, many workers use an *ashing aid*, such as MgO, $Mg(NO_3)_2$, HNO_3, or H_2SO_4, which is added to the sample before or during ashing. These ashing acids act in various ways. HNO_3 acts as an oxidizing agent, speeding up the destruction of organics or allowing lower temperatures to be used. H_2SO_4 can convert chlorides into less volatile sulfates. $Mg(NO_3)_2$ has a double action. It decomposes to nitrogen oxides, which hasten oxidation, and MgO, which dilutes the ash and reduces the total area of contact between the sample ash and the container walls, thus reducing retention losses. Some elements, for example, arsenic, copper, and silver can be dry-ashed successfully with a $Mg(NO_3)_2$ ashing aid but suffer serious losses in dry-ashing without aids.

The success, or lack of success, of various dry-ashing procedures for each element is summarized in Table 3.1.

TABLE 3.1. Sample Preparation Methods for Elemental Analysis of Organic Matter[a]

Element	Dry-Ashing	Wet Digestion	Other Methods
Al	6, 11, 20, 21	14	18
Sb	6, 9, 10, 11, 21	1–3, 11	16, 17
As	9, 10, 11	1–3, 5, 11	16, 17
Ba	6	4	—
Be	11, 20	14	—
Bi	6, 8	14	—
B	11	11	—
Cd	10–12	1–3	16
Ca	6, 20	4	18
Cs	7, 14 ($<500°C$)	14	16, 17
Cr	6–8, 10, 11, 13, 21	1–3, 11	16, 17
Co	6, 7, 9, 10, 13	1–3	16, 17
Cu	9–11, 13, 20	1–3	15–17
Ga	11	—	—
Ge	11, 21	11, 21	17
Au	12	14	16
In	11	—	—
Ir	12, 9?	14	16
Fe	6–10, 13, 20, 21	1–3	15–18
Pb	$<500°C$ 6–11, 21	1, 4	15, 17
Li	7, 14 ($<500°C$)	2, 14	—
Mg	6	14	18
Mn	6	14	16, 17
Hg	12	11	15, 18
			17 (92% recovery)
Mo	6–11	1–3	16, 17
Ni	6, 7, 11	14	—
Nb	14, 20	14	—
Os	14	11, 12	—
Pd	12, 9?	14	—
Pt	12, 9?	14	—
K	7, 14 ($<500°C$)	14	18
Re	9?	11, 12	16
Rh	12, 9?	14	—
Rb	14	14	—
Ru	14	11, 12	16
Sc	14	11, 12	16
Se	12	1, 2, 5, 11	17, 19
Si	14, 20	14	15, 18
Ag	9, 13	1–3	16
Na	6, 7	14	16–18
Sr	6–11	1–4	16
Ta	14, 20	14	—
Te	11	14	—
Tl	11	3, 14	16

TABLE 3.1. (*Continued*)

Element	Dry-Ashing	Wet Digestion	Other Methods
Sn	6, 11, 20	3, 11, 21	—
Ti	—	14	18
Trans- uranium	6, 8	14	—
V	6, 7, 11	14	15
Zn	6–11, 21	1–3	16, 17

[a]NOTES:

1. With HNO_3 and $HClO_4$.
2. With HNO_3, $HClO_4$, and H_2SO_4.
3. With HNO_3 and H_2SO_4.
4. Avoid H_2SO_4, due to insolubility of sulfate.
5. Mo(VI) catalyst found useful.
6. Without ashing aids.
7. With H_2SO_4.
8. With HNO_3.
9. With $Mg(NO_3)_2$.
10. Repeated heating to dryness with HNO_3 at 350° C. G. Middleton and R. E. Stuckey, Analyst, **78**, 532 (1953); **79**, 138 (1954).
11. Losses reported by some workers.
12. Not recommended.
13. Retention on crucible walls has been reported.
14. Generally satisfactory with conventional techniques.
15. Oxygen bomb technique. S. Fujiwara and H. Narasaki, *Anal. Chem.*, **36**, 206 (1964).
16. Fusion with $NaNO_3$—KNO_3 at 390°C. H. J. M. Bowen, *Anal. Chem.*, **40**, 969 (1968).
17. Low-temperature oxygen plasma ashing technique. C. E. Gleit and W. D. Holland, *Anal. Chem.*, **34**, 1454 (1962).
18. Teflon bomb. B. Bernas, *Anal. Chem.*, **40**, 1683 (1968).
19. Schöniger flask.
20. Ash may not dissolve easily in some cases.
21. Presence of excess chloride tends to increase losses.

b. Wet Digestion

In wet digestion, the sample is treated with concentrated mineral acids and/or strong oxidizing agents in solution. Oxidizing conditions are maintained throughout the procedure. Most often, the mixture is heated to 100–200°C to aid the digestion process. Wet digestion is fast and is much less troubled with volatilization losses, because of the lower temperature. The major disadvantage is the possibility of contamination from the large excess of reagents employed.

The most widely used acids for wet digestion are HNO_3, H_2SO_4, and $HClO_4$. A $3:1:1$ mixture, respectively, dissolves its weight of most organic materials (11). H_2SO_4 is eliminated if the sample contains large amounts of calcium, because of the danger of the coprecipitation of trace elements on $CaSO_4$. The insolubility of certain other sulfates (e.g., of Pb^{2+}, Ag^+, Ba^{2+}) and chlorides (e.g., of Ag^+, Pb^{2+}) limits the choice of suitable digestion acids in certain cases, although the K_{sp} values of these compounds are sometimes not exceeded at trace levels. Other oxidizing agents, such as H_2O_2 and permanganate, are occasionally employed in wet digestions. Salts of molybdenum(VI) are sometimes used as catalysts to speed up the oxidation reactions. Table 3.1 also lists wet-ashing methods which have been used for various elements.

Although volatility losses are expected to be much less severe in wet digestion than in dry ashing, they are not entirely absent. Mercury is lost when wet digestions are performed in open beakers; an enclosed reflux system is recommended (5). Elements that form volatile oxides, such as ruthenium and osmium, are lost. If the organic material is allowed to char during the digestion, reducing conditions are temporarily induced, and losses of selenium, arsenic, and antimony, which form volatile hydrides, may occur. The presence of organically bound chlorine has been shown to lead to losses of germanium and arsenic, which form volatile chlorides. Thiers (6) lists data indicating that antimony, arsenic, boron, chromium, germanium, selenium, and tin can be lost when acid mixtures containing HCl and H_2SO_4 or $HClO_4$ are boiled.

The use of concentrated $HClO_4$ in wet digestion procedures requires several precautions (7). Its use is desirable because the hot, concentrated acid is an exceedingly strong oxidizing agent, capable of destroying the most resistant organic materials. However, this high reactivity can lead to violent explosions if the acid is mishandled. First, it must be realized that the acid is an oxidizing agent only if both hot *and* concentrated. The cold, concentrated (70%) acid, although a strong acid, is not an oxidizing agent and will not even oxidize iodide to iodine or iron(II) to iron(III). Basically, the rules of safe handling are:

1. Never bring undigested organic matter directly into contact with hot, concentrated $HClO_4$; a fire or explosion may result.
2. Always predigest organic matter with HNO_3 first to destroy the easily oxidized compounds.
3. If hot, concentrated $HClO_4$ is spilled, dilute *immediately* with quantities of water. This dilutes and cools the acid and effectively "turns off" its oxidizing power.
4. Never boil $HClO_4$ to dryness. If the digestion cannot be watched

carefully, H_2SO_4 should be incorporated into the digestion mixture; the H_2SO_4 will remain after the $HClO_4$ has boiled away.

5. At the first sign of charring, dilute and cool the digest immediately.

6. The use of a $HClO_4$ fume hood is recommended. Alternatively, the $HClO_4$ fumes can be condensed and collected.

c. Other Methods

A variety of other methods has been proposed for the destruction of organic matter prior to elemental analysis. Middleton and Stuckey (8) treated a sample with HNO_3, heated it to dryness on a hot plate at 350°C, and repeated the procedure until ashing was complete. Good recoveries for arsenic, antimony, and cadmium among others, were reported. For very volatile elements, such as mercury, the oxygen bomb technique (9) and the acid pressure decomposition vessel (4) have been used with success. The low-temperature oxygen plasma ashing technique of Gleit and Holland (10) is useful for arsenic, selenium, cadmium, germanium, tin and others.

d. Choice of Method

The most comprehensive comparisons of recoveries by different sample preparation techniques have been published by Gorsuch (5). Thiers (6) has reviewed the topic carefully. The results of these and many other workers are summarized in Table 3.1. With this table, it should be possible to select a satisfactory sample preparation technique for any element listed. The table is inevitably incomplete because of the lack of published data on all elements with all techniques. In some cases conflicting recommendations are made. No apology for this situation is made; it would be unwise to pretend that the matter of sample preparation is a closed issue.

B. PRECONCENTRATION

1. EVAPORATION (12)

Evaporation is a simple, but slow, preconcentration method for solutions. Its main advantage is that it does not involve the use of large amounts of reagents or of complicated glassware; thus contamination is minimized. The main disadvantage is that the total dissolved solids content of the solution is increased, so that precipitation losses, viscosity, burner clogging, scatter, and so on, may become a problem. Also, the procedure is very slow, requiring many hours even for modest concentration factors. It is necessary to use a dust cover over the evaporation dish to prevent contamination from dust. Clean, dry, filtered air should be passed over the dish to carry away the vapor. Polyethylene or Teflon evaporation dishes are best. The dust cover can also be fabricated from a large plastic beaker fitted with a side arm for the flushing air. The sample is usually acidified (e.g., with 1 or 2% concentrated HCl by volume) to prevent hydrolysis. Even so, precipitation losses from $CaSO_4$, $BaSO_4$, silica, and so on, may be expected in some cases. Volatility losses also occur, especially for mercury and to a lesser extent for arsenic and antimony. The lowest possible temperature should be used, consistent with reasonable evaporation times. Boiling should obviously be avoided to prevent spattering losses. Either a heat lamp or a hot plate may be used, or both.

Evaporation is mainly a technique to be used with samples with low dissolved solids contents, such as water, mineral acids, organic liquids, and so on.

2. CHELATION AND SOLVENT EXTRACTION (11, 12)

Solvent extraction techniques are widely used in atomic and molecular absorption and fluorescence spectrometry for separation and preconcentration of the analyte. Selective extractions are often employed in molecular absorption and fluorescence spectrometry to avoid spectral interferences from other metals in the sample matrix. In atomic spectrometric techniques the inherent spectral selectivity is usually sufficient so that the selectivity of the extraction is unimportant. Nevertheless, solvent extraction is still widely used for preconcentration and for separation of transition metals from large amounts of alkali metals which could cause scattering, burner clogging, or other problems. Most commonly, metals in an aqueous solution are extracted into an immiscible organic solvent, usually with the use of a chelating agent, and the organic phase is analyzed directly without back-extracting into the aqueous phase. In atomic flame methods, aspiration of a metal dissolved in an organic solvent yields a three- to fivefold enhancement of the signal (refer to Chapter 6).

The most commonly used chelating agents for the solvent extraction of metals are listed in Table 3.2. The recommended pH ranges for the extraction of metals by dithizone, cupferron, oxine, sodium diethyldithiocarbamate (NaDDC), and ammonium pyrrolidine dithiocarbamate (APDC) are listed in Table 3.3. Note the APDC is the most comprehensive, that is, the least selective, reagent. Its use in colorimetry is therefore limited, but it is especially useful in atomic spectroscopic techniques, because several trace metals can be extracted simultaneously from an aqueous solution and determined in a single extract.

TABLE 3.2. Chelating Agents Commonly Used for Preconcentration in Atomic Spectroscopy

Common Name	Chemical Name	Commonly Used for
APDC	Ammonium pyrrolidine dithiocarbamate	Cu, Pb, Cd, Co, Mn, Fe, Ni, Bi, Zn, As, Ir, Pd, Pt, Se, Te, Tl, Mo, V, Cr
Cupferron	Ammonium salt of N-nitrosophenylhydroxylamine	V, Ti, Cu, Mn, Fe, Ni
Oxine	8-Hydroxyquinoline	Al, alkaline earths, others
Dithizone	Diphenylthiocarbazone	Ag, Pb, Cd, Zn, Cr
ACAC	Acetylacetone; 2,4-pentanedione	Transition metals
NaDDC	Sodium diethyldithiocarbamate	Pb, Cu, Fe, Mn, Te

TABLE 3.3. Recommended pH Ranges for Solvent Extraction of Metal Chelates[a,b]

Element	Dithizone[c]	Cupferron	Oxine	NaDDC	APDC
Ag	0–7 (1)	—	3–6 (5)	8–11 (1)[d]	1–10 (11)[d]
Al	—	3.5–9.0[d] (3)	5–6 (5)	—	—
As	—	—	—	6 (1)[d]	2–6 (4)
Be	—	—	6 (5)	—	—
Bi	>2 (1)	—	—	8–11 (1)[d]	1–10 (11)[d]
Cd	6–14 (1)	—	6 (5)	8–11 (1)[d]	1–6 (4); 1–10 (11)[d]
Co	6–8 (1)	—	5–6 (5)	3.6 (8); 8–11 (1)[d]	2–4 (4); 1–10 (11)[d]
Cr	0.5 (2)	—	—	3.6 (8)	3–9 (4)
Cu	2–5 (1)	7 (2); 1 (4)	2–6 (7)	3.6 (8); 8–11 (1)[d]	0.1–8 (4); 1–10 (11)[d]
Ga	—	—	—	—	3–7 (11)[d]
In	—	—	—	—	1–10 (11)[d]
Fe	—	7 (2)	6 (5)	3.6 (8)	2–5 (4); 5±0.3 (11)[d]
Hg	0–4 (1)	—	—	8–11 (1)[d]	2–4 (9); 1–10 (11)[d]
Pb	7–10 (1)	—	6 (5)	3.6 (8); 8–11 (1)[d]	0.1–6 (4); 1–10 (11)[d]
Mn	—	7 (2)	—	3.6 (8)	2–4 (4); 5±0.3 (11)[d]
Mo	—	—	1.8–2.6 (6)	—	3–4 (4); 3–6 (11)[d]
Ni	6–8 (1)	7 (2)	6 (5)	3.6 (8); 8–11 (1)[d]	2–4 (4); 1–10 (11)[d]

TABLE 3.3. (*Continued*)

Element	Dithizone[c]	Cupferron	Oxine	NaDDC	APDC
Se	—	—	—	$6 (1)^d$	3–6 (4)
Sb	—	—	—	—	$1 (11)^d$; 3.7 (12)
Sn	—	—	2.5–6 (13)	—	$3–6 (11)^d$
Tl	—	—	—	$8–11 (1)^d$	$3–10 (4)$; $3–10 (11)^d$
V	—	1 (4)	—	$6 (1)^d$	1–2 (10); 4 (4); $3–6 (11)^d$
W	—	—	—	—	$1–3 (11)^d$
Zn	6–9 (1)	—	6 (5)	3.6 (8); $8–11 (1)^d$	2–6 (4); $1–10 (11)^d$

^{a}References (given in parentheses) are:
1. E. Sandell, *Colorimetric Metal Analysis*, 3rd ed., Interscience, New York, 1959.
2. B. Delaughter, *At. Absorpt. Newslett.* **4**, 273 (1965).
3. J. Stary, *The Solvent Extraction of Metal Chelates*, H. Irving, ed., Pergamon Press, Oxford, 1964.
4. C. E. Mulford, *At. Absorpt. Newslett.* **5**, 88, 1966.
5. S. L. Sachder and P. W. West, *Envir. Sci. Tech.* **4** (9), 749 (1970).
6. Y. Chau and K. Lum-Shue-Chan. *Anal. Chim. Acta*, **48**, 205 (1969).
7. P. Cooke, *Atomic Absorption Spectrophotometry*, Pye Unicam, Cambridge, 1969.
8. J. Nix and T. Goodwin, *At. Absorpt. Newslett.*, **9** (6), 119 (1970).
9. B. B. Mesman and B. S. Smith, *At. Absorpt. Newslett.* **9** (4), 81 (1969).
10. Y. Chau and K. Lum-Shue-Chan, *Anal. Chim. Acta*, **50**, 201 (1970).
11. E. Lakanen, *At. Absorpt. Newslett.*, **5**, 17 (1965).
12. M. Yanagisawa, M. Suzuki, and T. Takeuchi, *Anal. Chim. Acta*, **47**, 121 (1969).
13. H. Malissa and S. Gomicek, *Z. Anal. Chem.*, **169**, 402 (1959).
^{b}Solvent is MIBK unless otherwise noted.
^{c}Solvent is CCl_4.
^{d}Solvent is $CHCl_3$.

The most commonly used extracting solvents are $CHCl_3$, CCl_4, methyl isobutyl ketone (MIBK), esters such as ethyl acetate or propionate, and ethers such as ethyl ether. For direct aspiration in atomic flame techniques, chlorinated solvents or very volatile solvents are not satisfactory; rather, MIBK is recommended. The solubility of MIBK in water (20 ml liter^{-1} at 25°C) is too great to allow very large concentration factors. Methyamyl ketone (MAK) is much less soluble in water and is a suitable alternative. However, MIBK is less expensive and more widely available.

In atomic absorption work, the APDC–MIBK system is overwhelmingly the most popular. An example of a typical extraction procedure, suitable for many of the metals listed in Table 3.3, is:

Adjust 100 ml of the aqueous solution to the appropriate pH range (usually between 3 and 4) by the addition of HCl or NH_3. Transfer to a separatory funnel. Add 5 ml of a freshly prepared 1% solution of APDC in water and shake to mix. Add 10 ml MIBK, shake 2 minutes, and allow the phases to separate. The organic (upper) phase may be sprayed directly into the flame.

It is sometimes convenient to use a narrow-necked volumetric flask or similar vessel instead of a separatory funnel. After the phases have separated, additional water can be poured in to raise the level of the organic phase into the neck, where it can be aspirated directly into the flame.

In solvent extraction work, it is always necessary to carry the blank and the standards through the same extraction procedure. The purpose of this is to (1) allow for less than 100% extraction, (2) correct for the concentration of reagent impurities, (3) eliminate calculation of the exact concentration ratio, and (4) provide standard solutions in the appropriate chemical form and matrix.

Not all metal extractions are based on chelation formation. For example, a few metals can be extracted directly as a halide complex from strongly acid solutions. Table 3.4 lists the details. Extractions of this type

TABLE 3.4. Halide Extraction Systems

Metal extracted	Aqueous phase	Organic phase
Fe(III)	6 M HCl	Diethyl ether–isobutyl acetate
Ga(III)	6–8 M HCl	Diisopropyl ether
U	HCl	Tributyl phosphate–chloroform
In	5 M HBr	Isopropyl ether
Au	3M HBr 6 M HCl	Diisopropyl ether–ethyl ether
Hg	I$^-$	Cyclohexanone
Cd	1.5 M I$^-$ 1.5 N H$_2$SO$_4$	Diethyl ether
Cr(III)	2 M HCl	MIBK
Sb(V)	2–8 M HCl	MIBK
Ti(III)	6 M HCl	Isopropyl ether
Mo(VI)	6 M HCl	Ether

are unusually selective, since many common elements are not at all extracted under these conditions. It is possible to use halide extractions either to concentrate the analyte metal or to remove a large excess of an interfering matrix element. Thus the determination of trace metals in iron metal can be aided by extracting most (about 99%) of the iron from a 6 M HCl solution of the metal into ether. Many metals, for example, aluminum, bismuth, cadmium, cobalt, beryllium, lead, manganese, nickel, and silver, are not extracted and can be analyzed in the remaining aqueous phase.

3. ION EXCHANGE (12)

Ion exchange is becoming an increasingly popular method of separation and preconcentration in trace element analysis, especially for water analysis at ultratrace levels. Basically three different types of ion exchange resins are used for these purposes: cation exchange, anion exchange, and chelating-type resins. Cation (acid) exchange resins are those that exchange cations with the solution, replacing all the cations in solution with H^+ or Na^+. Anion exchange resins replace the anions in solution with OH^- or Cl^-. Chelating resins contain functional groups similar to those in conventional chelating agents. These resins remove from solution any ions with which the functional groups can form a chelate bond. They are somewhat more selective than cation or anion exchange resins. All three types are widely used for trace element concentration. The large ion exchange capacity of many resins means that large volumes of dilute solutions can be passed through the ion exchange column with nearly complete retention of the ionic species. The retained ions are then eluted with a relatively small volume of a strong acid (e.g., 2 M HNO_3) or, for anion exchange resins, an alkali (e.g., 4 M NH_4OH). Nearly 100% recovery can be obtained for many metals. Very large concentration factors are practical. If the sample contains large amounts of major cations (e.g., Na^+, K^+, Mg^{2+}, Ca^{2+}), a cation exchange column may not be practical because it may be saturated by the major cations, thus preventing complete retention of trace cations. In such cases, a chelating resin that does not retain alkali or alkaline earth elements is preferred.

Riley and Taylor (13) separated a large number of metals from sea water on a 1.2×6 cm column of Chelex-100, a chelating resin (Bio-Rad Laboratories, Richmond, Calif.). The optimum sample pH was between 5 and 9, and the column was eluted with 2 M HNO_3, HCl, H_2SO_4, or $HClO_4$. Cesium, vanadium(VI), phosphate, and arsenate were not retained on the column, and poor recoveries for thallium(I), tin, and tungsten(VI) were found. Recoveries of 90–100% were found for silver, bismuth, cadmium, cerium(III), cobalt, copper, indium, molybdenum(VI), nickel, lead, rhenium(VII), scandium, thorium, vanadium(V), yttrium, and zinc.

Biechler (14) separated cadmium, copper, iron, lead, nickel, and zinc from industrial wastewater using a 1×10 cm column of a chelating resin.

McCraken et al (15) used an anion exchange resin in the Cl^- form to separate tin, cadmium, and zinc from excess copper. The sample was made 2 M in HCl, under which conditions only copper is not retained on the column.

C. LOSSES AND CONTAMINATION (6)

1. STABILITY OF DILUTE SOLUTIONS

Very dilute (<1 ppm) solutions of many elements are known to be subject to significant losses because of adsorption of the elements onto the walls of the containers in which they are stored. The extent and rate of this adsorption depends on the element, its concentration, the pH of the solution, and the type and history of the container. Serious losses of silver, molybdenum, manganese, vanadium and nickel have been noted in glass containers. However, these elements are stable in polyethylene containers, except for silver. Losses of 1 to 10% in 2 weeks have been reported for strontium, zinc, copper, iron, lead, and aluminum in Pyrex containers. Silver is the most notorious case of all; losses of 50% or more occur in only a few days.

In general, wall adsorption losses may be minimized by using polyethylene containers and keeping the solution acidic (pH <2).

Losses of mercury at 0.1 ppm levels or less may occur through volatilization or by adsorption. It is recommended that mercury solutions at this level be made up in 1 M HCl.

2. CONTAMINATION

a. Glassware

(1) Selection

In general, for laboratory ware, Teflon, polyethylene, platinum, and fused silica are considered superior to soft glass, borosilicate glass, porcelain, and stainless steel from the standpoint of contamination. Nevertheless, conventional borosilicate volumetric flasks and pipets can be safely used if well cleaned and if the period of contact with the sample or solution is not prolonged.

(2) Cleaning

Chromic acid cleaning baths, although widely used in chemical laboratories, are not recommended for trace analysis work, because of the difficulty of removing adsorbed chromium from the surfaces. A 1:1 mixture of concentrated HNO_3 and H_2SO_4 is a much more suitable alternative.

Thiers (6) recommends the following cleaning procedure for glass and plastic apparatus:

1. Soak and scrub with brushes in hot detergent solution.
2. Rinse completely in tap water; drain.
3. Soak 15 min in a 1:1 mixture of concentrated HNO_3 and H_2SO_4.
4. Rinse four times in tap water.
5. Rinse four times in deionized water.
6. Drain and dry in a dust-free location.

Large apparatus need not be completely immersed in the acid mixture. It is sufficient to coat their surfaces with a film of the acid and allow to stand 30 min before rinsing. Teflon and polyethylene can be cleaned by the above procedure. One should avoid *prolonged* soaking of polyethylene in concentrated acids, however.

Platinum dishes and crucibles are cleaned by filling them with molten $NaHSO_3$, keeping the salt fused over a Meeker flame for 10 min, cooling, removing the solidified block, storing in 6 M HCl, and rinsing four times in deionized water before use (6).

(3) Storage

Even the most carefully cleaned glassware can be contaminated in storage by the accumulation of dust, and so on. An easy way to store cleaned apparatus is in large envelopes made by folding and stapling polyethylene sheets or polyethylene-coated (freezer wrap) paper. Alternatively, flasks may be stored with their mouths covered with tightly stretched plastic wrapping film.

b. Reagents

(1) Water

Water is commonly purified either by distillation or by ion exchange, although the data collected by Thiers (6) shows that a mixed-bed ion exchange column produces water of equal or better quality than that produced by repeated distillation. Certainly, ion exchange is more convenient. However, ion exchange columns may not remove all traces of nitrogenous compounds, which would be a problem in the trace determination of nitrogen, ammonia, or proteins. Distillation of the ion exchange water is recommended for these purposes. Particulate matter can be removed by filtration through a microporous cellulose ester membrane.

(2) Mineral Acids, Bases, and Other Reagents

HCl of very high purity is readily prepared by dissolving the anhydrous gas in pure water. The HCl gas, available commercially in tanks, is first bubbled through concentrated H_2SO_4, passed through a 1 m column packed with borosilicate glass wool, bubbled through HCl-saturated water, and then dissolved in highly purified water at $0°C$. (The receiving vessel is cooled in an ice bath.) Concentrations up to 12 M can be prepared in this way.

HNO_3 can be purified by the repeated distillation of the azeotrope (65% HNO_3), but it is much more convenient to purchase acid of the required purity from commercial sources. The same is true of H_2SO_4 and $HClO_4$.

Ammonium hydroxide is best prepared by the dissolution of NH_3 gas in water. A very pure product can be obtained in this way. The alakli hydroxides and salts can usually be obtained commercially at sufficient purity, except with respect to contamination by other alkali metals. If necessary, heavy metals may be extracted from a concentrated aqueous solution with dithizone and oxine in chloroform. Solutions of heavy metal salts are best prepared by dissolving the pure metal in purified acid.

Organic solvents of the "spectro" grade are usually sufficiently pure to be used as received in trace metal work. If necessary, they can of course be distilled.

It is worthwhile to reiterate that it is *always* advisable to prepare a reagent blank, carried through the *entire* analytical procedure, in order to evaluate and correct for reagent contamination. If an excessively high blank is obtained, the purity of the reagents and/or the cleanliness of the glassware should be questioned. Although in principle the blank solution should correct for any amount of contamination, in practice the lack of "perfect reproducibility" of blank readings causes large analytical errors if the blank reading is too high.

BIBLIOGRAPHY

I. M. Korenman, *Analytical Chemistry of Low Concentrations*, J. Schmorak, Transl., Israel Program for Scientific Translations, Jersualem, 1968.

G. Tolg, *Ultramicro Elemental Analysis*, C. G. Thalmayer, Transl., John Wiley, New York, 1970.

REFERENCES

1. N. H. Suhr and C. O. Ingamells, *Anal. Chem.*, **38**, 730 (1966).
2. W. Slavin, *Atomic Absorption Spectroscopy*, Interscience, New York, 1968.

3. G. K. Billings, *The Preparation of Geological Samples for Analysis by Atomic Absorption*, Varian-Techtron Publication T-1001, Palo Alto, Ca.

4. B. Bernas, *Anal. Chem.*, **40**, 1682 (1968).

5. T. T. Gorsuch, *The Destruction of Organic Matter*, Pergamon Press, Oxford, 1970.

6. R. E. Thiers, "Contamination in Trace Element Analysis and Its Control," in *Methods of Biochemical Analysis*, D. Glick, Ed., Vol. V, Interscience, New York, 1957.

7. Analytical Methods Committee, *Analyst*, **84**, 214 (1959).

8. G. Middleton and R. E. Stuckey, *Analyst*, **78**, 532 (1953).

9. S. Fujiwara and H. Narasaki, *Anal. Chem.*, **36**, 206 (1964).

10. C. E. Gleit and W. D. Holland, *Anal. Chem.*, **34**, 1454 (1962).

11. G. D. Christian and F. J. Feldman, *Atomic Absorption Spectroscopy*, Wiley-Interscience, New York, 1970.

12. C. R. Parker, *Water Analysis by Atomic Absorption Spectroscopy*, Varian Techtron, Springvale, Australia, 1972.

13. J. P. Riley and D. Taylor, *Anal. Chim. Acta*, **40**, 479 (1968).

14. D. G. Biechler, *Anal. Chem.*, **37**, 1054 (1965).

15. J. D. McCraken, M. C. Vecchione, and S. L. Longo, *At. Absorpt. Newslett.*, **8**, 102 (1969).

4

SPECTROSCOPIC METHODS

J. D. WINEFORDNER

Department of Chemistry
University of Florida
Gainesville, Florida

A. ELECTROMAGNETIC RADIATION (1a, 2a, 3a)

1. LIGHT

Light or electromagnetic radiation can be described in two ways, one based on the *wavelike nature*, which is required to explain such optical phenomena as reflection, diffraction, interference, refraction, and scattering, and one based on the *particle nature*, which is required to explain the processes of emission and absorption of radiation by atoms and molecules. For a thorough discussion of physical and geometric optics refer to the appropriate references listed at the end of this chapter.

2. FIGURES OF MERIT FOR LIGHT

According to Maxwell, an electromagnetic wave can be represented as an alternating current with an associated magnetic field (see Figure 4.1). Interaction of the wave with the surroundings can be discussed in terms of the electric and magnetic vectors. Some figures of merit of light based on the wavelike nature of light are listed below.

Plane of polarization is the plane (xy) determined by the electric vector.

Direction of polarization is the direction of the electric (y) or magnetic (z) vector. If the radiation is circularly polarized, the electric vector has constant amplitude but rotates about the axis of propagation (x).

Direction of propagation is the direction of motion (x). The electromagnetic field varies periodically perpendicularly to the direction of propagation.

Intensity of light wave is indicative of the strength of the wave and is proportional to the square of the amplitude.

Wavelength λ is the distance covered by one wave in one complete oscillation (see Figure 4.1) (in m, cm, μm, nm, etc.).

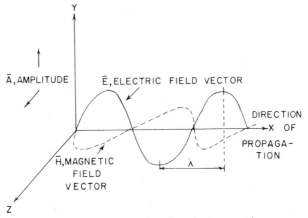

Fig. 4.1 Schematic representation of an electromagnetic wave.

Frequency ν is the number of times per second the electric (or magnetic) field reaches a maximum positive value (in Hz, i.e., in s^{-1}).

Wavenumber $\bar{\nu}$ (or σ) is the number of waves per unit length [generally in cm^{-1}, or Kaysers (K)].

Velocity of propagation v_q is the velocity of light in the medium q and is related to λ_q and ν by:

$$v_q = \lambda_q \nu \qquad (\text{in m s}^{-1}) \tag{1}$$

If the medium q is a vacuum,

$$v_{q=\text{vac}} = c = \lambda \nu \tag{2}$$

where c is the speed of light ($c = 3 \times 10^8$ m s^{-1}).

Refractive index $n_{q\lambda}$ is defined by:

$$n_{q\lambda} = \frac{c}{v_{q\lambda}} \tag{3}$$

where the refractive index is for a given wavelength λ and material q. The variation in refractive index with wavelength is the basis of operation of prisms as dispersers of radiation and is the basis of chromatic aberration in lenses.

Diffraction of radiation is a result of *scattering* of radiation from narrow apertures and *interference* (*superposition* of waves) of the resulting wavefronts. Diffraction is the basis of the operation of gratings. Interference is also the basis of operation of *Fabry–Perot* (multiple beams) and

Michelson (dual-beam) interferometers which are finding considerable use in the study of line shapes and wide-wavelength multispectral component coverage, respectively.

Some additional figures of merit of light based on the particle nature of light are listed below.

Emission of energy results when an elementary system (nuclear, atomic, or molecular) undergoes a radiational deactivation resulting in a loss of energy as light, for example, R_2 and R_3 in Figure 4.2.

Absorption of radiational energy results in excitation of an elementary system from a lower to an upper level, for example, R_1 in Figure 4.2.

If the emitting species has been excited by light, the emission is called *photoluminescence*. In the case of molecules, spin-allowed photoluminescence (emission) transitions are called *fluorescence*, and spin-unallowed photoluminescence (emission) transitions are called *phosphorescence*. In the case of atoms, unfortunately, all transitions, whether spin-allowed or not, are called fluorescence.

Nonradiational excitation of an elementary system results in excitation of the species from a lower to an upper level, for example, *D*, in Figure 4.2. Such excitation can be via chemical reactions (if the species then undergoes emission, it is called *chemiluminescence* and, if the species is biologically excited, it is called (*bioluminescence*), via electrical energy (called *electroluminescence* if the species then undergoes emission), via an electron beam (called *cathodoluminescence* if the species then undergoes emission), and via thermal energy (called *thermoluminescence* if the species then undergoes emission). It should be noted that excitation of elementary systems in high-temperature gases and plasmas, for example, flames, arcs, and sparks, occurs predominately via collisions with energetically excited species; this mode of excitation is generally called *collisional*, and a system is often said to be in *thermal equilibrium* if it can be described by a single temperature T; however, this mode of excitation, sometimes called *thermal excitation*, should not be confused with thermoluminescence which

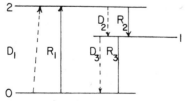

Fig. 4.2 Schematic representation of the energy levels of an elementary system with radiational (solid lines), nonradiational (dashed lines), excitation (upward pointing arrows), and deexcitation (downward pointing arrows) energy indicated.

generally refers to molecular species heated to relatively low temperatures, for example, 100–400°C following radiational excitation.

The link between the wavelike and particle nature of light is the Planck equation:

$$\Delta E = h\nu = \frac{hc}{\lambda} = hc\bar{\nu} \tag{4}$$

where ΔE is the energy change (absorption or emission) in the elementary system that results when a photon of energy $h\nu$ has been absorbed or emitted. The appropriate units of the terms in equation 4 are $h = 6.6 \times 10^{27}$ ergs or 6.6×10^{-34} J s (since 1 J $= 10^7$ erg $= 1$ VC). Also note that since 1 eV $= 1.6 \times 10^{-19}$ C $\times$ 1 V $= 1.6 \times 10^{-19}$ J, $h = 4.1 \times 10^{-15}$ eV s.

Contrary to the implications originating from the energy level diagram (Figure 4.2), transitions involve a finite spread of energies, and so a spectral transition has an energy width of δE. To convert δE into the width in frequency $\delta\nu$ and the width in wavelength $\delta\lambda$, equations 2 and 4 are used, and so:

$$\delta E = h\delta\lambda = \frac{hc}{\lambda_0^2}\delta\lambda = hc\delta\bar{\nu} \tag{5}$$

where λ_0 is the peak wavelength of the transition. The spread of energies involved in a spectral transition is a result of the following fundamental limits: *natural line broadening* due to the time the species remains at the energy level—the shorter the time at a level, the greater the energy spread of that level (Heisenberg uncertainty principle); *Doppler broadening* due to thermal motion of emitting (and absorbing) species relative to the detector —this broadening depends directly on $\sqrt{T}$, where T is the temperature of the species; *collisional broadening* due to collisions of the excited species with similar or other ground-state species results in either small energy perturbations (*adiabatic collisions*) or in complete quenching of the energy (*quenching collisions*); and other fundamental types of broadening including *Stark* and *Zeeman* effects—these effects are seldom of significance in most types of spectroscopy used to determine elemental species.

In addition to the above fundamental broadening effects, various fine and hyperfine effects cause atomic and molecular transitions to be broader than predicted. In the gas state, molecular electronic transitions are broadened by the vibrational and rotational levels. In the liquid phase, molecular rotations are restricted, resulting in a loss of the rotational fine structure and in a general broadening of the molecular spectral band. In fact, in the liquid phase, much of the vibrational fine structure is also lost as a result of solvent interactions with the molecular species. Finally, even if the spectral line of an atomic transition is narrow and has hyperfine structure, and if the spectral band of a molecular species in the gas state has considerable fine structure, the actual recorded spectral profile may be far more characteristic of the spectrometer than the elementary system. For example, the recorded spectrometric signal S_λ at any wavelength λ depends on the spectral radiant flux Φ_λ reaching the detector, and on the spectrometer (or apparatus) profile function F_λ, and so:

$$S_\lambda \alpha \int_{\lambda_l}^{\lambda_u} \Phi_\lambda F_\lambda \, d\lambda \tag{6}$$

where λ_u and λ_l are the upper and lower wavelength limits determined by the spectrometer spectral bandpass. If the spectral component being recorded has a width of $\Delta\lambda$ and if the spectral bandpass of the spectrometer is $\Delta\lambda_s$, it should be evident from equation 6 that the recorded profile will be indicative only of the spectral component profile of $\Delta\lambda_s \ll \Delta\lambda$. If $\Delta\lambda_s \gtrsim \Delta\lambda$, the recorded profile will be indicative of the spectrometric function. Of course, the influence of $\Delta\lambda_s$ on absorption and emission profiles can be exactly found by substituting for Φ_λ in terms of the source profile and the absorption or emission profile. If the detector sensitivity and optics transmittance also vary greatly with wavelength, equation 6 can be appropriately modified to contain them in the product within the integral (called a *convolution integral*).

3.   THE ELECTROMAGNETIC SPECTRUM

In Figure 4.3, the electromagnetic spectrum is given. Included in Figure 4.3 are the types of transition (the most energetic ones possible in a given region) and the wavelength, wave number, frequency, and energies for each region. The regions are not very distinct, that is, the ranges are only approximate and may vary considerably from book to book. Some useful *conversion factors* and *constants* are listed in Table 4.1.

In order to "speak the language of spectroscopy," the reader must also be familiar with the terms, symbols, and units of *radiant energy* approved for use by IUPAC. These are listed in Table 4.2.

Several basic relationships of use in spectroscopy are also given. The Boltzmann expression is given by:

$$\frac{n_j}{\sum_i n_i} = \frac{g_j e^{-E_j/kT}}{\sum_i g_i e^{-E_i/kT}} \tag{7}$$

Fig. 4.3 The electromagnetic spectrum.

Type of Transition	Region Names	Wavelength (m)	Wavelength (Å)	Wavenumber (K or cm⁻¹)	Frequency (Hz)	Photon (eV)	Energy (J)	Type of Spectroscopy in Region
						-10^{-6}		
Nuclear spin	Radio frequency	-10^{-1}	-10^{9}	-10^{-1}	-10^{9}		-10^{-24}	NMR
						-10^{-5}		
					-10^{10}			
Electron spin	Microwave	-10^{-2}	10^{8}	-1		-10^{-4}	-10^{-23}	Microwave spectroscopy, EPR
					-10^{11}			
		-10^{-3}	-10^{7}	-10^{1}		-10^{-3}	-10^{-22}	
Molecular rotations	Far IR				-10^{12}			IR spectroscopy (absorption-emission)
		-10^{-4}	-10^{6}	-10^{2}		-10^{-2}	-10^{-21}	
					-10^{13}			
Molecular vibrations	IR	-10^{-5}	-10^{5}	-10^{3}		-10^{-1}	-10^{-20}	IR absorption spectroscopy
					-10^{14}			

84

Electromagnetic spectrum chart (rotated):

Process	Region						Energy	Technique
	Near IR	−25,000 −10⁴	−10⁻⁶ −10⁴	−10⁴			−10⁻¹⁹	Near-IR absorption spectroscopy
Valence and bonding electron	Visible	−8,000 −3,500						UV-visible absorption emission spectroscopy
	UV	−2,000 −1,650	−10⁻⁷ −10³	−10⁵	−10¹⁵	−1	−10⁻¹⁸	
		(N₂ purge)			−10¹⁶	−10¹		
	Far UV		−10⁻⁸ −10²	−10⁶		−10²	−10⁻¹⁷	Far-UV absorption spectroscopy, photoelectron spectroscopy
	X-ray		−10⁻⁹ −10¹	−10⁷	−10¹⁷	−10³	−10⁻¹⁶	X-ray emission-absorption fluorescence-diffraction spectroscopy
Inner shell electrons			−10⁻¹⁰ −1	−10⁸	−10¹⁸	−10⁴	−10⁻¹⁵	ESCA, Auger spectroscopy
Nuclear	γ-ray		−10⁻¹¹ −10⁻¹	−10⁹	−10¹⁹		−10⁻¹⁴	Neutron activation analysis, Mössbauer spectroscopy

85

TABLE 4.1   Terms, Symbols and Units of Radiant Energy*

	Symbol	Unit
Radiant flux (or power)	Φ	W
Radiant Energy	$Q = \int_0^{\pm} \Phi\, dt$	J (or Ws)
Radiant intensity[a]	$I = \Phi/\Omega$	$W\ sr^{-1}$
Radiance[b]	$B = \Phi/S\Omega$	$W\ sr^{-1}\ m^{-2}$
Radiant emissivity	$J = \Phi/V\Omega$	$W\ sr^{-1}\ m^{-3}$
Radiant energy density	$u = Q/V$	$J\ m^{-3}$
Irradiance	$E = \Phi/S$	$W\ m^{-2}$
Radiant exposure	$H = \int_0^t E\, dt$	$J\ m^{-2}$ (or $W\ s\ m^{-2}$)

[a]Ω is the solid angle of radiation.
[b]S is the area of the source.
*From IUPAC Commission V-4, Part I.

where n_j is the number of atomic or molecular species in state j, q_j is the statistical weight of state j, E_j is the energy (in J) of state j, and the other terms are defined for the general state i. The Boltzmann expression is useful in predicting the fractional number of species in a given state (here j) as compared to those in all other states or to those in another specific state.

Radiational transitions between energy levels occur with certain probabilities. The probability of spontaneous emission A_{ul} (in s^{-1}) of radiant energy from an upper level u resulting in conversion of a species to level l is related to the probability of induced emission B_{ul} (in $s^{-1}/J\ s\ m^{-3}$, i.e., induced photons per spectral energy density) by:

$$A_{ul} = \left(\frac{8\pi h\nu^3}{c^3} \right) B_{ul} \tag{8}$$

The induced emission probability (for induced emission like induced absorption, that is, radiational excitation, depends on the presence of external radiation) is related to the induced absorption probability B_{lu} by:

$$B_{lu} g_l = B_{ul} g_u \tag{9}$$

where the g's are the respective statistical weights. According to equation 8, spontaneous emission greatly exceeds induced emission in the UV-visible region, since $A_{ul}/B_{ul} \alpha \nu^3$, and so for UV-visible radiation $A_{ul} >>> B_{ul}$. In combustion flames and in high-temperature plasmas, such as arcs, sparks, plasma jets, and so on, where the source temperature is much greater than the surrounding (atmospheric) temperature, the rate of spontaneous emission is considerably greater than that of induced emission, that is, $A_{ul}n_u \gg B_{ul}n_u$. Induced emission becomes important only if extremely high source intensities (spectral densities) are available (e.g., lasers). However, spontaneous emission will be similar in magnitude to induced emission at longer wavelengths, such as the far IR and microwave regions. In addition, since absorption spectroscopy depends on the difference in population in the lower and upper levels, the absorption signal is large in the optical region (UV-visible-IR) and low in the microwave and radio frequency regions of the electromagnetic spectrum; as a result, instrumental problems are significantly greater for long-wavelength spectroscopy.

The final term worthy of definition is *oscillator strength f*, which arose in the early days of spectroscopy to describe the number of electrons per atom undergoing radiational transitions. Although this term has lost much of its meaning with the advent of the quantum theory of radiation, it is still used by spectroscopists, and tables of f values exist. The relationship between f_{ul} and A_{ul} is given by:

$$f_{ul} = A_{ul}\lambda_{ul}^2 \left(\frac{mc}{8\pi^2 e^2} \right) \tag{10}$$

and also

$$f_{ul}g_u = f_{lu}g_l \tag{11}$$

B. SIGNAL EXPRESSIONS IN SPECTROSCOPY

1. GENERAL COMMENTS (REFER TO FIGURE 4.4)

Signal expression is the expression that relates the measured signal to the analyte concentration or amount. No attempt is made to derive in detail the exact expressions, but rather the reader is referred to the literature.

2. OPTICAL SPECTROSCOPY (1–5)

a. General Expression for Radiance (See Table 4.1)

The basic radiance expressions in atomic and molecular optical (UV, visible, and IR regions) spectroscopy are given in Table 4.2. The reader should note the extensive footnotes. The resulting spectrometer signal expression for emission and luminescence spectrometry in all cases depends directly on the radiance values (assuming, as in Table 4.1, that resonance luminescence is being measured or modifying the expressions

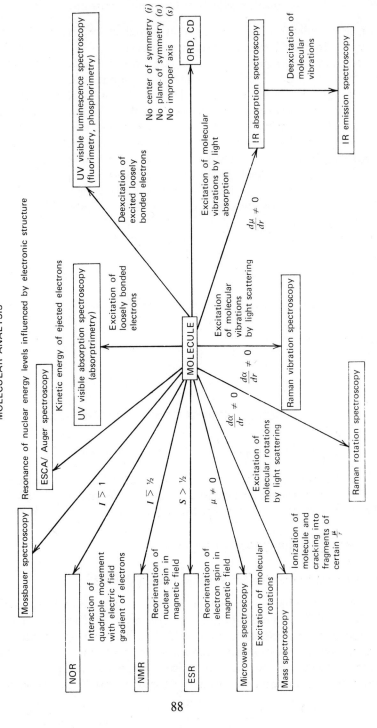

MOLECULAR ANALYSIS

ATOMIC ANALYSIS

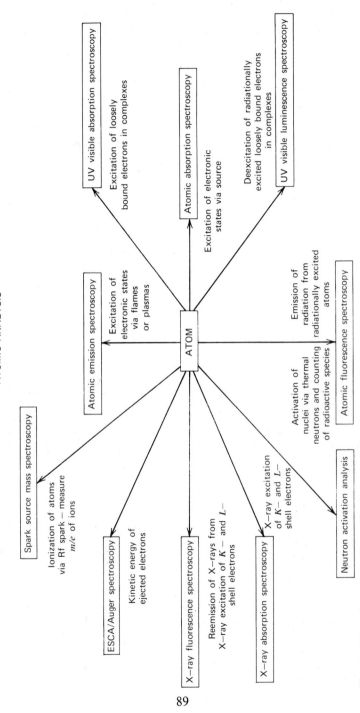

Fig. 4.4 Basis of and general uses of various types of spectroscopy.

TABLE 4.2. Radiance Expressions in Atomic and Molecular Optical Spectroscopy[a,b,c]

Method[d]	Low Optical Density (Low n_l)[e,f,g]	High Optical Density (High n_l)[e,g]
AES, MES	$B_{AES,MES} = C_2 B_{B\lambda_0} K_0 n_l fl\delta\lambda_D$	$B_{AES} = B_{B\lambda_0}\sqrt{C_3 K_0 n_l fla\delta\lambda_D^2}$
AAC	$B_{AAC} = C_2 B_{C\lambda_0} K_0 n_l fl\delta\lambda_D$ $\alpha_{AAC} = \dfrac{C_2 K_0 n_l fl\delta\lambda_D}{\delta\lambda_s}$	$B_{AAC} = B_{C\lambda_0}\sqrt{C_3 K_0 n_l fla\delta\lambda_D^2}$ $\alpha_{AAC} = \dfrac{1}{\delta\lambda_s}\sqrt{C_3 K_0 n_l fla\delta\lambda_D^2}$
AAL, MAL	$B_{AAL,MAL} = B_L K_0 n_l fl\delta_0$ $\alpha_{AAL,MAL} = K_0 n_l fl\delta_0$	$B_{AAL,MAL} = B_L$ $\alpha_{AAL,MAL} = 1$
AFC, MLL	$B_{AFC,MLC} = C_2 B_{C\lambda_0} K_0 n_l fl\delta\lambda_D\, Y\!\left(\dfrac{\Omega_A}{4\pi}\right)$	$B_{AFC} = \dfrac{C_3}{C_2} B_{C\lambda_0} a\delta\lambda_D\sqrt{\dfrac{L}{l}}\; Y\!\left(\dfrac{\Omega_A}{4\pi}\right)$
AFL, MLL	$B_{AFL,MLC} = C_2 B_{C\lambda_0} K_0 n_l fL\delta\lambda_D\, Y\!\left(\dfrac{\Omega_A}{4\pi}\right)$	$B_{AFL} = B_L\sqrt{\dfrac{4aL}{\pi K_0 n_l fl^2}}\; Y\!\left(\dfrac{\Omega_A}{4\pi}\right)$

[a] All radiance values are evaluated at λ_s, the spectrometer setting (see text) that is generally the peak of the transition. All expressions given for luminescence spectrometry are for resonance transitions. To change the expressions to nonresonance luminescence, the self-absorption factor must be for the reabsorption process rather than the initial exciting process, and the quantum yield must be changed to an energy or flux yield. Also, it should be stressed that the luminescence expressions are given for the case in which there are *no* prefilter or *no* postfilter effects, since these are the only analytically useful cases. The expressions in the table are taken primarily from Winefordner, Svoboda, and Cline (4) and from Winefordner, Schulman, and O'Haver (5). For absorption spectrometry, α is the fraction absorbed.

[b] Definitions of symbols:

$B_{B\lambda_0}$ = blackbody spectral radiance at temperature of atomizer (cell) and at wavelength λ_0 (in $\mathrm{W\ m^{-2}\ sr^{-1}\ m^{-1}}$)

$B_{C\lambda_0}$ = continuum source spectral radiance at temperature of source and at wavelength λ_0 (in $\mathrm{W\ m^{-2}\ sr^{-1}\ m^{-1}}$)

B_L = radiance of line source, $\int B_{L\lambda}\,d\lambda$ (in $\mathrm{W\ m^{-2}\ sr^{-1}}$)

$C_2 = \sqrt{\pi}\,/2\sqrt{\ln 2}$ (no units)

$C_3 = \sqrt{\pi}\,\sqrt{\ln 2}$ (no units)

a = damping constant, $\sqrt{\ln 2}\,\delta\lambda_C/\delta\lambda_D$ (no units)

f = absorption oscillator strength (no units)

l = absorption or emission path length (in m)

L = luminescence path length (in cm)

Y = luminescence quantum efficiency (no units)

n_l = concentration of species at lower level (in $\mathrm{m^{-3}}$)

$\delta\lambda_D$ = Doppler half-width (in m)
$\delta\lambda_C$ = Lorentzian (collisional) half-width (in m)
$\Delta\lambda_s$ = spectrometer bandpass (in m)

δ_0 = Voigt profile factor for $a = \sqrt{\ln 2}\ \delta\lambda_C/\delta\lambda_D$, where $\delta\lambda_C$ and $\delta\lambda_D$ refer to the absorption line or band, and for $\bar{V} \cong \delta\lambda_s/\delta\lambda_A$ where $\delta\lambda_A$ is the absorption line or band half-width and $\Delta\lambda_s$ is the source (or spectrometer) half-width, that is, δ_0 is evaluated from the Voight integral of

$$\delta_0 = \frac{a}{\pi} \int_{-\infty}^{+\infty} \frac{e^{-y^2}\,dy}{a^2 + (\bar{v} - y)^2}$$

where a and $\bar{v}$ are as defined above and y is $2\sqrt{\ln 2}\ (\lambda - \lambda_0)/\delta\lambda_D$ (where λ is any wavelength; λ_0 is the peak absorption wavelength, and $\delta\lambda_D$ is the Doppler half-width of the absorption line).

Ω_A = excitation source solid angle impinging in absorption cell (in sr)
K_0 = modified absorption coefficient, $k_0 n_1 f$ (in m^2)
k_0 = absorption coefficient for pure Doppler (gaussian) broadening (in m^{-1})

$$k_0 = K_0 n_l f = \frac{2\sqrt{\ln 2}\ \times \lambda_0^2 F_s}{\sqrt{\pi}\ C\delta\lambda_D}$$

$$x = \frac{\pi e^2}{mc} = 3.0 \times 10^{-4}\ m^2\ s^{-1}$$

λ_0 = peak absorption or emission wavelength (in m)
c = speed of light (in m s^{-1})
m = mass of electron (in kg)
F_s = source factor to account for saturation of energy levels (no units)
F_{sl} = source factor for line source, $E^s/(E + E_s)$ (no units)
F_{sc} = source factor for continuum source, $E_\lambda^s/(E_\lambda + E_s^2)$ (no units)
E = integrated spectral irradiance of line source, $B\Omega_A$ (in W m^{-2})
E_λ = spectral irradiance of continuum source, $B_\lambda \Omega_A$ (in W m^{-2} m^{-v})
E^s = saturation irradiance of line source (applies only to lasers) corresponding to source irradiance resulting in a decrease in the peak absorption coefficient by a factor of 2 over that observed with a low-intensity source (nonlaser source) (in W m^{-2})
E_λ^s = saturation spectral irradiance of continuum source (applies only to lasers) corresponding to source spectral irradiance resulting in a decrease in the integrated absorption coefficient, $\int_0^\infty k_\lambda d\lambda$ for the line or band, by a factor of 2 over that measured with a low-intensity source.

[c]A low-intensity source is any line source with an integrated spectral irradiance of $\lesssim 10^{-2}$ W cm^{-2} sr^{-1}, for example, electrodeless discharge lamps, hollow cathode lamps, or any continuum source with a spectral irradiance of $\lesssim 1$ W cm^{-2} sr^{-1} nm^{-1}. High-intensity (laser) sources correspond to $\gtrsim 100$ W cm^{-2} or $\gtrsim 10^4$ kW cm^{-2} nm^{-1}.

The quantum yield of a transition is defined by $Y = A/(A + K)$, where A is the emission transition probability, and K is the probability of radiationless transitions.

[d]AES, Atomic emission spectrometry (excitation of atoms via combustion flames and high-temperature plasmas such as arcs, sparks, RF plasmas, etc.); MES, molecular emission spectrometry (excitation of molecules via combustion flames and high-temperature plasmas); AAC, atomic absorption spectrometry with a continuum source and with a flame or nonflame

atomizer; AAL, atomic absorption spectrometry with a line source and with a flame or nonflame atomizer; MAL, molecular absorption spectrometry with a line source; AFC, atomic fluorescence spectrometry with a continuum source and with a flame or nonflame atomizer; AFL, atomic fluorescence spectrometry with a line source and with a flame or nonflame atomizer; MLC, molecular luminescence spectrometry with a continuum source; MLL, molecular luminescence spectrometry with a line source.

Continuum source refers to a source that has broad spectral output, that is, $B_{C\lambda}$ is constant over the spectral absorption line or band. Line source refers to a source that has a spectral output narrower than the absorption line or band. A line source can be a continuum source with a monochromator having a spectral bandpass $\Delta\lambda_s$ less than the absorption line or bandwidth. Fluorescence for atoms refers to any photoluminescence transition. Luminescence for molecules refers to a photoluminescence transition; however, fluorescence specifically refers to spin-allowed transitions and phosphorescence to spin-unallowed transitions. Emission for atoms and molecules refers to collisionally activated species in hot plasmas and not to chemiluminescence, bioluminescence, cathodoluminescence, and so on.

[e]Low optical density means $k_\lambda x \lesssim 0.05$, where k_λ is the spectral absorption coefficient and x is the absorption path length.

[f]The expressions for molecular (M) spectrometry given for low optical density also apply approximately to high optical densities, because self-absorption should be negligible, that is, most molecular emission is Stokes' non-resonance emission (longer-wavelength emission).

[g]The B's are radiances for the processes of concern; that is, they refer to the integrated spectral radiances or the line or band areas $\int B_\lambda d\lambda$. For example, for a molecular electronic band B is the area under the entire band as determined by the vibrational-rotational fine structures and, in IR absorption spectrometry, B is the area under the vibrational band as determined by the rotational fine structure.

---

for non-resonance luminescence). The measured emission or luminescence signal:

$$S_{\text{em or lum}})_{\lambda_s} = K_0 \int_{\lambda_l}^{\lambda_u} B_{E\lambda}(\text{or } B_{L\lambda}) T_\lambda F_\lambda \gamma_\lambda \, d\lambda \tag{12}$$

where K_0 is an optics (entrance and spectrometer) factor which includes the spectrometer slit width and height, or aperture area in the case of a filter system, and the measurement solid angle generally is determined by the spectrometer or filter aperture, T_λ is the transmission factor of all optics, F_λ is the spectrometer function (often triangular), γ_λ is the detector response factor, λ_u and λ_e are the wavelength limits of the spectrometer, that is, $\lambda_u \cong \lambda_s + \Delta\lambda_s/2$ and $\lambda_1 \cong \lambda_s - \Delta\lambda_s/2$ [where $\Delta\lambda_s$ is the spectral bandpass of the spectrometer of the optical system and λ_s is the wavelength setting (peak transmittance) of the spectrometer or optical system], and $B_{E\lambda}$ and $B_{L\lambda}$ are the spectral radiances of emission and luminescence, respectively (in W m^{-2} sr^{-1} m^{-1}) and can be found approximately from the expressions in Table 4.1 by dividing by the line or band half-widths. The emission or luminescence signal is of course evaluated at λ_s. For atomic transitions measured with medium or low-resolution spectroscopic devices, λ_u can be replaced by ∞ and λ_l by 0 to simplify the integration. For broad band molecular transitions or for atomic lines measured with high-resolution spectrometers, λ_u and λ_l are finite, as discussed above. The specific factors influencing the measured emission or luminescence signal (or absorption signal) are discussed in the sections that follow.

The measured absorption signal is seldom $B_{A_{\lambda,s}}$ but is rather $\alpha\lambda_s$, the fraction absorbed, or A_{λ_s}, the absorbance. These terms are related as follows:

$$\alpha_{\lambda_s} = \frac{B_{A_{\lambda,s}}}{\int B_\lambda \, d\lambda} \tag{13}$$

where $\int_{\lambda_l}^{\lambda_u} B_\lambda \, d_\lambda$ is the integrated source radiance reaching the sample. The absorbance A_{λ_s} is given by:

$$A_{\lambda_s} = -\log(1-\alpha) \tag{14}$$

The relationship between the absorption coefficient k (in cm^{-1}) and the molar extinction coefficient ϵ (in cm^{-1} M^{-1}) is:

$$k_0 = 2.3 C \epsilon_0 \tag{15}$$

where the subscript 0 refers to peak values, and C is the concentration of analyte (in mole liter^{-1} or M) in solution. Also,

$$k_\lambda = 2.3 C \epsilon_\lambda \tag{16}$$

where the subscripts refer to any wavelength λ.

b. Concentration of Analyte in Gaseous State—Flames, Arcs, and Such

The concentration of analyte in the lower atomic or molecular states of a gaseous species n_l is related to the concentration of analyte in solution C (in mole liter^{-1}). The relationships, however, depend on the manner of converting the sample solution to atomic vapor. For a flame atomizer or any high-temperature plasma when the sample is introduced as an aerosol, it can be shown (5) that:

$$n_l = \frac{N_A F \epsilon \beta C g_l}{Q_t e_f Z_T} \tag{17}$$

where N_A is Avogadro's number (6×10^{23}), ϵ is the efficiency of conversion of sample solution to submicroscopic species (atoms, molecules, and ions) at the height of measurement (no units), β is the efficiency of conversion of submicroscopic species to atoms (no units), C is the analyte solution concentration (in mole liter^{-1}), g_l is the statistical weight of the lower state involved in the transition (no units), Q_t is the flow rate of unburnt gases into the flame (or any other plasma where the sample is introduced as an aerosol (in m^3 s^{-1}), $\epsilon_f = (T/T_R)(n_T/n_R)$ [where T is the flame (or plasma) temperature, T_R is the temperature of the gases in the tanks, and n_T is the number of moles of flame gas products resulting from n_R moles of room temperature reactant gases], and Z_T is the electronic partition function (no units) and is given by:

$$Z_T = \sum_i g_i \epsilon^{-E_i/kT} \tag{18}$$

where g_i is the statistical weight of level i, E_i is the energy of level i, (in eV) k is the Boltzmann constant (8.31×10^{-5} eV K^{-1}), and T is the flame (or plasma) temperature (in K).

For the case of nonflame cells such as furnace and filament atomizers, and for high-temperature arc and spark plasmas where the sample is introduced as a so via thermal or electrical means (i.e., in the case of furnaces and filaments, a solution sample is placed in or on the atomizer and dried prior to atomization and, in the case of high-temperature plasmas, the sample is usually part of the electrode system or is thermally vaporized or electrically sputtered from a filament into the plasma), the relationship between n_1 and the volatilization rate Q (in atoms s^{-1}), the velocity v at which the particles travel upward from the surface of the atomizer (in m s^{-1}), and A the cross-sectional area of the source (in m^2) is given by:

$$n_1 = \frac{Q}{vA} = \frac{Q}{\psi} \tag{19}$$

where ψ is the transport rate of sample into the plasma or gas volume being studied (in m^3 s^{-1}). It is assumed in the above expression that lateral diffusion is negligible and that the processes involved in volatilization occur at a constant rate. The velocity v may be a result of either a gas flowing past the atomizer, as in many nonflame atomic absorption systems, or the combined effect of convection and an electrical field (on ions) in the case in which the sample is directly atomized into the plasma or gas volume being studied via electrodes, such as in the DC arc. The amount of analyte q (in kg) is related to the above in parameters by:

$$q = \frac{M_A}{N_A} \int_{t_Q} \frac{V_p n_l}{\tau} dt \cong \frac{M_A V_p n_l t_Q}{N_A \tau} \tag{20}$$

where M_A is the molecular weight of the analyte (in kg mole^{-1}), N_A is Avogadro's number, n_l is the number density of atoms in the lower state l, t_Q is the total volatilization time (in s), τ is the mean time of any species in a plasma volume or gas volume being studied V_p (in m^3). The plasma volume is related to vA or ψ by $V_p = \tau\psi$ or τvA, and so:

$$q \cong \frac{M_A \psi n_l t_Q}{N_A} \tag{21}$$

If the volatilization rate is not completely efficient in producing atoms in the lower state l, the above equations must be multiplied by $\epsilon\beta/Z_T$, where ϵ is now the efficiency of converting solid sample to submicroscopic species and β is the same as before. If an ion line is being measured, β is the efficiency of producing the appropriate ions rather than atoms. The above expression for q is valid for any discrete-type atomizer, that is, one in which a solid is atomized for a period of time t_Q, as long as the atomization is assumed uniform. The difficulty in using equations 19 and 21 is that the terms ψ (or vA) and Q may be extremely difficult to evaluate; especially the latter parameter. Boumans (6) has discussed evaluation of these terms for an arc atomizer. L'vov (7) has discussed evaluation of similar parameters for nonflame atomic absorption atomizers. In the case of electrical sputtering, as in the case of sparks, evaluation is even more difficult.

c. Concentration of Analyte in Liquid or Solid State

If the sample is assumed to be homogeneous, the concentration n (in m^{-3}) is given by:

$$n = \frac{N_A q}{M_A V} \tag{22}$$

where N_A is Avogadro's number, g is the mass of analyte (in kg) in the (solid or liquid) volume V (in m^3), and M_A is the molecular weight of the analyte. If the analyte reacts with itself or with other species, undergoes photodecomposition, or dissociates, associates, or undergoes some equilibrium process to change the concentration of the measured species, equation 19 must be multiplied by a fraction γ to account for the loss of analyte. Of course, it is assumed that the analyte does not react or interact in such a manner that the signal of the measured species is swamped by a reaction product; generally such interferences can be avoided by judicious choices of chemical and spectroscopic conditions. If V is in liters rather than cubic meters,

$$n = 10^3 N_A C \quad (\text{in m}^{-3}) \tag{23}$$

If the chemical system is in thermodynamic equilibrium, it is generally a simple matter to calculate γ, assuming all the important chemical equilibria and their equilibrium constants are known.

3. X-RAY SPECTROSCOPY

The basic expressions for X-ray absorption spectroscopy with either monochromatic or polychromatic X-rays is basically (7) the same as for optical absorption spectroscopy, that is,

$$A = \log\left(\frac{S_s}{S_u}\right) = k'(q_u - q_s) \tag{24}$$

where q_u is the amount of analyte in the unknown (in kg), q_s is the amount of analyte in a standard similar in concentration to the unknown (in kg), S_u is the signal (number of counts in a specified time) due to the unknown, and S_s is similarly defined for the standard. Equation 24 is very similar to the expression for absorbance for optical spectroscopy with a continuum (or line) source. The effective absorption coefficient k' (in kg^{-1}) is related to the mass absorption coefficient μ (in m^2 kg^{-1}) and the linear absorption coefficient k (in m^{-1}) by:

$$\mu = \frac{k}{\rho} = 2.3 S_c k' \tag{25}$$

where S_c is the cross-sectional sample area (in m^2) perpendicular to the exciting beam, and ρ is the density of the sample (in kg m^{-3}).

The basic irradiance expression for X-ray fluorescence with a line source for excitation is given (9) by:

$$E_{XF} = E_L G_l (\csc\phi_1)(\sigma_A \beta n_A l)\left(1 - \frac{1}{J_K}\right) Y_K \theta \alpha_a \frac{[1 - e^{-\mu^* \rho l}]}{\mu^* \rho l} \tag{26}$$

assuming front surface (90°) excitation-fluorescence measurement, assuming no enhancement effect, and assuming a sample of thickness l (in m). The terms in equation 26 are:

E_{XF} = X-ray fluorescence from sample (in W m^{-2})

E_L = line radiance of exciting source incident on the sample, generally being the K_α doublet, for the source (in W m^{-2} sr^{-1})

G_1 = a geometric factor concerning the collimation and impingement of X-rays from the source onto the sample (no units)

ϕ_1 = angle of incidence of X-ray exciting beam with respect to surface (in deg.)

σ_A = total photoelectric cross section for the exciting radiation (in m^2)

n_A = concentration of analyte in sample (in cm^{-3}) (q_A = analyte amount in sample = $M_A n_A V / N_A$, where V is the sample volume; it is assumed that the analyte is homogeneously distributed)

M_A = molecular weight of analyte (in kg mole^{-1})

N_A = Avogadro's number (in mole^{-1})

β = geometric factor accounting for the area over which X-rays are absorbed and over which X-rays are emitted (no units)

l = absorption path length (in m)

J_K = ratio between photoelectric mass absorption coefficient at the top and bottom of the K-absorption edge, assuming K-α-line used for excitation (no units)

$1 - 1/J_K$ = fraction of photoelectric events occurring in the K-shell (no units)

Y_K = fluorescence quantum yield for X-rays from analyte (no units)

θ = fraction of K-fluorescence X-rays of energy E_j (energy of K-α X-rays as compared to total K-X-rays emitted (no units)

α_a = fraction of exciting X-rays lost as a result of air (atmosphere) absorption (no units)

μ^* = mass absorption coefficient for exciting X-rays (in m^2 kg^{-1})

$\mu^* = \mu_1 \csc \phi_1 + \mu_2 \csc \phi_2$ (in m^2 kg^{-1})

ϕ_2 = angle of measurement of emitted fluorescence X-ray taken with respect to surface of sample (in deg)

μ_1 = total mass absorption coefficient of sample for exciting X-rays (in m^2 kg^{-1})

μ_2 = total mass absorption coefficient of sample for emitted X-rays (in m^2 kg^{-1})

ρ = density of sample (in kg m^{-3})

It should be noted that the expression for X-ray fluorescence contains terms accounting for matrix absorption of exciting and fluorescence X-rays, whereas luminescence optical spectroscopy expressions are given only for the analyte. Because absorption by the matrix is *always* important and generally impossible or difficult to separate in X-ray fluorescence spectroscopy, but only rarely important in atomic fluorescence spectroscopy and often avoidable (correction or separation) in molecular luminescence spectroscopy; it was felt unnecessary to account for matrix interference in optical spectroscopy.

Several limiting cases of equation 26 are possible. If the sample is very thin (thin film),

$$E_{XF} = E_L G_1 (\csc\phi_1)(\sigma_A n_A l)\left(1 - \frac{1}{J_K}\right) Y_K \theta \alpha_a \tag{27}$$

If the sample is infinitely thick,

$$E_{XF} = E_L G_1 (\csc\phi_1)\left(\frac{\sigma_A n_A}{\mu^* \rho}\right)\left(1 - \frac{1}{J_K}\right) Y_K \theta \alpha_u \tag{28}$$

The above expressions (equations 27 and 28) apply to X-ray fluorescence spectrometry, whatever the detection device, that is, whether crystal spectrometers and scintillation detectors are used or whether the newer semiconductive detectors with no spectrometers are used.

The electron microprobe is a special type of X-ray spectrometric system in that X-rays from a surface are produced via a focused electron beam. The general signal irradiance expression for the (10, 11) electron microprobe signal resulting from a metallic film of thickness l (in m) is:

$$E_{XFE} = \frac{K\rho_A N_A \Phi_e}{M_A} \int_0^l e^{-\alpha\rho z} e^{-(\mu^*\rho z)\csc\theta} dz \tag{29}$$

where
   K = constant depending on V/V_K (no units)
   V = energy of incident electrons (in keV)
   V_K = critical excitation potential of primary radiation (in keV)
   ρ_A = density of analyte (in kg m^{-3})
   M_A = atomic weight of analyte (in kg mole^{-1})
   N_A = Avogadro's number (in mole^{-1})
   Φ_e = electron flux
   α = constant dependent on K/V^2 (in m^2 kg^{-1})
   ρ = density of sample (in kg m^{-3})
   z = absorption path length for electrons (in m)
   θ = collection angle from sample taken with respect to exciting electron beam (in deg)
   l = thickness of sample (in m)
   μ^* = mass absorption coefficient of sample for emitted X-rays (in m^2 kg^{-1})

If the above integration is performed and n_A, the concentration of analyte (in m^{-3}) is substituted for $\rho_A N_A / M_A$,

$$E_{XFE} = K n_A \Phi_e l \left(\frac{1 - e^{-(\mu^* + \alpha)\rho l}}{(\mu^* + \alpha)\rho l}\right) \tag{30}$$

and, if the sample is infinitely thick,

$$E_{XFE} = Kn_A \Phi_e \left(\frac{1}{\mu^* + \alpha} \right) \left(\frac{1}{\rho} \right)$$ (31)

and, if the sample is very thin ($l \sim 0$),

$$E_{XFE} = Kn_A \Phi_e$$ (32)

Of course, n_A is related to the amount of analyte g_A in the same manner as for X-ray fluorescence (see equation 22).

4. NEUTRON ACTIVATION (12–14)

The basic expression for the counting rate in neutron activation analysis is well-known and is given (Chapter 2, Section B) by:

$$R_R = \epsilon p N_Q \Phi \sigma_Q (1 - e^{-0.693 t_i / t_{1/2,R}})$$ (33)

where R_R is the counting rate for a radioactive nucleus R produced by irradiation of a sample with element Q (in s^{-1}), ϵ is the fraction of the total counting rate in the measured photopeak (no unit), p is the fraction of the isotopes of Q that can be converted to radioactive isotopes R via an (n, γ) readout (no units), N_Q is the number of analyte Q atoms in the irradiated sample (no units), Φ is the neutron flux incident on the sample (in m^{-2} s^{-1}), σ_Q is the cross section of the nuclide for reaction with thermal neutrons of a given energy (in m^2), t_i is the irradiation time (in s), and $t_{1/2,R}$ is the half-life of the radioactive isotope R (in s). Equation 33 can be modified in terms of sample amount of Q, q_Q, by:

$$R_R = \epsilon p g_Q \frac{N_A}{M_Q} \Phi \sigma_Q (1 - e^{-0.693 t_i / t} 1/2, R)$$ (34)

where N_A is Avogadro's number, and M_A is the atomic weight of Q (in kg mole^{-1}). In the case of neutron activation analysis, the source intensity is given in terms of a flux rather than a radiance or irradiance, since the sample is within the neutron reactor.

If the irradiation time t_i is *much smaller* than the half-life $t_{1/2,R}$,

$$R_R = \frac{0.693 \epsilon p q_Q N_A \Phi \sigma_Q t_i}{M_Q t_{1/2,R}}$$ (35)

and if the irradiation time t_i is much larger than the half-life $t_{1/2,R}$,

$$R_R = \frac{\epsilon p q_Q N_A \Phi \sigma_Q}{M_A}$$ (36)

It should be pointed out that, although the major source of activation is thermal neutrons, there are several other means of activation of nuclei (15, 16). These means include *fast neutrons* for example, 14 MeV neutron generators are used commercially, *gamma photons*, for example, 7 to 20 MeV gamma rays [also called *photonactivation* (15, 16)], and *charged particles*, for example, *alpha particles* of 44 MeV, *protons* of 6 to 15 MeV, and *deuterons*. The expressions given above for thermal neutron activation analysis also apply to activation by means of these other means, as long as σ_Q and Φ are changed to correspond to the appropriate reaction and activation processes, respectively.

Although *isotopic* and *nonisotopic* methods are generally not considered instrumental methods of analysis, they involve the measurement of low concentrations of elements via radioactivity. Radioactive methods (16, 17) involve either the use of a radioactive tracer of the element (analyte) to be determined (isotopic labeling) or the use of a radioactive tracer of an element different from the one being measured (nonisotopic labeling). In isotopic labeling methods, the principle is that, if the fraction of the element to be determined can be isolated (the fraction being determined by measuring the radioactivity) and if the mass of the element in that fraction can be determined either by some conventional measurement technique or by reacting the element with an equivalent amount of a reagent, the original mass of the analyte species can be calculated. In nonisotopic labeling methods, the principle simply involves reacting the element to be determined with an excess of a second material which is radioactively labeled and, if the fraction of the second material thus released is measured via its radioactivity, the original mass of the analyte element can be determined.

In addition to the above methods involving radioactive tracers, there are also *radiometric titrations* which involve use of a labeled titrant, a labeled titrand, or a labeled indicator. The change in radioactivity is measured as the titrant is added.

Single isotope dilution involves measuring the radioactivity R_0 (in counts s^{-1}) of the tracer of weight q_0 (in kg) and then the activity R_x (in counts s^{-1}) of the tracer after addition to a weight q_x (in kg) of the analyte. The basic equation is:

$$q_x = q_0 \left(\frac{R_0}{R_x} \right) \tag{37}$$

as long as $q_0 \ll q_x$.

Substoichiometric isotopic dilution involves removal of equal partial amounts of the analyte element from the tracer and from the tracer plus sample solution. If these activators are again R_0 and R_x, respectively,

$$q_x = q_0\left(\frac{R_0}{R_x} - 1\right) \tag{38}$$

where q_0 and q_x are as defined above.

Isotopic exchange has also been used for trace element analysis. Consider:

$$M^*X + MY = (M, M^*)X + (M, M^*)Y$$

$$MX + M^*Y = (M, M^*)X + (M, M^*)Y \tag{39}$$

where MX and MY are complexes of M, and M* is the activated metal ion; also, MX and MY must be easily separable by solvent extraction and the most strongly complexing liquid must be associated stoichiometrically with one portion of M prior to exchange. Thus, if the amount of M associated with Y (q_{MY}) is unknown, and that with X (q_{MX}) is known,

$$q_{MY} = q_{MX}\left(\frac{1}{D_{M^*}}\right) \tag{40}$$

where D_{M^*} is the distribution of M* between the organic solvent and the aqueous phase, assuming MX is the extractable species.

Nonisotopic methods include *isotopic derivatives* and *derivative dilution, isotopic replacement,* and *radiorelease.* In the former, the element to be determined is made to react quantitatively with a labeled reagent of known specific activity S_R, (where $S_R = R_R/q_R$ [where R_R is the radioactivity (in counts s^{-1}), and q_R is the weight of the labeled reagent], to form a compound which can be separated and counted. If the recovery is quantitative, the amount of the element q_x can be determined:

$$q_x = \frac{R_X}{R_R}(q_R \cdot \epsilon) \tag{41}$$

where R_X is the activity (in counts s^{-1}) of the compound formed, and ϵ is the ratio of equivalent weights of the element and reagent. Isotopic displacement involves forming a complex of high stability, a certain reagent, and measurement of the released activity from a labeled complex of that reagent of lower stability. In radiorelease methods, the released radioactivity of a labeled material when the labeled material reacts with the trace element (the element may or may not be isotopic with the label); the reaction must be highly selective.

5. SPARK SOURCE MASS SPECTROMETRY (17, 18)

There does not appear to be any formal relationship for the signal expression in spark source mass spectrometry that is comparable to those for the other methods discussed in this book (19). The major reason for the lack of a formal signal expressions is probably a direct result of the complexity of operating and controlling and the lack of knowledge concerning the Rf spark source for producing ions. Other more theoretically sound but less sensitive sources of producing ions of solid species for quantitative analysis have not been used in commercial mass spectrometers for routine trace analysis of elements. Other difficulties include plate fogging and calibration (assuming the photographic emulsion detector—the common one for spark source mass spectrometry—is used), diffuse plate background due to scatter of ions by collisions with residual gas molecules and to secondary products resulting in a high background near the major component, line background due to traces of hydrocarbon in the vacuum system due to overlap of mass lines of similar m/e due to production of many multiply charged ions by the spark, and difference in composition of the ion sample produced by the Rf spark and the composition of the sample.

The overall signal for the element j is given (21) by:

$$S_j = S_{sj} f_t f_p n_s \left(\frac{1}{t_A} \right) \qquad (42)$$

where S_{sj} is the ratio of ions produced per sample atom vaporized, f_t is an instrumental transmission factor, f_p is a plate response, and n_s is the total number of sample atoms vaporized in a unit time t_A (in s) factor. The term S_j is simply the number of ions of the analyte of given m/e recorded with the double-focusing mass spectrographs (it lies between 10^{-7} and 10^{-8}). The factor S_{sj} depends on the type of solid sample and the vaporization conditions (i.e., local and bulk temperatures, the diffusion coefficient of analyte in the solid and vapor phases, the ionization work function, and the sputtering yield), on the type and ionization conditions of the vaporized analyte atoms (i.e., the kinetic energy of the species, the ionization potential and cross section, the partial pressure of the species, and the condensation coefficient), and on the energy and density of the ionizing electrons in the Rf plasma. The term f_t depends on the energy and charge distributions of the ions produced in the Rf plasma and the instrumental geometry. The term f_p depends on the mass and charge of the ions and the type of emulsion.

The plate exposure H from a ion beam can be expressed (21) by the empirical expression:

$$H = f_i n S_j = f_i n S_{sj} f_s f_p \tag{43}$$

where all terms have been defined above, except n which is the analyte concentration and f_i which is the isotope fraction of the analyte element isotope being measured in the sample. If an electrical detection system is used, f_p is defined for the electrical detector.

6. ELECTRON SPECTROSCOPY (22–24)

The basic electron flux (in m s^{-1}) expression in electron spectroscopy is given by:

$$\phi_{ES} = \sigma_A n_A \alpha l \Phi^0 (1 - e^{-d/l}) K \tag{44}$$

where

σ_A = photoelectron cross section (for X-ray or vacuum UV excitation) for a given shell of element A for incident photons of energy $h\nu$ (in m^2)

n_A = concentration of element A in homogeneous sample (in m^{-3})

l = mean escape depth for electrons of energy E (in m)

Φ^0 = photon flux of incident monochromatic line source beam of x-rays at surface of sample (in s^{-1})

α = sample area being illuminated (in m^2)

K = instrumental and geometric factor, containing parameters such as solid angle of measurement (no units)

d = effective sample thickness (in m)

The electron flux ϕ_{ES} (in counts s^{-1}) is dependent on the electron energy E, that is, ϕ_{ES} depends on l which depends on E. Actually, equation 41 should also apply to electron impact spectroscopy and to Auger spectroscopy as long as Φ^0 and σ_A refer to the appropriate processes, that is, for electron impact spectroscopy, Φ^0 refers to the electron flux incident on the sample, and σ_A refers to the cross section for electron impact resulting in excitation-ionization of the analyte; for Auger spectroscopy, Φ^0 has the same meaning as in equation 41 if the source of excitation-ionization is an X-ray or vacuum UV source, or has the same meaning as for electron impact spectroscopy if the source of excitation-ionization is the electron beam and σ_A is defined for the appropriate source of excitation-ionization.

Perhaps the four basic types of spectroscopy should be reviewed briefly: *electron impact spectroscopy* refers to the use of an electron beam source which causes excitation of valence shell electrons to higher electronic states (the sample is a gas) and so gives information about electronic transition energies via the loss in energy of the electron beam; *photoelectron vacuum ultraviolet (VUV) spectroscopy* refers to the use of a VUV source which causes ionization of the valence electrons and so gives information about ionization energies of atoms and molecules via the kinetic energy of the ejected electron (the sample is a gas); *ESCA* (electron spectroscopy for chemical analysis) refers to the use of an X-ray beam source which causes ionization of inner shell electrons (the sample is a solid) and so gives ionization energy information via the kinetic energy of the ejected electrons; *Auger spectroscopy* refers to the use of either high-energy electrons or X-rays to eject the inner shell electrons and the measurement of the kinetic energy of a secondary (Auger) electron (25) resulting when the ionized atom relaxes (the sample is a solid). The absolute signal levels of electron impact spectrometers are often several orders of magnitude larger than the other electron spectroscopic methods.

BIBLIOGRAPHY

F. A. Jenkins and H. E. White, *Fundamentals of Optics*, McGraw-Hill, New York, 1957.

R. W. Wood, *Physical Optics*, Dover, New York, 1934.

H. D. Young, *Fundamentals of Optics and Modern Physics*, McGraw-Hill, New York, 1968.

REFERENCES

1. G. T. J. Alkemade, "A Theoretical Discussion of Some Aspects of Atomic Fluorescence Spectroscopy in Flames," in *Proceedings of the International Atomic Absorption Conference, Shefield, England, July 1969*, Butterworth, London, 1970.

2. H. P. Hooymayers, *Spectrochim. Acta*, **23B**, 567 (1968).

3. P. J. T. Zeegers, R. Smith, and J. D. Winefordner, *Anal. Chem.*, **40** (13), 26A (1970).

4. J. D. Winefordner, V. Svoboda, and L. Cline, *Crit. Rev. Anal. Chem.* **1**, 233 (1970).

5. J. D. Winefordner, S. G. Schulman, and T. C. O'Haver, *Luminescence Spectrometry in Analytical Chemistry*, John Wiley, New York, 1972.

6. P. W. J. M. Boumans, *Theory of Spectrochemical Excitation*, Adam Hilger, London, 1966.

7. B. V. L'vov, *Atomic Absorption Spectrochemical Analysis*, Adam Hilger, London 1970.

8. H. A. Liebhafsky, H. G. Heiffer, E. H. Winslow, and Y. D. Zemany, *X-Ray Absorption and Emission in Analytical Chemistry*, John Wiley, New York, 1960.

9. L. S. Birks, *X-Ray Spectrochemical Analysis*, Interscience, New York, 1969.

10. L. S. Birks, *Electron Probe Microanalysis*, John Wiley, New York, 1962.

11. R. Theisen, *Quantitative Electron Microprobe Analysis*, Springer Verlag, New York, 1965.

12. J. M. A. Lenihan and S. T. Thomson, Ed., *Activation Analysis*, Academic Press, London, 1965.

13. M. Rakovk, *Activation Analysis*, Iliffe, London, 1970.

14. R. F. Coleman and T. B. Pierce, *Analyst*, **92**, 1 (1967).

15. G. J. Lutz, *Anal. Chem.*, **43**, 93 (1973).

16. A. A. Smales, "Radioactivity Techniques in Trace Characterization," in *Trace Characterization, Chemical and Physical*, NBS, Monograph 100, U. S. Govt. Printing Office, Washington, D. C., 1967.

17. A. J. Ahearn, Ed., *Mass Spectrometric Analysis of Solids*, Interscience, New York, 1965.

18. G. H. Morrison, Ed., *Trace Analysis, Physical Methods*, Elsevier, Amsterdam, 1966.

19. A. J. Ahearn, "Spark Source Mass Spectrometric Analysis of Solids," in *Trace Characterization, Chemical and Physical*, NBS Monograph 100, U. S. Govt. Printing Office, Washington, D. C., 1967.

20. R. E. Honig, "Analysis of Solids by Mass Spectrometry," in *Advances in Mass Spectrometry*, U of F., W. L. Meady, Ed., Institute of Petroleum, London, 1966.

21. N. W. H. Addink, "Quantitative Determination of Impurities by Means of a Spark Source Mass Spectrometer," in *Mass Spectrometry*, R. I. Reed, Ed., Academic Press, London, 1965.

22. C. A. Lucchesi and J. E. Lester, *J. Chem. Ed.*, **50** (4), A205; **50** (5) A269 (1973).

23. D. M. Hercules, *Anal. Chem.*, **42** (1) 13A (1966).

24. J. F. Redina and R. E. Grojean, *Appl. Spectrosc.*, **25**, 24 (1971).

25. L. A. Harris, *Anal. Chem.*, **40** (14), 24A (1968).

OPTICAL INSTRUMENTATION

R. C. ELSER

Department of Pathology
York Hospital
York, Pennsylvania

In the preceding pages, the equations that account for the phenomena observable in optical systems were derived and explained. Their application requires optical instrumentation capable of measuring the phenomena these expressions predict. The component parts of such instrumental arrangements and the parameters used to characterize them are discussed in the following sections. Although sources of radiant energy necessarily are included as integral parts of commercial spectrometers, their characteristics of output radiance with respect to wavelength are not reviewed. It is assumed that appropriate sources will be employed.

A. CHARACTERISTICS OF SPECTROMETERS (1, 2)

1. PARAMETERS

Spectrometers as a general class of instruments include some device to transmit radiation emanating from a source to a reservoir containing the sample to be irradiated. If the device is a dispersive type, the following parameters may be used to reflect its qualitites: (a) dispersion, (b) spectral bandwidth, (c) resolving power, and (d) luminosity. A generalized diagram of a monochromator is given in Figure 5.1.

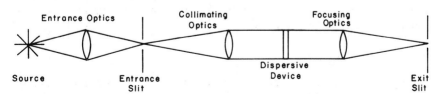

Fig. 5.1 Schematic diagram of typical optical monochromator.

Radiation emanating from a source is focused on the entrance slit by an appropriate optical arrangement which may include lenses or mirrors. Radiation passing through the slit that lies in the focal plane of the monochromator optical system is directed onto the dispersive element, which may be a prism or a grating, by means of collimating optics (lenses or mirrors) and finally focused onto an exit slit by means of appropriate lenses or mirrors. In practice, the collimating and focusing optics within the monochromator are often of the same focal length and may be combined into a single optical element.

a. Dispersion

One of the important characteristics of a monochromator is its ability to separate the component wavelengths of radiation incident on it into a spectrum. This is achieved by the dispersive element of the monochromator, either a grating or a prism, through its capacity for dispersing radiation of differing wavelengths at differing angles. The *angular dispersion* of the element is given by $d\theta/d\lambda$ (in rad nm^{-1}) where $d\theta$ is the angular separation of two dispersed beams differing in wavelength by $d\lambda$. The value of $d\theta/d\lambda$ depends on the construction of the prism or grating. For prisms, $d\theta/d\lambda$ is proportional to $1/\lambda^3$ and is also dependent on the material of which the prism is constructed. For gratings, $d\theta/d\lambda$ is inversely proportional to the distance between ruled lines. The linear dispersion $dx/d\lambda$ is the separation dx (in cm) between different wavelengths $d\lambda$ (in nm) at the focal plane. The relationship between angular dispersion and linear dispersion is given by:

$$dx/d\lambda = f d\theta/d\lambda \tag{1}$$

where f is the focal length of the monochromator.

A more commonly quoted (and more easily utilized) parameter of dispersion is the *reciprocal linear dispersion*:

$$d\lambda/dx = \frac{1}{f}\frac{d\lambda}{d\theta} \tag{2}$$

The reciprocal linear dispersion is usually quoted by manufacturers' literature on UV-visible spectrometers in angstroms per millimeter. Typical values range from 6 to 100 Å mm^{-1}.

b. Spectral Bandwidth

The spectral purity of radiation emerging from a spectrometer is indicated by its *spectral bandpass* or *spectral bandwidth*. This defines the range of wavelengths transmitted through the exit slit when the entrance slit is illuminated by a source of polychromatic radiation. To define accurately the spectral bandwidth of a spectrometer, the *slit function* of the monochromator must be known. This is equivalent to the plot of intensity versus wavelength when the monochromator is scanned through a monochromatic line emanating from a line source which evenly and completely illuminates the entrance slit. It is experimentally difficult to measure the true slit function of a spectrometer, because line sources of infinitesimal width are not available and thus the natural width of the source contributes to the slit function of the spectrometer. The theoretical slit function for a spectrometer having equal entrance and exit slits and in perfect optical alignment is a triangle (Figure 5.2) centered on the nominal wavelength setting of the monochromator and having a width at half maximum which is defined as the spectral bandwidth $\Delta\lambda_s$. Practically, the spectral bandwidth is given by:

$$\Delta\lambda_s = \frac{d\lambda}{dx} w \qquad (3)$$

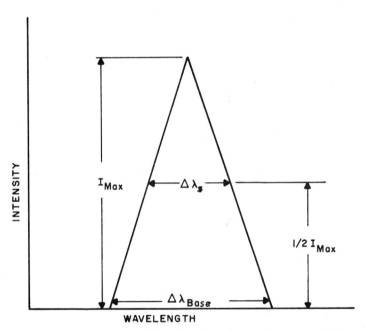

Fig. 5.2   Variation in intensity of radiation exiting from a monochromator exit slit, assuming a monochromatic source of radiation evenly illuminates the entrance slit. I_{max}, Maximum intensity exiting from monochromator exit slit; $\Delta\lambda_s$, spectral bandwidth of monochromator; $\Delta\lambda_{base}$ = base width of triangular function. In reality, the slit function may be much more gaussian or lorentzian in shape, but either profile can be approximated by a triangle.

where $d\lambda/dx$ is the reciprocal linear dispersion (in Å mm^{-1}), and w is the width (in mm) of the entrance or exit slit if they are equal or of the larger if they are unequal. The base width of the slit function $\Delta\lambda$ is given by:

$$\Delta\lambda_{base} = \frac{d\lambda}{dx}(w_s + w_e) \tag{4}$$

where w_s and w_e are the widths of the entrance and exit slits, respectively. When $w_s = w_e = w$,

$$\Delta\lambda_{base} = 2\frac{d\lambda}{dx}w \tag{5}$$

c. Resolution and Resolving Power

"Resolution" and "resolving power" are terms used to describe the quality of an optical system with respect to its capacity to distinguish between two adjacent resolution elements. Resolution may be defined as the minimum difference between two wavelengths that can be unequivocally differentiated. Conceptually, for complete resolution, this corresponds to nonoverlapping slit functions (see Figure 5.2), wherein the bases of the two slit functions are continuous but not overlapping. In this case, the resolution is equal to the base width of one of the slit functions given by equation 4 or 5, if perfect optical alignment and aberration-free components are assumed.

From equation 5 one might assume that $\Delta\lambda_s$ can be made infinitesimally small for a given system by decreasing the slit width w to infinitesimally small values, that is, $w \to 0$. Because of Fraunhofer diffraction, however, the geometric slit image can not become infinitely narrow, but turns into a diffraction pattern at a critical value of the slit width, which is determined by the focal length of the monochromator. Fraunhofer diffraction predicts that the width of the central fringe of the diffraction pattern at some wavelength λ is:

$$x_d = \frac{\lambda f}{a} \tag{6}$$

where f is the focal length of the optical system, and a is the diameter of the parallel beam of radiation reaching the disperser. For diffraction effects to result in a negligible influence on the recorded slit image, that is, for the slit image to be determined primarily by the geometric slit width, w

is given by:

$$w \geqslant \sqrt{\lambda f} \qquad (7)$$

When w is less than $\sqrt{\lambda f}$, the effective aperture or slit width x must be used to calculate spectral bandwidth. Thus, as slit widths approach zero, the spectral bandwidth $\Delta\lambda_d$ becomes:

$$\lim_{w \to 0} \Delta\lambda_d = \Delta\lambda_d = \frac{d\lambda}{dx} x_d = \left(\frac{d\lambda}{dx}\right)\left(\frac{\lambda f}{a}\right) \qquad (8)$$

The minimum effective slit width $\Delta\lambda_{\text{min}}$ of a real monochromator is also affected by optical imperfections in the system consisting of spherical aberration, coma, and astigmatism which result in entrance slit image mismatch at the exit slit. The result is that $\Delta\lambda_{\text{min}}$ is only a theoretical value and must always be greater in practice than $\Delta\lambda_d$. The actual spectral bandwidth at a given wavelength is approximately the sum of the dispersion bandwidth given by equation 3, the diffraction bandwidth given by equation 8, and slit image mismatch. Since the dispersion bandwidth $\Delta\lambda_s$ and diffraction bandwidth $\Delta\lambda_d$ can be calculated, a comparison of the sum of these components with an experimentally determined spectral bandwidth is an indication of the quality of optical components contributing to slit image mismatch. Slit image mismatch effects should not contribute more than 0.05 nm to total spectral bandwidth.

The resolving power R of a spectrometer is the ratio of the mean wavelength of two closely spaced wavelength elements to their wavelength separation when they are just resolved:

$$R = \frac{\bar{\lambda}}{\Delta\lambda_s} \qquad (9)$$

The theoretical resolving power of an optically perfect spectrometer at zero slit width is defined by the Rayleigh criterion for the resolution of two point sources. When the first minimum of one diffraction pattern coincides with the central (peak) image of the second, the two sources are considered to be resolved according to the Rayleigh criterion. Their condition corresponds to a separation in wavelength given by equation 8 and by:

$$\Delta\lambda_d = \left(\frac{d\lambda}{dx}\right)\left(\frac{\lambda f}{a}\right) \qquad (10)$$

where $\Delta\lambda_d$ is the theoretical diffraction limit of separation between two wavelength elements. The theoretical resolution* R_0 becomes

$$R_0 = \left(\frac{dx}{d\lambda}\right)\left(\frac{\bar{\lambda}a}{\lambda f}\right) = \left(\frac{dx}{d\lambda}\right)\left(\frac{1}{f}\right)a = a\left(\frac{d\theta}{d\lambda}\right) \qquad (11)$$

The degree to which R approximates R_0 is another indication of spectrometer quality. The experimental resolving power R is given by combination of equations 3 and 9:

$$R = \left(\frac{dx}{d\lambda}\right)_s\left(\frac{\bar{\lambda}}{w}\right) = \left(\frac{d\theta}{d\lambda}\right)\left(\frac{\bar{\lambda}}{fw}\right) \qquad (12)$$

The experimental resolving power depends on the angular dispersion of the grating or prism, the wavelength setting, the focal length of the optical system, and the slit width, whereas the theoretical resolving power depends only on the dispersion itself. In most cases in analytical spectrometry, $R \ll R_0$, and so $\Delta\lambda_s \gg \Delta\lambda_{\min}$ (or $\Delta\lambda_d$). Only when the geometric slit width w of the monochromator is ~ 0.01 mm or less does $R \to R_0$ and $\Delta\lambda_s \to \Delta\lambda_d$ (or $\Delta\lambda_{\min}$).

Finally, it should be stated that gratings are generally blazed at a certain angle, and so the incident radiation is then preferentially reflected in the direction that is symmetric to the normal to the oblique groove (the blaze angle is the angle of the oblique groove with the grating base). Because reflected radiation always obeys the grating equation, the blaze angle is perfect *only* for the reflection of one wavelength. However, the blaze angle extends from about two-thirds of the blaze wavelength λ_β to about three-halves of λ_β, for example, a grating blaze for 600 nm extends from $600 \times \frac{2}{3} = 400$ nm to $600 \times \frac{3}{2} = 900$ nm. Also, it should be stressed that blazed and unblazed gratings both have order overlap, that is, radiation in different orders is superimposed at the same angle of diffraction. For example, at 800 nm for $m = 1$, 400 nm ($m = 2$), 267 nm ($m = 3$), 200 nm ($m = 4$), and so on, spectral components also exist.

Despite the difficulties with order overlap in using grating spectrometers, gratings are the most commonly used dispersers in optical spectrometers.

---

*For a prism monochromator, $R_0 = b(dn/d\lambda)$, where b is the base of the prism and $dn/d\lambda$ is the variation in refractive index with wavelength for the prism material.

For a grating monochromator, $R_0 = ma/(d\cos\theta) = mN$, where m is the order of diffraction, d is the spacing between grating grooves, θ is the angle between the diffracted ray and the grating normal, and N is the number of illuminated grooves in the grating. The value of R_0 for a grating unlike a prism is independent of the material and the wavelength. Thus a grating has nearly constant angular dispersion and constant spectral bandwidth over the entire spectrum of interest.

The reasons for this are: (1) the angular dispersion and spectral bandpass are nearly constant over the wavelength region of interest; (2) gratings are much less expensive, and so several gratings can be utilized to cover optimally the entire spectral region of interest; (3) stray light can be made small with gratings, although prisms are generally better; (4) gratings can be used for a much wider wavelength range as a result of the lack of transmittance of prism materials for some wavelengths; (5) the grating has a potentially much larger dispersion than the prism.

d. Luminosity

The term "luminosity" is used here to refer to the light gathering and transmission qualities of a spectrometer. In general parlance, the light-gathering capacity of optical systems is referred to as *speed* and is expressed by the *f-number*:

$$f\text{-}number = f/D \qquad (13)$$

where D is the diameter of the light (aperture) optical component (lens or mirror), and f is its focal length. The smaller the f number, the greater the light collecting power and the faster the speed of the optics. The solid angle Ω (in sr), of light collected by an optical system from a point source is given by:

$$\Omega = \frac{A}{f^2} \qquad (14)$$

where A is the limiting useful area of the mirror or lens, and f is its focal length. Thus fast components have large useful areas and/or short focal lengths and collect large solid angles of radiation.

Radiation passing through a spectrometer encounters many optical components. The transmission factor τ_f of a spectrometer indicates the efficiency with which the optical components permit light at wavelength λ to reach the exit slit, that is, the ratio of light flux of wavelength λ emanating from the spectrometer to that collected by the entrance slit and associated optics. Losses of light may occur in spectrometers as a result of reflection, absorption, and scattering by lenses, mirrors, prisms, or gratings. It is obvious that spectrometers should be used only for wavelength regions in which the transmission characteristics of lens and prism materials is high and in which the reflectivity of mirrors and gratings is also maximal. Light losses are proportional to the number of optical components in the optical train. Thus spectrometers having complicated folded optical paths are likely to have lower transmission efficiency than simpler systems.

Reflection losses by lenses and prisms may be minimized by coating with low reflectance loss coatings. These are usually MgF_2 coatings and are of a thickness equal to one-quarter of the wavelength of maximal transmission. Overall reflection may be reduced to a fraction of 1% by the use of such coatings. Absorption losses by reflective surfaces may also be minimized by coating the surface with multiple layers of one-quarter wavelength coats. Reflectivity can be increased to virtually 100% by this technique.

The luminosity $wh\tau\Omega$ of the spectrometer determines the amount of radiant flux (in W) emanating from the exit slit, assuming the entrance slit is fully illuminated by a continuum source:

$$\Phi = B_{C\lambda}^0 wh\tau_f \Omega \Delta\lambda_s \qquad (15)$$

where Φ = total radiant flux at exit slit (in W)

$B_{C\lambda}^0$ = spectral radiance of continuum source (in W cm^{-2} sr^{-1} nm^{-1})

w = illuminated width of slit (in cm) (entrance slit = exit slit)

h = illuminated height of slit (in cm) (entrance slit = exit slit)

τ_f = transmission factor of optical system (no units)

Ω = solid angle of radiation collected by optical system (in sr)

$\Delta\lambda_s$ = spectral bandpass of spectrometer (in nm)

In Appendix D, a more complete treatment of the factors affecting the measured flux is given.

2. NONDISPERSIVE SPECTROMETERS

Nondispersive spectrometers utilize filters instead of gratings and prisms as the spectral isolation device. Filters commonly used are either of the Wratten type (a glass sandwich of a dye-containing layer of gelatin) or of the Fabry–Perot interference type.

Wratten filters can be obtained with a variety of spectral characteristics (Figure 5.3); Curve a depicts a low-pass filter, while curve b depicts a high-pass filter. These types find use as low- and high-blocking filters. A combination of the two results in a bandpass filter. Wratten bandpass filters exhibit wide spectral bandwidths (on the order of 20 nm and greater), which render them unsuitable for accurate quantitative analytical use.

Fabry–Perot interference filters offer spectrally purer transmission characteristics. Typical half-bandwidths* (see Figure 5.4) $\Delta\lambda_{0.5}$ range from less than 1 nm to approximately 10 nm, depending on the center wavelength and construction. Transmission of light by interference filters is ordinarily

---

*Half-bandwidth is defined as the spectral width (in nm) of the filter characteristic spectrum at one-half the peak transmission.

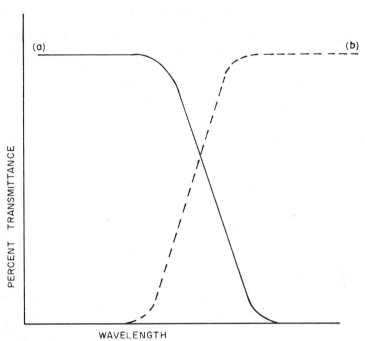

Fig. 5.3 Schematic representation of transmittance wavelength spectral plots for (*a*) low-pass and (*b*) high-pass filters.

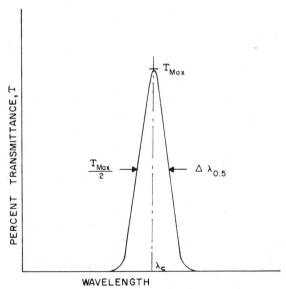

Fig. 5.4 Schematic representation of a transmittance wavelength spectral plot for an interference filter with definition of filter bandpass.

better than 50% in the near-UV region. Interference filters are blocked to radiation outside their spectral pass downward to far UV and upward to near IR. Filters are available to cover the UV and visible spectrum from 300 to 650 nm at approximately 10 to 20 nm intervals.

Nondispersive spectrometers offer an economic advantage over dispersive spectrometers, as well as increased luminosity. However, filters must be kept in stock for all wavelengths of measurement. If an assay requires a measurement at a wavelength for which no filter is available, less than maximal sensitivity must be accepted.

Dispersive spectrometers are expensive but offer the advantage of continuous wavelength selectability and, as a direct result, the ability to generate continuum spectra when necessary.

3.   SINGLE-BEAM VERSUS DOUBLE-BEAM SPECTROMETERS

The instrumental response of a single-beam spectrometer is the resultant of all the individual component responses, most of which vary nonlinearly as a function of wavelength. As a consequence, the instrumental output must be normalized each time a measurement is made at a wavelength different from the previous measurement. Single-beam spectrometers therefore are most useful for analytical measurements requiring observation at single wavelengths.

Double-beam spectrometers are required when a spectral scan of the optical characteristics of an analyte is desired. Since I_0, the intensity of light passing through the reference (blank), varies with source energy, monochromator transmission, reference material transmission, and detector response, all of which vary with wavelength, some means must be provided to maintain the phototube signal corresponding to I_0 at a constant level. This may be accomplished by one of several alternative methods. A common means is to provide feedback control to regulate photodetector sensitivity via dynode voltage. A second method is to control the monochromator slit width by means of servomotors and mechanical slit drives. A third alternative is to position an optical wedge in the light path automatically to increase or attenuate radiation reaching the detector.

Automatic gain control is the least expensive of the three modes of operation, since it involves only electronic circuitry and no mechanical components. It has the advantage of providing constant slit width and thus constant resolving power during the scan when a grating monochromator is used. However, the noise level of the photodetector varies with gain and thus is not constant throughout the scan.

Automatic slit control is much more costly, since mechanical slit drives must be incorporated. However, photodetector gain and therefore noise remain constant. As the slit width varies, resolution does likewise. Automatic slit controls appear to have utility only where a constant noise level is required.

Optical wedge systems seem to have approximately intermediate utility. A mechanical drive system, although not nearly as complex as needed for slit drives, is required. Phototube gain as well as resolution remains constant throughout the scan.

Figure 5.5 illustrates schematically the arrangement of a typical double-beam spectrometer employing automatic photodetector gain adjustment.

4. MULTIPLEXING TECHNIQUES (REFER TO APPENDICES B AND C)

5. CHARACTERISTICS OF DETECTORS (3)

Photodetectors can be arranged functionally into several groups. The slow, insensitive, rugged, inexpensive barrier layer cell serves admirably in filter photometers and applications requiring ruggedness as a principal quality. Barrier layer cells respond slowly and possess a spectral response characteristic similar to that of the human eye. They are not useful in the UV portion of the spectrum.

The more sensitive photodiode requires low-voltage biasing of the photocathode on the order of 250 V DC. This type can be either a gas-filled or a vacuum tube and is available with photocathodes having an S-1, S-3, S-4, S-5, or S-9 spectral response (plots of detector response versus wavelength). Applications of photodiodes are generally under relatively high light level conditions which require fast response times.

Multiplier phototubes* are sensitive to very low levels of radiant energy. Typical gain obtainable from multipliers ranges from $\sim 10^5$ to $\sim 5 \times 10^7$ in special cases. The need to measure very low light levels necessitates the use of multiplier phototubes. Operating voltage may range from several hundred to several thousand volts. When dealing with high voltage, it is imperative that the detectors and their mounting sockets be kept scrupulously clean and free from dirt or grease. The presence of either may provide leakage paths to ground and render the measurement system unusable.

A device has recently become available which may provide the electronic equivalent of the photographic plate. Self-scanning arrays of silicon photodiodes are obtainable which contain up to 256 elements on a single silicon chip. Two recent reports by Horlick and Codding (4, 5) discuss their characteristics and applications. The advantage over conventional

*Multiplier phototubes are also called *photomultiplier tubes*.

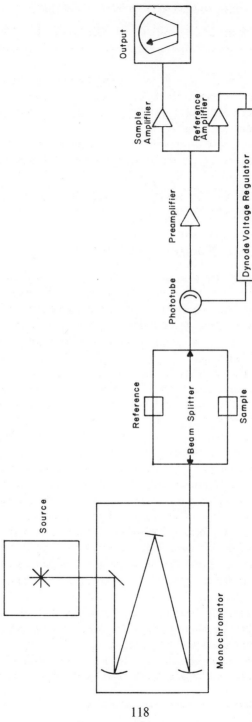

Fig. 5.5 Schematic diagram of typical double-beam absorption spectrometer employing automatic photodetector gain adjustment for beam balancing.

118

multiplier phototubes lies primarily in the capability of resolving spatially close spectral elements in the spectrometer focal plane, thus obviating the need for narrow exit slits.

B. TYPICAL SPECTROMETRIC SYSTEMS

1. ABSORPTION SPECTROMETER

Figure 5.6 is a block diagram of a typical single-beam absorption spectrometer.

Such an arrangement is employed in virtually all absorption spectrometric systems whether they be UV, visible, or IR spectrophotometers, atomic absorption spectrophotometers, or X-ray absorption spectrometers. Sources and detectors must be appropriate for the analytical technique and be spectrally matched. Spectral isolation devices may be grating or prism monochromators or interference filters. Sample reservoirs may be cuvets containing absorbing solutions (UV, visible), sample holders containing solid pelletized samples (IR, X-ray), or flames or furnaces containing clouds of atoms (UV, visible).

The analytical signal in all absorption spectrometers behaves according to the equations described in the preceding chapter, that is, it is dependent on the length of the sample reservoir, and so on. Signal-to-noise considerations were discussed previously in Chapter 2.

2. EMISSION SPECTROMETER

In Figure 5.7 the general arrangement of an emission spectrometer is given. It differs from the absorption spectrometer only in that the sample reservoir serves also as the source.

3. LUMINESCENCE SPECTROMETER

A luminescence spectrometer is essentially the combination of an absorption and an emission spectrometer with a sample reservoir common to both systems. The geometric arrangement is usually at right angles about the sample reservoir (see Figure 5.8).

A 90° configuration is often chosen to minimize any source radiation transmission through the emission portion of the spectrometer, which may be scattered by the sample or optical components. Scattering of radiation is discussed in Chapter 4, Section A.3.d.

The generalized luminescence spectrometer exemplifies systems used in measuring molecular or atomic fluorescence, phosphorescence, and X-ray fluorescence.

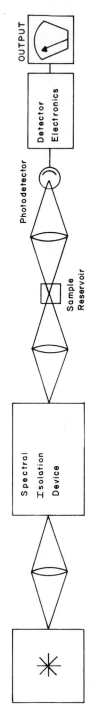

Fig. 5.6 Schematic diagram of single-beam absorption spectrometer.

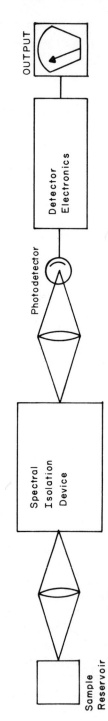

Fig. 5.7 Schematic diagram of single-beam emission spectrometer.

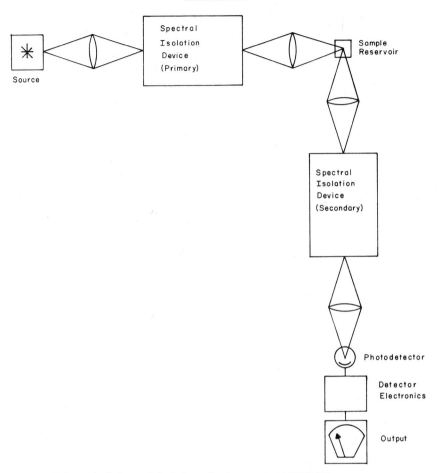

Fig. 5.8 Schematic diagram of single-beam luminescence spectrometer.

REFERENCES

1. Williams, C. S. and Becklund, O. A., *Optics: A Short Course for Engineers and Scientists*. Wiley-Interscience, New York, 1972.

2. Jenkins, F. A. and White, H. E., *Fundamentals of Optics*, McGraw Hill, New York, 1957.

3. Radio Corporation of America, *RCA Photomultiplier Manual*, Harrison, New Jersey, 1970.

4. Horlick, G. and Codding, E. G. *Anal. Chem.*, **45**, 1490 (1973).

5. Horlick, G. and Codding, E. G. *Anal. Chem.*, **45**, 1749 (1973).

OPTICAL ATOMIC SPECTROSCOPIC METHODS

CLAUDE VEILLON

Biophysics Research Laboratory
Harvard Medical School
Peter B. Brigham Hospital
Boston, Massachusetts

A. INTRODUCTION

Optical atomic spectroscopic methods can be grouped into three techniques: atomic emission, atomic absorption, and atomic fluorescence spectroscopy. Atomic emission spectrometry or spectrography is usually considered in two categories: flame and nonflame. That is, one might employ a flame for atomization and excitation, usually of aqueous sample solutions, or a nonflame device, for example, some type of electrical discharge (DC arc, AC spark, RF plasma, etc.). With the arc and spark, solid samples are usually excited directly and measured in a multielement, spectrographic instrument. Atomic absorption and fluorescence spectrometry are instrumentally similar to flame emission spectrometry, but here the flame (or nonflame device) serves only as an atomizer. Nonflame atomization is becoming more widespread in use in the latter two techniques, usually in the form of high-temperature tube furnaces.

Instrumentally, all three techniques have many similarities. Each method has unique advantages and disadvantages, which are dependent in many cases on the application, type of sample, number of samples, amount of sample, sensitivity required, and information desired.

All three techniques are widely used, predominantly in trace elemental determinations (mostly metals) for virtually every type of sample. Some recent developments, particularly the nonflame atomization systems for atomic absorption and fluorescence, have greatly lowered the minimum detectable amounts of elements, opening up many new application and research areas. These include trace elements in the environment, the role of trace elements in biological systems, and many other previously unattainable applications—previously unattainable because of limited sensitivity and/or limited quantity of sample. This greatly increased sensitivity is not without a price, however, because sampling, contamination, and stability (samples and standards) become increasingly troublesome as lower elemental concentrations need to be measured.

Let us now look, in some detail, at these three techniques, and hopefully give enough detail about them that the reader can decide which one(s) best fits his needs for a particular analytical problem.

B. ATOMIC EMISSION SPECTROMETRY

1. BASIS OF THE METHOD

The basis of atomic emission can be illustrated:

$$M + energy \rightarrow M^* \rightarrow M + h\nu \qquad (1)$$

where M represents the metal atom in the gas phase (and usually in its ground electronic state). The *energy* is imparted to the atom, usually by collisions with high-temperature (i.e., high kinetic energy) atoms and molecules in the atomization-excitation source (flame, arc, etc.), resulting in an electronic transition within the atom. This results in an excited atom M* which can then (as one possible means) lose the energy acquired by a radiational process (emission of a photon $h\nu$) and return to the original state. The energy of the emitted photon $h\nu$, is equal to the energy difference between the electron energy levels involved in the transition. Naturally, emission is not the only possible means of energy loss, but it is the only one of importance in this case. Also, other transitions from higher to lower energy levels and not necessarily ending in the ground state are possible, resulting in more than one emission wavelength for an element (ν is the frequency of the emitted radiation and is related to the wavelength λ, by $\nu = c/\lambda$, where c is the velocity of light). Usually one or more of the transitions ending in the ground electronic state is the most probable, resulting in one of the characteristic emission wavelengths (lines) of greatest intensity (so-called resonance lines).

The following requirements for atomic emission spectroscopic methods are pertinent: (a) the sample must be atomized and the resulting atoms excited to emission (in atomic emission methods, these two functions are accomplished simultaneously in the atomization-excitation source); (b) the resulting characteristic line emissions must be spectrally (wavelength) separated and their relative intensities measured, employing a suitable dispersion*-detection system (spectrometer or spectrograph); and (c) the

---

*Nondispersive rather than dispersive systems, for example, Hadamard or Fourier transform systems, can be used (see Appendices A and B) but so far have found no analytical applications.

resulting intensities must then be compared with standards of known elemental concentration and the elemental content of the desired element(s) determined.

The fraction of sample atoms excited under any given source conditions varies exponentially with the source temperature, the Boltzmann distribution being a good approximation where the source attains or approaches thermodynamic equilibrium. Thus the higher the temperature of the excitation source, the greater the emission intensity for a given analyte atomic concentration in the source. Coupled to the increased intensity with increased excitiation temperature, the sample atomic concentration also usually increases in cases in which atomization is not complete at lower temperatures. One might at first expect that the analytical sensitivity (decreased detection limit) can be increased almost without limit by simply going to higher-temperature excitation sources, but there are several limitations to this approach. At higher temperatures, ionization of the sample atoms becomes important; the spectrum becomes more complex as more upper-level lines are excited; and, perhaps most important, the source background emission also increases rapidly.

The principal advantages of atomic emission are low analytical detection limits for many elements, relatively simple instrumentation, good specificity and speed of analysis, and adaptation to simultaneous, multielemental analysis. The principal limitations are centered around the type of excitation source used and the inseparability of the atomization and excitation processes. With relatively low-temperature sources, many elements with high excitation energies are not adequately excited, while with relatively high-temperature sources high background and complex spectra may require a high-resolution spectrometric system.

2. SOURCES OF EXCITATION

In atomic emission spectroscopic methods the source is responsible for both analyte atomization and for excitation of the resulting analyte atoms. These sources can be placed in two general categories: flame and furnace. The former employ chemical flames formed from the combustion of various fuels in various oxidants. Furnace sources usually are electrical discharges of some kind, such as arcs, sparks, and plasmas. The primary function of each—atomization and excitation—is the same, but the mode of operation of each may have significant differences.

a. Flames

The chemical flame is one of the oldest emission sources, dating back to the time of Kirchoff and Bunsen, and it is still in wide use today. Various types of chemical flames and burner designs have evolved, and these have all been quite successful from an analytical standpoint. As emission sources, flames have much to offer. They are simple, inexpensive to operate, and tens of elements can be determined in the micrograms-per-milliliter concentration range or below. The temperatures of most commonly used flames are in the 2000 to 3000 K region, which results in good analytical sensitivity for elements having excitation energies (resonance lines) of about 5 eV or less and being atomized to an appreciable extent in the flame (i.e., having no strong tendency toward compound formation).

Flames also have some important limitations for use as emission sources: (a) Temperatures much above 3000 K cannot be obtained with the usual fuel–oxidant combinations, resulting in relatively poor analytical sensitivity for many elements with high excitation energies; (b) the chemical environment of the flame is quite reactive, and many elements exhibit strong compound formation (usually metal monoxide) which effectively removes them from the atomic emission process; and (c) considerable background emission may be present in certain spectral regions (e.g., the OH band emission between 3000 and 3500 Å and the C_2, CN, and CH band emissions from hydrocarbon-fueled flames). Despite these limitations, flame atomic emission remains one of the simplest and most sensitive analytical methods for easily excited elements that do not form highly stable compounds (at high temperatures), such as alkali, alkaline earth, and several transition elements.

Some commonly used flame gas combinations are shown in Table 6.1. The temperature range indicated are only approximate, since the flame temperature depends to no small extent on the stoichiometry, region observed and method of measurement, extent of premixing of gases, sample (solvent) introduction rate, desolvation efficiency, and so on. Of more significance is the fact that most common flames are in the 2000 to 3000 K region. Temperatures in excess of 4000 K can be obtained in flames like C_2N_2–O_2 and H_2–F_2, but these have not found great acceptance. C_2N_2 is relatively expensive, and the combustion product of the H_2–F_2 flame presents an obvious problem.

TABLE 6.1. Common Fuel-Oxidant Combinations for Flame

Fuel	Oxidant	Approximate temperature range (K)
H_2	O_2	2600–3000
H_2	Air	2100–2300
C_2H_2	O_2	2800–3300
C_2H_2	Air	2300–2500
C_2H_2	N_2O	2700–3100
Propane	O_2	2600–3000
Propane	Air	2000–2200

Burners used in flame atomic emission spectrometry generally fall into two general categories: direct injection (produces turbulent flames) and premixed (or "chamber-type"). The former are the familiar Beckman-type burners, designed by P. T. Gilbert, Jr., of the Beckman Instrument Company (Figure 6.1). Burners similar in principle are also available from Zeiss and perhaps other manufacturers. These burners produce a turbulent flame, since the fuel and oxidant mix external to the burner, and all of the sample solution is sprayed directly into the flame center. These burners are quite safe to operate because of the external fuel mixing, although they are somewhat noisy (audibly). The most popular types employ $C_2H_2-O_2$ or H_2-O_2, and both aqueous and organic sample solutions can be readily handled.

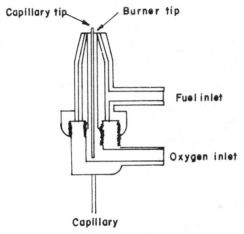

Capillary tip   Burner tip

Fuel inlet

Oxygen inlet

Capillary

Fig. 6.1 "Total-consumption" burner.

Premix burners (Figure 6.2) usually employ premixed fuel and oxidant and a Meker-, Bunsen- or slot-type burner head. Fine aerosol droplets of the sample solution are generated by spraying the sample into the gas mixing chamber (usually pneumatically, using the oxidant gas), and the larger droplets drain away. The remaining sample aerosol is swept into the flame along with the premixed gases. Premixing the fuel and oxidant results in a much quieter flame (both audibly and electrically in terms of signal flicker components). This feature, plus usually lower background emission, would be expected to improve the detection limits, but much of this hoped-for gain is lost because of the relative inefficiency of the sample introduction system.

The slot-type burner heads frequently used in atomic absorption spectrometry have also been used with considerable success for flame

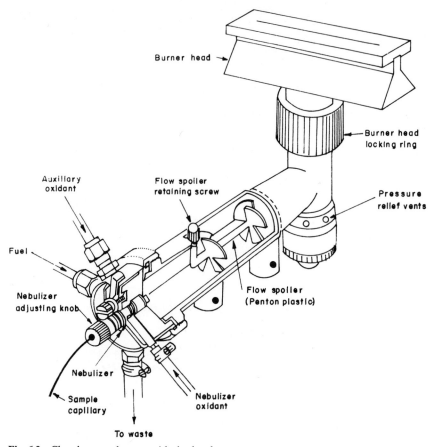

Fig. 6.2   Chamber-type burner, with slot head.

atomic emission, because of their favorable geometry. For example, Pickett and Koirtyohann (1) employed this configuration in a thorough investigation of the analytical capabilities of the $N_2O–C_2H_2$ flame.

Rather than go into the many details and nuances of burner design, nebulizer design, burning velocity considerations, flame composition, and so on, we refer the reader to the excellent treatise on emission flame spectroscopy by Mavrodineanu and Boiteux (2).

Regardless of the nebulizer burner system used, the atomization-excitation process is basically the same. One begins with a sample solution (usually aqueous) which is converted to a fine mist and introduced into a high-temperature flame cell. Here several processes occur in rapid sequence: nebulization, desolvation, dissociation (atomization), and excitation. The various steps are illustrated in Figure 6.3. The absorption and fluorescence processes are also shown, since the process through atomization is the same for these techniques also; only the mode of excitation and/or measurement is different.

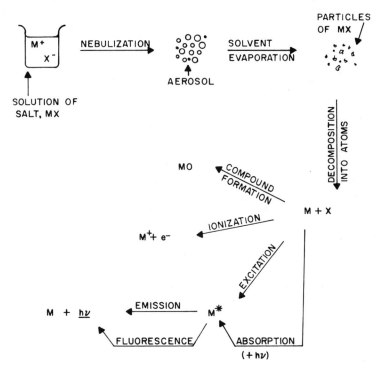

Fig. 6.3   Atomization-excitation process.

The production of gas phase atoms of the sample metal depends initially on the thermodynamics and kinetics of the volatilization-atomization processes, that is, those processes involved in converting the dried sample salt particle to gaseous atoms. The various ways in which analyte atoms, once produced, can be further lost include compound formation with flame gas products and ionization (obviously, once an atom is ionized or becomes part of the compound, it no longer contributes to the *atomic* emission being measured). Ionization is not usually important, except in the case of easily ionized elements in relatively hot flames. By far the most important loss mechanism is compound formation of analyte atoms with flame gas radicals. Many elements form quite stable compounds in the flame, notably metal monoxides. If one considers the general equilbrium:

$$M + O \rightleftharpoons MO \qquad\qquad (2)$$

the reaction is exothermic, and an increase in temperature shifts the equilibrium toward the free metal atom, assuming the oxygen concentration remains the same or decreases. Since one normally has little control over the temperature in chemical flames, a more productive approach is to lower the free oxygen content, which has the same effect on the equilibrium shift. One means of doing this, especially in hydrocarbon flames, is to make the flame fuel-rich. The excess fuel reduces the partial pressure of oxygen atoms in the flame, shifting the equilibrium toward the free metal. Usually the flame temperature does not fall rapidly as one begins to operate at fuel-rich conditions, so a net gain is achieved. Extracting the sample into organic solvents can also help, by reducing the cooling effect of the aqueous solvent and by improving the nebulization and desolvation efficiencies.

Another way in which one might consider the reduction in compound formation to occur is as follows. As the flame conditions become more fuel-rich, the concentration of strongly reducing species, such as C and CO within the flame, increases. This causes a decrease in the monoxides via the following processes:

$$MO + C \rightleftarrows M + CO$$

$$MO + CO \rightleftarrows M + CO_2 \tag{3}$$

Naturally, this works only in cases in which, for example, CO is more stable than MO.

It is important to note at this point that the processes of ionization and compound formation of the analyte atoms in the flame result in a loss of atoms available for atomic emission, atomic absorption, or atomic fluorescence, *and this affects all three analytical techniques equally.*

As indicated earlier, excitation of atoms results primarily (in flames) from collisions with energetic atoms (kinetic) and molecules (kinetic and internal energy). If the observation region of the flame is in thermal equilibrium, or nearly so, the fraction of atoms excited will follow the Boltzmann distribution:

$$\frac{n_u}{n_1} = K \exp(-\Delta E / kT) \tag{4}$$

where n_u is the number of atoms in the upper (excited) state, n_1 is the number of atoms in the lower (usually ground) state, K is a proportionality constant involving the ratio of statistical weights for the upper and lower states of the particular transition in the particular atom, ΔE is the energy difference between the upper and lower states of the particular transition in the particular atom (in eV)(E = excitation energy for transition $= \Delta E -$

$h\nu = hc/\lambda$), k is the Boltzmann constant (in eV K^{-1}), and T is the absolute temperature (in K). Even at low values of ΔE and high values of T, the fraction of atoms excited, n_u/n_1, is quite small, that is, only a small fraction of the atoms resulting from the atomization process are ever excited. In addition to this, only a small fraction of the atoms excited emit, or lose energy in the form of radiation. An excited atom may have a lifetime in the excited state of $\cong 10^{-8}$ s before emission occurs. Because these excited atoms are produced in a hot gas at atmospheric pressure, they (statistically) undergo several collisions during this 10 ns period, and many transfer their excitation energy to the colliding species (especially if it is a molecule) and never emit.

Despite the fact that only a small number of the atoms are excited, and only a small number of these ever emit, one might wonder why the method works at all and is capable of determining trace levels of many elements. For good measure, one can also consider the facts that many, if not most, samples are *not completely atomized* in the flame, and that only a small fraction of the sample reaches the flame. This is further discussed later.

In employing flame atomic emission for trace analytical purposes, one frequently uses an experimental setup consisting of a flame (with an appropriate sample introduction system and gas handling equipment), a monochromator (preferably a model that can be wavelength scanned), a detector and an appropriate amplifier-readout system. Of all the instrumental parameters available to the analyst in optimizing the determination, *those associated with the flame are usually the most critical*. Selecting the proper analytical wavelength and optimizing the variables associated with detection and readout are usually straightforward. A slight change in flame source condition may change the sensitivity significantly, and optimum conditions for one determination may not be optimum for another element.

Depending on the particular nebulizer-burner design, the analyst may or may not have a great deal of control over the sample introduction system. Increasing the *gas pressure* to the nebulizer increases the *solution aspiration rate* and increases the *flow rate of the nebulizing gas* (usually the oxidant), which requires an increased *fuel gas flow rate* to maintain the fuel/oxidant flow ratio (stoichiometry). This can affect the flame size, which might affect the *optimum viewing region*. Changing the flame stoichiometry can affect the temperature, chemical conditions within the flame, atomization efficiency, and the optimum viewing region. Since the analytical detection limits can vary to a greater or lesser extent with any or all of these parameters (flow rates of oxidant and fuel, pressure of oxidant, solution flow rate, and region of observation), and since the optimum flame parameters may be different for each element (or even sample type, for the same element), these are extremely critical parameters. Complete optimization of these variables (some are dependent on others) is certainly not always important, unless one wishes or needs to obtain the lowest detection limits. Where such a degree of fine-tuning is necessary, it can be approached systematically (statistically) and based on the best available theories. An excellent discussion of this has been published by Winefordner and Vickers (3).

In choosing an analytical technique for trace elemental analysis, one very important criterion is the smallest amount of an element that can be determined, the so-called detection limit. In many techniques one frequently hears the terms "sensitivity" and "detection limit". Both are important criteria in evaluating the potential of an instrument or technique for a particular determination, and they frequently mean different things to different people (refer to Chapter 1 and Appendix A for a thorough discussion of these concepts).

In atomic emission spectrometry, flame or otherwise, the sensitivity is a measure of the increase in the analytical signal as the concentration of the analyte element increases. In essence, if one plots the emission signal [relative intensity (in arbitrary units) of the atomic line above the background or blank intensity] *versus* the concentration of the element in the sample or standards, the slope of the linear portion of the curve will be a measure of the sensitivity for that line of that element under those analytical conditions. It is not a "hard" number, because a change in the source operating conditions (e.g., temperature) causes it to change. As another example, if the line emission is superimposed on a continuum background emission from the source, changing the monochromator spectral bandwidth will alter the sensitivity. Narrowing the spectral bandwidth causes the line intensity to decrease, and the continuum background to decrease, but not at the same rate. The background decreases faster than the line, by the square root of the spectral bandwidth. Thus reducing the spectral bandwidth by a factor of 4 improves the sensitivity by a factor of $\sqrt{4}$ or 2. This may or may not improve the detection limit, because decreasing the spectral bandwidth decreases the detector signal and, since the detection limit is determined by noise as well as signal, it may not improve.

The term "detection limit" is somewhat of an enigma. The name implies the smallest amount of an element that can be reliably *detected*. In practice, it refers to the smallest concentration or amount of analyte that can be measured (i.e., concentration or the amount present). Obviously, there is no sharp boundary here between purely qualitative detection and quantitative determination; the analytical precision, of course, improves rapidly as one moves up from the detection limit. Some criterion had to be established and generally accepted. Today most workers use a signal-to-rms noise ratio of 2 to establish this point. In the case of emission, this usually amounts to "that concentration or quantity of analyte where the signal (above background or blank) is equal to twice the root-mean-square (rms) fluctuations of the background or blank."* Other criteria and definitions have been proposed and are used, but this is perhaps the most widely used one. It too is not a "hard" number, because it may depend on several factors, such as sensitivity, type of signal processing used, and time constant of the electronic measurement system.

Frequently, one can trade off measurement time for an improved signal-to-noise ratio; it is not a linear trade-off, but rather a square root, like the spectral bandwidth–sensitivity trade mentioned earlier.

While on the subject of detection limit, it might be appropriate to point out that these are usually expressed in concentration units. This is acceptable in cases in which one is not sample-limited, by reasons of sensitivity or whatever. A more objective approach, allowing a more direct comparison of data and analytical techniques, is to express the detection limits on an absolute basis such as weight. In other words, what is the smallest amount of an element that can be determined? There is a trend toward this, particularly since the advent of the furnace atomization systems used in atomic absorption and fluorescence spectrometry.

The preceeding is intended to point out that sensitivity and detection limit values are good indications of the trace analysis capabilities of an instrument or technique, but small differences may not be meaningful. An order of magnitude is certainly significant, while a factor of 2 probably is not.

Detection limits for a number of elements by flame atomic emission spectrometry are shown in Table 6.2. These are for the N_2O–C_2H_2 flame (slot burner) and are taken from the work of Pickett and Koirtyohann (1). Their system consumed about 3 ml min^{-1} of sample solution, so if 20 s (i.e., 1 ml) is sufficient for a measurement, the minimum detectable amounts (absolute basis) of each element (in μg) will be as shown in Table 6.2. About a dozen of these elements can be determined at levels of 10 ng or less, about 30 can be determined at levels of 1 μg or less, and only a few have detection limits above the 1 μg ml^{-1}. While this study did not report on all the elements, it gives one a good idea of the capabilities of flame atomic emission spectrometry, namely, the ability to determine perhaps half of the elements in the periodic table at microgram levels and below.

*Standard deviation—see Appendix A1.

TABLE 6.2. Detection Limits by Emission in the N_2O–C_2H_2 Flame

Element	Wavelength (Å)	Detection limit ($\mu g\ ml^{-1}$)
Al	3962	0.01
Ag	3281	0.02
Au	2676	0.5
Ba	5536	0.002
Be	2349	40.
Ca	4227	0.0001
Cd	3261	2.
Co	3454	0.05
Cr	4254	0.005
Cu	3274	0.01
Fe	3720	0.05
Ga	4172	0.01
Ge	2651	0.5
In	4511	0.005
La	5791	2.
Li	6708	0.00003
Mg	2852	0.005
Mn	4031	0.005
Mo	3903	0.1
Nb	4059	1.
Ni	3415	0.03
Pb	4058	0.2
Pd	3635	0.05
Pt	2660	2.
Re	3461	0.2
Sc	4020	0.03
Sn	2840	0.3
Sr	4607	0.0002
Ti	3999	0.2
Tl	5351	0.02
V	4379	0.01
W	4009	0.5
Y	4077	1.
Zr	3601	3.

b. DC Arc

The DC arc discharge is a widely used spectrochemical excitation source and is almost always employed with a spectrograph or multichannel spectrometer. A relatively high-current, relatively low-voltage discharge is maintained between two electrodes (usually graphite), one containing the sample (usually the anode), and operating in air (free-burning) or some other gas mixture at atmospheric pressure. Electrodes are most often made from high-purity carbon or graphite, because of its high-temperature stability, ease of fabrication, and ease of purification. One popular electrode configuration is shown in Figure 6.4; the lower-cup electrode contains the sample to be analyzed. An arc plasma is formed between the electrodes, heating the sample electrode and vaporizing the sample into the high-temperature discharge where atomization and excitation take place.

The temperature observed within the arc plasma depends to some extent on the sample composition [the major cation(s)], on the operating (electrical) parameters, and *even* on the method of temperature measurement. However, the temperatures experienced by the atoms are considerably higher than those attainable in chemical flames, leading to improved sensitivity in many cases. In his thesis, de Galan (University of Amsterdam, 1965) reported arc temperatures under various conditions and observed values in the 5000 to 6000 K region using the zinc 3076/3282 Å line pair as the thermometric species.

The term "temperature" usually implies that the system is in thermal equilibrium. In other words, the *electron temperature* (kinetic energy of the electrons), the *gas temperature* (kinetic energy of the neutral atoms), the *excitation (electronic) temperature* (determines population of the various energy levels), and the *ionization temperature* (determines the ionization equilbria) are all numerically equal when the system is in thermal equilibrium. For molecules, analogous temperatures associated with dissociation equilbria and vibrational-rotational states are also important. A *gaseous system in thermal equilibrium* meets the following conditions (4):

1. Velocity distribution of all free particles (molecules, atoms, ions, electrons) at all energy levels satisifies Maxwell's equation.

Fig. 6.4 A popular type of graphite electrode arrangement for DC arc spectrography.

2.   For each kind of particle, the relative population of energy levels conforms to Boltzmann's distribution law.

3.   Ionization (atoms, molecules) is described by Saha's equation, and dissociation of molecules follows the general rules for chemical equilibria (mass action law).

4.   Radiation density is consistent with Planck's law.

Since light quanta are lost from the arc plasma, the radiation density does not satisfy Planck's law. However, this energy loss by radiation amounts to only a few percent of the total energy (4), and thermal equilibrium is approximated.

In an arc, there is a radial decrease in temperature, from the central core of the plasma outward, and this fundamentally interferes with the achievement of thermal equilibrium. However, the change in temperature over a distance equal to one mean free path is usually small, and this gradient has little effect on the equilibrium conditions. In general, one may consider each volume element separately in a nonhomogeneous source, and the system can be described as being in *local thermodynamic equilibrium*.

With a high arc temperature, the analytical sensitivity is also quite high, and low detection limits for most elements result. However, the nature of the discharge is such that the amount of radiation reaching the detector is not very reproducible from sample to sample. This is caused primarily by wandering of the arc on the electrode surfaces during the discharge. Consequently, reproducibilities better than about $\pm 20\%$ are difficult to achieve. This makes the DC arc source better suited to qualitative or semiquantitative analyses, rather than quantitative work. This is not as great a disadvantage as it might at first appear. The tremendous sensitivity of this high-temperature source, combined with the simultaneous multielement capabilities of the spectrograph, make the combination a very powerful and useful analytical tool.

The atomization-excitation process in the DC arc is similar to that of the flame, except that one usually employs solid samples rather than aqueous solutions. Solid samples, usually in the form of powders, are placed in the graphite cup electrode, alone or mixed with powered graphite, and the arc is struck. Liquid samples may be placed in the electrode and evaporated to dryness, or a rotating disc electrode used (see Section 2.c). The choice of operating conditions is straightforward; arc currents in the 3 to 30 A region are used, with 5 to 20 A being the most common. Rather than go into the many fine points of the operation of the arc (electrodes, sample handling, techniques, etc.) here, we refer the reader to the excellent books by Pinta (5) and by Slavin (6).

Detection limits, as before, can be expressed in terms of concentration (e.g., $\mu g \ ml^{-1}$ for solutions, $\mu g/g^{-1}$ for solids) when the amount of sample is unlimited, or in terms of amount (e.g., μg, ng,) when one is sample-limited. Micrograms per milliliter (μg ml^{-1}) is referred to as parts-per-million (ppm) and is usually applied to dilute aqueous solutions. For samples with densities other than unity, micrograms per gram (μg g^{-1}) is preferred. In spectrography, the concentration may also be expressed on a weight percentage basis (0.0001% = 1 ppm). Because the detection limits in arc spectrography depend on so many variables, such as sample composition (matrix), arc conditions, and optical arrangement, only approximate ranges can be given. Table 6.3 illustrates the various elements and the approximate concentrations at which they can be detected. Here the usefulness of the DC arc source in qualitative and semiquantitative determinations is evident; virtually every metal in the periodic chart can be detected at concentration levels of 100 ppm (0.01%) or below, with many in the 1 to 10 ppm region (note that these detection limits are wt % for solid samples rather that ppm for solution samples).

TABLE 6.3.   Elements Detectable with the DC Arc Source, and Approximate Detection Limits

Approximate detection limit*	Elements
Below 10^{-4}%	Li, Na, Cu, Ag
10^{-4}–10^{-3}%	K, Rb, Cs, Be, Mg, Ca, Sr, Ba, Sc, Y, La, Ti, Zr, V, Cr, Mo, Mn, Fe, Ru, Co, Rh, Ni, Pd, Au, Zn, Cd, B, Al, Ga, In, Tl, Ge, Sn, Pb, Pr, Nd, Eu, Tb, Dy, Ho, Er, Tm, Yb, Lu
10^{-3}–10^{-2}%	Hf, Nb, Ta, W, Re, Os, Ir, Pt, Hg, Si, P, As, Sb, Bi, F, Th, U
10^{-2}–10^{-1}%	Se, Te, Ce, Sm, Gd

*10^{-4}% = 1 μg analyte per 1 g solid sample

c.   AC Spark

The AC or RF spark discharge is another widely used emission source in spectrochemical analysis. It too is almost always employed with a spectrograph or direct-reading, multichannel spectrometer. While resulting in poorer detection limits than the DC arc, it has a relatively high degree of reproducibility and can be used in quantitative analysis.

The DC arc source employs a relatively simple current-regulated power supply, furnishing a relatively high current at a fairly low voltage. The AC spark, however, employs a high-voltage, low-current (average) short-duration, oscillatory type of discharge. A basic circuit arrangement for the spark source is shown in Figure 6.5. The step-up transformer (T) raises the mains (usually 220 V, 60 Hz in the United States) to perhaps 30 kV AC. The charging resistor (R_c) determines the rate at which the capacitor (C) is charged, and the inductance (L) and resistance (R) determine the characteristics of the electrical discharge. Once the voltage exceeds the breakdown voltage of the gaps, a spark discharge occurs until the voltage across C drops below the breakdown voltage. The auxiliary gap (B) has a constant breakdown voltage that is higher than the breakdown voltage of the analytical gap (A). Thus this gap determines the rate of sparking (breakdown), independently of the prehistory of the analytical gap, resulting in a well-controlled, reproducible discharge.

The individual discharges consist of a high-voltage pulse, followed by a rapid series of oscillatory, arclike discharges, rapidly decreasing until the breakdown voltage is no longer exceeded. The individual discharges usually occur at radiofrequencies and have a total duration on the order of 10^{-5} s; approximately 1000 arclike discharges occur per second. Thus the spark discharge only occurs for a relatively small fraction of the total time (for the above case, 10^{-5} s $\times 10^3$ s$^{-1} = 10^{-2}$ or 1% duty factor), resulting in little heating of the bulk sample.

For an oscillatory discharge, the Rf frequency of the individual discharges is given by:

$$f = \frac{1}{2\pi}\sqrt{\frac{1}{LC}} \tag{5}$$

and the current by:

$$i = V\sqrt{\frac{C}{L}} \tag{6}$$

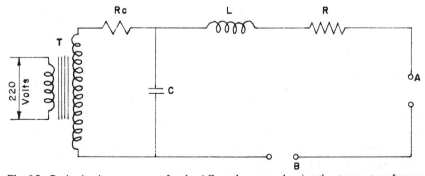

Fig. 6.5 Basic circuit arrangement for the AC spark source, showing the step-up transformer (T), the charging resistor (R_c), the capacitor (C), inductance (L), resistance (R), auxiliary gap (B), and analytical gap (A).

where V is the instantaneous voltage. By varying the values of V, L, and C, the nature of the discharge can be varied.

The fact that the spark occurs repeatedly over a fairly small area of the sample, and that little heating of the bulk sample occurs as a result of the long "off" time, is an advantage in that homogeneity and limited-area studies can be made and solutions can be analyzed directly. However, the relatively small amount of sample consumed leads to somewhat poorer analytical sensitivity and detection limits.

Conducting samples (e.g., metals) are usually ground flat and used as one electrode with a pointed graphite counterelectrode (point-to-plane technique). Powdered samples (conducting and nonconducting) are usually mixed with graphite powder and pressed into a pellet which is used as the plane electrode. Solutions are usually determined using a porous cup (graphite) electrode or a rotating disc electrode. The former consists of a porous-bottom graphite cup containing the sample solution and the counterelectrode beneath the cup, discharging to the wet bottom of the porous cup. The rotating disc electrode consists of a rotating graphite disc, the lower edge of which dips into the sample solution and carries it to the spark discharge region at the top of the disc. Numerous other electrode arrangements have been used, but these are the most popular.

The atomization-excitation process in the spark source is somewhat different than for the previously discussed sources. In the initial high-voltage pulse, mostly atom and ion lines of the gap atmosphere occur. In the more arclike discharge following the initial pulse, spectra of the sample atoms and ions appear. The *effective excitation temperature* in the discharge may be several times that in the DC arc. This means that virtually any element will be excited, but the spectra are somewhat more complex and ionization is greater. By using time-resolved spectroscopic techniques, the line-to-background ratio can be greatly improved (7).

Detection limits, in general, are somewhat poorer than with the DC arc, primarily because of the small amount of sample consumed. The 50-or-so elements most frequently analyzed for with an AC spark exhibit detection limits ranging from $10^{-7}\%$ (by weight) to $10^{-2}\%$ (by weight), with most being in the 10^{-4} to $10^{-2}\%$ range (about half or more are in the 10^{-3} to $10^{-2}\%$ range).

d. Other Sources of Excitation

Several newer types of emission sources have been developed in recent years. These include the DC arc plasma jet, the laser microprobe, the RF plasma torch, and the capillary arc.

(1)   DC Arc Plasma Jet

The DC arc *plasma jet*, as developed by Margoshes and Scribner (8, 9), is essentially a gas-stabilized, thermally pinched DC arc plasma adapted to the analysis of solutions. A schematic of the source is shown in Figure 6.6. The cathode is a thoriated tungsten rod, and the anode is a graphic disc through which the sample solution is sprayed. The arc plasma is forced to pass through a cooled graphite ring, which constricts the arc and results in high effective plasma temperatures. An improved design has been described by Owen (10), and a dual-jet, ecomonically operated version has been described by Valente and Schrenk (11). Merchant and Veillon (12) have characterized a DC arc plasma jet; their data indicate that only a small fraction of the sample aerosol particles actually enters the plasma, and that high analytical sensitivity requires the use of a high-resolution optical system because of the intense plasma background emission. Elliott (13) has described a commercially available DC arc plasma jet used with a high resolving power echelle spectrometer.

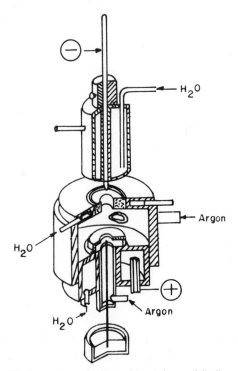

Fig. 6.6   Schematic of DC are plasma jet [from Margoshes and Scribner (9)].

The latter two plasma jets have apparent temperatures in the 8000 to 11000 K range in the plasma core. Some typical detection limits, as reported by Elliott (13) and by Valente and Schrenk (11), are shown in Table 6.4 (in μg ml^{-1} in aqueous solutions, and assuming 1 ml of sample is required for a determination).

(2) Laser Microprobe

The *laser microprobe* is a novel type of emission source well suited to the analysis of very small samples or very small areas of a sample. A schematic of the device is shown in Figure 6.7. A Q-switched, pulsed ruby laser is

TABLE 6.4. Approximate Detection Limits for Aqueous Samples with DC Arc Plasma Jets

Element	Wavelength (Å)	Detection limits (μg ml^{-1}) Valente and Schrenk (11)	Elliot (13)
Ag	3280	—	0.002
Al	3944	—	0.008
B	2498	0.4	0.04
Ba	5535	—	0.005
Be	3321	—	0.0005
Ca	4226	—	0.0005
	3934	0.002	—
Cd	2288	0.03	0.0005
Co	3405	—	0.001
Cr	5208	—	0.001
	4254	0.003	—
Cu	3247	—	0.002
Fe	3720	0.005	0.03
K	7665	—	0.0003
La	3996	0.07	—
Li	6708	0.001	—
Mg	2795	—	0.001
Mn	4030	—	0.0005
Mo	3864	—	0.0005
Na	5890	—	0.00005
Ni	3492	—	0.005
	3415	0.003	—
Pb	4057	—	0.02
	2833	0.03	—
Sb	2598	—	0.003
Ti	3361	—	0.001
U	4242	0.5	—
Y	3710	0.008	—
Zn	4810	—	0.01
	2139	0.01	—

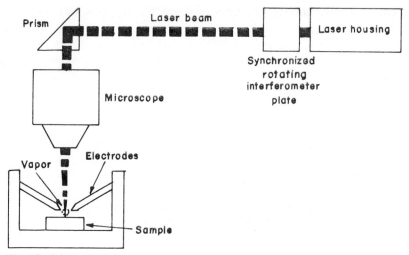

Fig. 6.7 Schematic of the laser microprobe (from literature of Jarrell-Ash Co.).

focused via a microscope onto a very small area of the sample. The intense laser pulse vaporizes a small amount of sample, leaving a hemispherical crater about 50 μm in diameter, and the vapor plume passes between the closely spaced electrodes. The electrodes are connected to a large capacitor charged to a high voltage, and the sample vapor causes the electrode gap to break down, resulting in excitation of the sample vapor.

(3) RF Plasma Torch

The RF *plasma torch* consists of an inductively coupled, RF plasma discharge in argon at atmospheric pressure. This is perhaps one of the more promising sources at the present time for simultaneous, multiele- mental analyses by emission spectroscopy. This source was first investi- gated for trace elemental determinations by Greenfield, Jones, and Berry (14). Fassel and co-workers (15–17) have studied this source for about a decade and have praised its virtues repeatedly at numerous scientific meetings. An excellent RF plasma torch system has recently been described by Boumans and de Boer (18), which appears to be very well suited to simultaneous, multielement analysis. The instrumentation is relatively sim- ple, no tuning of the circuit is necessary, an ultrasonic nebulization system is used, and excellent analytical sensitivity and detection limits are achieved. A schematic of the Boumans and de Boer (18) torch is shown in Figure 6.8. Another interesting RF plasma emission device has been described by Morrison and Talmi (19). This source employs a plasma

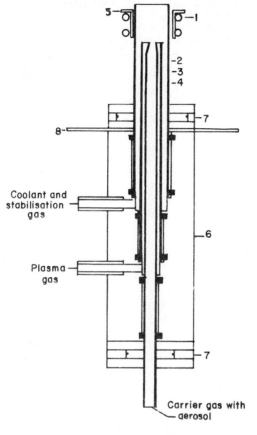

5 ─
─ 1

─ 2
─ 3
─ 4

8 ─
─ 7

Coolant and stabilisation gas

─ 6

Plasma gas

─ 7

Carrier gas with
aerosol

1. Two-turn, water-cooled induction coil
2. Coolant tube
3. Plasma tube
4. Aerosol tube
5. Quartz collar
6. Plexiglass base
7. Centering screws
8. Bottom of plasma torch chamber

Fig. 6.8 RF plasma torch [from Boumans and de Boer (18)].

TABLE 6.5. Detection Limits Reported for Several Rf Plasma Torches

	Detection Limit ($\mu g\ ml^{-1}$)		
Element	Reference 17	Reference 18	Reference 19[a]
Ag	—	0.002	0.0001
Al	0.001	0.003	0.001
Au	—	—	0.001
Ba	0.0001[b]	0.001	—
Bi	—	—	0.0005
Ca	0.0005[b]	0.0001	0.001
Cd	—	—	0.00001
Co	0.004	—	—
Cr	0.001	0.002	0.001
Eu	—	0.0002	—
Fe	—	0.005	0.01
Ga	—	—	0.001
Hg	—	—	0.00001
I	—	—	0.05
In	—	0.0004	0.001
Li	—	0.00005	—
Mg	—	—	0.0001
Mn	—	0.001	—
Ni	0.003	0.02	—
P	—	—	0.01
Pb	0.008	—	0.001
S	—	—	0.01[c]
Si	—	—	0.05
Sr	—	0.0002	—
Te	—	—	0.001
Ti	0.005[b]	0.03	—
Tm	—	0.004	—
V	0.002	0.007	—
W	—	0.7	—
Y	0.0006[b]	0.1	—
Yb	—	0.0002	—
Zn	0.01	—	0.00001

[a]These values are in micrograms and are comparable to the other values if a 1 ml minimum volume is assumed.
[b]Ion line used.
[c]Molecular band used.

discharge in helium and is designed especially to handle solid samples. Detection limits reported for these various devices are listed in Table 6.5.

(4) Capillary Arc

The *capillary arc* source, developed by Jones, Dahlquist, and Hoyt (20), is a unique and interesting sampling-excitation system for atomic emission spectrography on solid, conducting samples. The sampling and excitation processes have been separated; an aerosol generator produces fine particles directly from the surface of a solid metal sample, and these particles are conducted into a DC capillary arc operating in argon at atmospheric pressure. The aerosol generator is a simple device in which the sample acts as the cathode of a DC arc discharge. Rapid movement of the cathode spot produces uniform sampling over a well-defined area, and a flowing gas stream transports the aerosol particles to the capillary arc discharge. The aerosol generator and capillary arc devices are shown in Figure 6.9. These investigators applied the devices to the analysis of steel samples and, while detection limits were not reported, determinations in the 0.01 to 1% (wt%) range were made. Relative standard deviations of a few percent were observed near the 0.01% region, improving to less than 1% in the region of 0.1% and above. Good precision, ability to analyze large samples, and several other features make this a promising sampling-excitation system.

3. EMISSION SPECTROMETERS AND SPECTROGRAPHS

a. General

Once a suitable emission source has been selected, a means of isolating or separating the atomic lines and measuring their intensities is required. This is normally achieved with a spectrograph or spectrometer of some type. A spectrometer and a spectrograph are the same type of optical device, differing in the means of detection and measurement of the source radiation. Both consist of a dispersing instrument (monochromator) to separate the various emissions according to wavelength, and a detector (or detectors) of some type. A *spectrometer* normally employs a photoelectric detector (or detectors)(e.g., a photomultiplier), which converts the radiation to an electrical signal, while a *spectrograph* employs photographic detection, such as a photographic emulsion (film or plate). An image vidicon or similar device has advantages representative of both spectrometers and spectrographs (see Section B.3.e). Spectrometers may be further divided into single-channel or multichannel types, if they (respectively) isolate and measure one spectral interval at a time or more than one. (See Appendices B and C.)

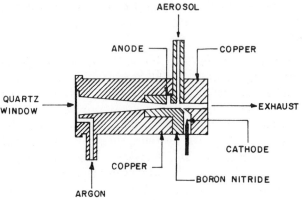

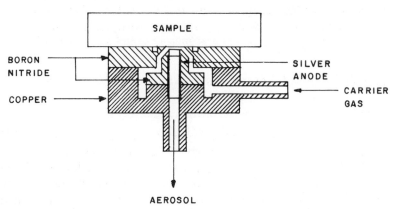

Fig. 6.9 Capillary arc source and aerosol generator for direct solid sampling of metals [from Jones, Dahlquist and Hoyt (20)].

b. Spectrograph

Prior to the availability of high-quality, relatively inexpensive diffraction gratings, spectrographs employed either a glass or quartz prism as the dispersing device. Prisms have largely been replaced by gratings, resulting in spectral systems with linear dispersion. A variety of optical arrangements or mountings are used, such as the Paschen–Runge and Eagle mountings (concave grating, based on the Rowland circle), the Wadsworth mounting (concave grating, *not* based on the Rowland circle), and the

Elbert and Czerny–Turner mountings (plane grating). A thorough description of these, as well as gratings, optics, and so on, is available in various texts, one of the better ones being the paperback by Sawyer (21).

The use of photographic emulsions as the detection system in spectrographs has several advantages and disadvantages. The photographic detector permits the collection of all the spectral information over a wide wavelength range at one time. It is basically an integrating device, in that the radiation detection is cumulative. It provides a permanent record of the data, which may be investigated and reinvestigated as often as one likes. All these make it an exceptional tool for qualitative analysis. The only serious disadvantages of this detector are with respect to its use in quantitative determinations. It takes perhaps 20 to 30 min to develop, fix, and wash the film or plate. The blackness of the images produced must be related to the intensity of the light that produced them. This requires a device to measure the blackness of the images, such as a densitometer (microphotometer), and is complicated by the fact that the blackening (e.g., expressed as $\% \, T$) is not linear with intensity *and* varies with wavelength. This requires that the emulsion be calibrated for accurate quantitative work. Details for performing this calibration are available from the manufacturers of spectrographs and spectrographic supplies, and from various texts (e.g., reference 21). Despite these limitations, when a versatile, sensitive, specific, simultaneous multielement analytical capability is required, the emission spectrograph is hard to beat. It should be stressed that the use of electronic image devices, such as photodiode arrays, image vidicons, and SIT tubes, which are now state-of-the-art devices with certain severe limitations (e.g., rather poor sensitivity and/or resolution), may eventually replace photographic devices as well as conventional multichannel spectrometers when the limitations are minimized and the cost of such devices is reduced (also see Section B.3.e).

c. Single-Channel Spectrometer

The single-channel spectrometer is by far the most widely used in flame atomic emisssion spectrometry, atomic absorption, and atomic fluorescence spectrometry. These spectrometers usually consist of a plane grating monochromator in an Ebert or Czerny–Turner mounting, a mirror focal length between 0.25 and 1 m, equal and bilaterally adjustable entrance and exit slits, a photomultiplier detector, and possibly provisions for electrical wavelength scanning. Signal processing electronics vary from simple electrometer circuits to sophisticated lock-in amplifiers, photon counting or integration, and readout systems vary from simple meters to direct concentration printout. Similar instrumentation is also used with certain high-energy excitation sources, for example, plasmas.

d. Multichannel Spectrometers

These instruments are usually spectrographs in which the photographic detector has been removed and several exit slit–photoelectric detector units substituted; the slits are placed at appropriate points on the focal curve of the instrument to monitor specific wavelengths (elements). These direct-reading, multichannel instruments are usually set up to analyze simultaneously for several to as many as 20 to 30 elements (in the larger instruments). The excitation sources are usually of the electrical discharge variety (arc, spark, plasma, etc.). While not very versatile in terms of being able to easily change the elements determined, they are fast, reproducible and very well suited to the analysis of large numbers of samples of a given type. Some of the larger instruments are quite sophisticated electronically, with built-in computers to process the data, control the source, and print out the analytical results. The only real limit to the number of elements that can be determined simultaneously is the physical space behind the focal curve for the mounting of detectors. Despite the level of sophistication of modern multichannel emission spectrometers, they still represent a "brute force" approach. Some promising new approaches to simultaneous multielement analysis by atomic emission spectrometry are mentioned in the following section.

e. New Developments

Several new instrument designs, dispersing devices, and detectors have appeared in recent years and show promise of improving the sensitivity, versatility, and multielement capabilities of emission spectrochemical analysis.

In the area of dispersing devices the development of holographically ruled gratings and the resurgence of interest in the echelle grating has recently occurred. Holographic gratings, while not yet perfected in terms of groove shape, are essentially perfect with respect to groove spacing, freedom from periodic errors, and so on, resulting in spectra of exceptional purity. Stray light, scatter, ghosts, and so on, are essentially absent, and very high resolving powers can be achieved. This should aid in the development of high-resolution, high-aperture monochromators which are important in emission.

Renewed interest in the echelle grating has also occurred in recent years. These are relatively coarse gratings which depend on use at high diffraction orders to achieve high dispersion. The problem of overlapping orders can be solved by cross-dispersion with another device, such as a prism. The result is a more-or-less square array of wavelengths (i.e., spectral lines), with an order-overlap-free region (free spectral range). One commercially available instrument (Spectrametrics, Inc., Andover, Mass.) employs a series of slits at appropriate locations on the spectral grid, followed by a chopper with appropriately located aperature circles so that a different frequency is established for each slit (wavelength). A single detector is used, and the signals are separated by a multiplex technique. The instrument has high dispersion (spectral bandwidths ~0.01 Å), has a high aperture, and is compact (focal length = 0.75 m). A plasma jet emission source has been employed in this system (13).

Perhaps the greatest interest in recent years in multielement emission techniques has centered on multichannel detectors. These include such devices as television camera tubes and arrays of photodiodes, phototransistors, or photoresistors (e.g., vidicons). Naturally, the photographic emulsion is a multichannel detector, but we are here referring to electronic devices to replace the emusion. The use of these devices was perhaps first suggested by Margoshes (22). Photodiode and phototransistor arrays were studied as detection devices for multichannel emission spectrometry by Boumans and Brouwer (23). They found phototransistors to be superior to photodiodes. Busch, Howell, and Morrison (24), and Mitchell, Jackson, and Aldous (25) used silicon vidicon tubes for simultaneous multielement flame emission spectrometric determinations. The advantages of these systems are that a compact, low-dispersion monochromator can be used to cover a wide spectral region, closely spaced lines can be measured, and they are fast. The main disadvantages, at least at present, is the low sensitivity relative to a photomultiplier. This appears to be a state-of-the-art limitation, and future developments with intensified tubes may overcome this limitation.

f. Choice of a Spectroscopic Instrument

The selection of an emission instrument depends almost entirely on the analytical problem (and perhaps one's resources). In fact, the choice of atomic emission over other techniques also depends on the particular problem to be handled. Factors that have to be considered in choosing an instrument (assuming adequate resources) include: detection limits required; elements to be determined; types of samples; number of samples; number of elements to be determined in each sample; quantity of sample; accuracy and precision required; skill of the operator; time and cost per determination; cost of instrument acquisition, operation, and maintenance; versatility required; and perhaps others. It is impossible to make suggestions covering more than a few clear cases, so the user must arm himself with facts and consider his specific problems and how well the techniques or instrument under consideration meets his requirements. No single instrument or technique does all jobs well, although one may be able to adequately handle several specific problems.

4. QUANTITATIVE ANALYSIS

a. Use of Standards

For a quantitative determination of an element or elements present in a sample, the instrument (actually the analytical procedure) must be calibrated. The measured intensity of a line (or lines) of the elements must be related to the concentration of the elements producing the emission lines. For this purpose, standards of known elemental concentration must be used.

Emission sources like the DC arc and AC spark, which employ solid samples directly, generally require analyzed samples (standards) which closely approximate the unknown samples in composition. This is because the matrix can affect the intensities observed. For example, the emission intensity of the copper lines at a given copper concentration may differ for samples such as steels, aluminum alloys, and soils. Different responses of a given concentration of a given element for different matricies are due primarily to different rates of vaporization and different conditions within the source during the determination. For this reason, it is important for highest accuracy that the standards and samples be as similar in composition as possible.

Other emission sources like the chemical flame, plasma jet, and plasma torch, and so on, which usually employ aqueous solutions of samples, are somewhat less affected by matrix effects. The reason is that solutions containing much above 1% total solids are rarely worked with, so the samples are in a more uniform matrix. For very dilute solutions, matrix effects are rarely a serious problem. This is not to imply that matrix effects do not occur or are nor serious in some cases. Anything that affects the rate at which analyte atoms are introduced into the source, such as changes in viscosity, surface tension, and density, affects the results. However, the most important cause of the observed matrix effects in solutions is solute volatilization; the *rate* at which sample aerosol particles are desolvated, volatilized, and atomized is altered. Compound formation may also be important (e.g., anion interference effects). These matrix effects in solutions, when they occur, can usually be compensated for by use of the standard addition technique.

b. Analytical Range

The useful analytical range of concentration (or amount) over which determinations can be made extends from (or near) the detection limit, up to some arbitrary upper limit. In theory, the upper limit is that of the pure element, while in practice determinations are rarely made for major constituents (e.g., $\gtrsim 10\%$) in the sample. Note that this is not the range obtainable on a single sample; at higher concentrations, samples can be diluted or a less sensitive emission line used. Perhaps the most useful description of the analytical range is the portion of the analytical curve (intensity versus concentration) that is linear. This usually ranges from the detection limit to the point where the analytical curve begins to depart from a straight line (because of self-absorption, reduced atomization efficiency, and so on). For many elements and sources, this may cover several decades of concentration (or amount).

c. Interferences

Anything causing the apparent concentration of an element in a sample to differ from the actual concentration can be termed an *interference*. Interferences in emission spectroscopy usually fall into three general categories: chemical, spectral, and physical.

Physical interferences refer to problems occurring during an analysis, such as changes in the source parameters (current, gas flow rates, sample introduction rate, etc.), changes in the monochromator wavelength setting, and changes in detector response (e.g., as a result of voltage change) or amplifier gain.

Spectral interference refer to such problems as the presence of an emission line not due to the analyte element within the monochromator spectral bandwidth (or an unresolved line in the case of spectrographs), changes in the background emission level during the determination or between samples, and so forth.

Chemical interferences usually refer to problems such as matrix effects (solute-vaporization effects), anion interference effects (compound formation), changes in the degree of ionization, excitation, and atomization, and so forth.

In most applications, these interferences are of little concern but, for quantitative determinations, the analyst must be aware of their existence and take precautions to minimize them when they do occur (refer to Chapter 2, Sections A.2 and A.3, and Appendix A).

d. Applications

To list all the analytical problems to which atomic emission spectroscopy has been applied would indeed be a challenging endeavor. These techniques have been applied to the determination of virtually every element in the periodic table in almost every imaginable matrix, primarily for trace or low concentration levels. Complete summaries of these various applications can be obtained by sifting through the various reviews on this subject. Such review are published in *Analytical Chemistry* every 2 years (see, e.g., reference 26).

C. ATOMIC ABSORPTION SPECTROMETRY

1. BASIS OF METHOD

Just as atoms can be thermally excited and emit radiation of discrete energies, unexcited atoms can also absorb radiation at these (and other) discrete energies. The measurement requires an external source emitting the narrow line spectrum of the desired element. The process can be illustrated as:

$$M + h\nu(\text{from external source}) \rightarrow M^* \tag{7}$$

Here one employs some type of atomization system, such as a flame, to atomize the analyte, and the decrease in the intensity of the appropriate source emission line due to absorption by sample atoms is measured. The measurement is usually made under conditions in which Beer's law is followed (e.g., low concentration).

It was mentioned earlier in the discussion of atomic emission that, at temperatures of a few thousand degrees, the Boltzmann equation predicts that only a small fraction of the atoms are excited. In addition, only a fraction of the atoms excited lose energy by radiative means, the remainder transferring energy in collisions with other species. This means that most of the atoms are in the ground state in the atomization cell and, in the early days of atomic absorption spectrometry, led to the erroneous conclusion that absorption would be inherently more sensitive than emission because more of the atoms are available to absorb radiation than are available to emit radiation. This is not the situation. Rather than go into a theoretical discussion of the reasons here, we consider the following: (a) In emission, the signal becomes progressively smaller as the analyte concentration decreases, while in absorption the reverse is true (i.e., the blank and analyte signals approach each other; the difference in these signals is small and of the order of the emission signals); (b) in practice it may be easier to measure a very small signal (emission) than to measure a very small *difference* in two large signals (absorption). A rough rule of thumb for absorption versus emission using flame atomizers is that elements whose principal resonance lines are below about 3000 Å can usually be determined with lower detection limits by atomic absorption, while those with principal lines above 3500 Å usually result in lower detection limits by emission. The region between about 3000 and 3500 Å contains resonance lines of elements having similar detection limits by both absorption and emission methods. It should be stressed that this comparison is valid only for flames and low-intensity light sources for atomic absorption, that is, nonlaser sources. With high-temperature sources, such as plasmas, this limitation no longer exists, but high dispersion is then needed to contend with the high background emission from these sources. Only for certain cases is there a clear superiority of absorption over emission. Atomic absorption can be used to determine most elements with good analytical sensitivity and low detection limits, although atomic emission is far more readily adaptable to simultaneous multielement determinations and has a wider dynamic range of concentrations determinable under given conditions.

2. ATOMIZERS

a. Flames

When atomic absorption spectrometry was proposed as a means of trace elemental analysis by Walsh (27) in 1955, the chemical flame was a logical choice for sample atomization. For many years, virtually every atomic absorption instrument employed a premixed air–acetylene flame and a slot-head, chamber-type nebulizer burner. This atomization system was quite successful for many elements, but for the more refractory elements there was a clear need for higher atomization temperatures than that provided by the air–acetylene flame. Mixtures like oxygen–hydrogen and oxygen–acetylene have significantly higher temperatures, but the high burning velocity of these mixtures essentially eliminated them because of burner design considerations. A major development was that of the nitrous oxide-acetylene flame (28); this mixture has both a high temperature and a low burning velocity, and greatly extended the obtainable sensitivity and number of elements that could be determined by atomic absorption spectrometry.

Higher sensitivity for each element is often obtained under different flame conditions, observation region, and instrumental parameters. The flame conditions strongly affect the sensitivity for many elements, so it is important that the various parameters be optimized for each determination. Optimum or near-optimum conditions for various measurements can be obtained from the various equipment manufacturers for their particular instruments and burners, or from books on atomic absorption spectrometry, such as the one by Christian and Feldman (29).

b. Furnace Devices

Several furnace atomization systems have been developed or investigated in recent years for atomic absorption spectrometry. Gatehouse and Walsh (30) developed a sputtering chamber atomization system for analysis of solid samples directly. The device operates like a hollow cathode discharge lamp and is shown in Figure 6.10.

One of the most interesting and successful furnace atomization systems is the high-temperature furnace. These are basically graphite tube furnaces; the development of these devices for analytical studies was pioneered by L'vov (31). Many similar devices consisting of heated graphite tubes, graphite rods, tantalum ribbons, platinum wires, tungsten wires, and so on, have been investigated recently. The most promising devices thus far appear to be heated graphite tube furnaces. Two popular versions are available commercially from Perkin-Elmer Corporation and from Varian Instruments, Techtron Division (Figure 6.11). These are essentially the same, namely, graphite tube furnaces, and the primary difference is one of size.

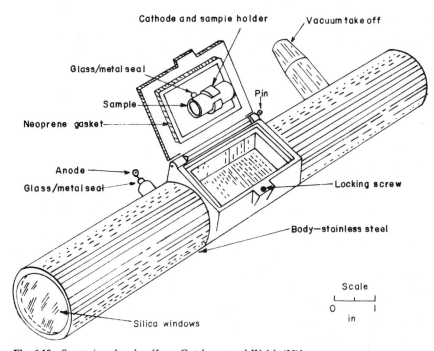

Fig. 6.10 Sputtering chamber [from Gatehouse and Walsh (30)].

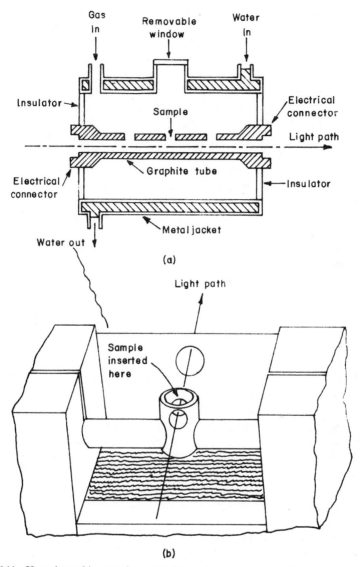

Fig. 6.11 Heated graphite atomizers. (*a*) Schematic of the Perkin-Elmer furnace, and (*b*) Schematic of the Varian Techtron system (the cup shown can be replaced with a small tube).

Graphite filaments (rods) and tantalum ribbon devices have also been used, in which the sample is placed directly on the graphite or tantalum and atomized. These very simple devices are somewhat less popular and have the disadvantage that, as soon as the sample is vaporized off the heated surface, it experiences a rapid temperature drop, which can lead to somewhat greater matrix effects in some cases. The tantalum device, however, has an advantage in the case of elements that form stable carbides in graphite or carbon furnaces.

In graphite tube devices, small samples ($\sim$1 to 100 μl) are usually employed. The sample is pipetted into the tube, a relatively low current is passed to warm the tube and evaporate the solvent, and then the temperature is rapidly raised to vaporize and atomize the sample. Organic matrices (e.g., biological samples) can also be destroyed in a higher-temperature "ash" mode prior to atomization. The two principal advantages of these heated graphite atomization devices are their very high sensitivity and low detection limits (concentrational and absolute), and their ability to utilize very small samples. Murphy and Veillon (32) showed that, even for easily atomized elements, absolute detection limits with heated graphite atomizers are at least two orders lower than with flame atomization. The principal reasons for the superior sensitivity achieved by graphite tube atomizers are: (a) a nonreactive (nitrogen or argon) atmosphere, (b) a strongly reducing medium (hot carbon), (c) the analyte atoms are confined to a relatively small volume for a relatively long time; and (d) low background emission from the furnace gases (assuming of course that the incandescent walls are not imaged on the entrance slit). Temperatures comparable to those in the hottest flames are achieved ($\sim$3000 K), at least by the furnace walls, although the gas temperatures reached are probably lower. Thus these devices overcome several of the limitations of flames and achieve very high analytical sensitivity and low detection limits with very small samples.

The last can also be considered a disadvantage when one is not sample-limited, because present designs *must* be used with small samples. If one wishes to use peak height measurements of absorbance, all the analyte element must be vaporized before any is lost by diffusion out of the tube ends and out of the optical path. If this is not the case, one must use the peak area in the measurement; but even here, only relatively small samples can be accommodated. The relatively poor reproducibility of measurement with these atomizers was initially believed to be due to the reproducibility of the micropipets used to place the sample in the tube. Evidence now indicates that the major cause is the repeatability of sample location and the surface condition of the graphite. Despite this limitation, and the perhaps greater problems with matrix effects, these heated graphite furnaces are clearly superior to flames as atomization systems for many elements. Some comparative detection limits are given in Section 5.d.

3. LIGHT SOURCES

In atomic absorption spectrometry, an external source of radiation (to be absorbed) must be used. Usually these are line sources, emitting the spectral lines of the element to be determined, although a few attempts at employing continuum sources have been made.

a. Hollow Cathode Lamps

By far the most widely used line source is the hollow cathode discharge tube or lamp. A schematic diagram of a simple hollow cathode lamp is shown in Figure 6.12. These are sealed glass devices containing an inert gas (usually neon or argon) at a low pressure (a few torr). The cathode, in the shape of a hollow cylinder or cup, is made from or lined with the pure metal or an alloy of the desired element. A high-voltage, low-current glow discharge occurs in the lamp, and the gas pressure and cathode configuration are such that the discharge occurs primarily inside the cathode cavity. The neon or argon ions bombard the cathode surface and sputter some of the cathode material into the gas phase, (thermal atomization–ionization is also important) where the atoms are excited through ionization, ion collision, and/or electron collision, producing the emission spectrum of the desired element. Usually, the discharge is further confined to the cathode cavity by shielding of the electrodes (Figure 6.13).

Hollow cathode lamps of sufficient intensity can be made for almost any element. In a few cases, in which the lamps are of low intensity, improved detection limits can be obtained by increasing the source line brightness. Two successful ways of achieving this are the use of high-intensity (auxiliary electrodes) hollow cathode lamps and the use of microwave-powered electrodeless discharge lamps. Both of these are discussed in Section D.3.b. Note that it has been stated that the detection limit in the case of very

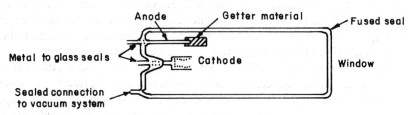

Fig. 6.12 Schematic of a hollow cathode lamp.

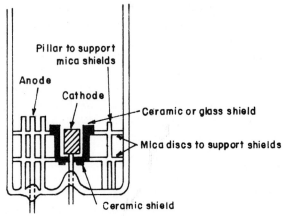

Fig. 6.13   Schematic of hollow cathode lamp, showing shielding usually employed to confine discharge to cathode cavity.

low-intensity sources may be improved by going to a higher-intensity line source. This is not to imply that the sensitivity will be improved. Since atomic absorption is based on Beer's law, an increase in source intensity does not *per se* affect the sensitivity. However, if source fluctuation is the limiting noise source, *and* if a brighter source is employed *without* a corresponding increase in fluctuation (or linewidth: see below), the signal-to-noise ratio will be improved, thus lowering the detection limit.

Most hollow cathode lamps operate at currents of 10 to 20 mA. Intensity generally increases with lamp current, but a maximum operating current is usually specified, based either on lifetime or linewidth considerations. As the lamp current increases, the emission linewidths also increase and may even begin to exhibit reversal at higher currents. For highest sensitivity, it is important that the source emission linewidth be narrower than the absorption linewidth of the element in the atomization cell. This is illustrated diagramatically in Figure 6.14. As long as essentially all of the source emission profile lies within the absorption profile of the sample atoms, little unabsorbed source radiation reaches the detector, and maximum analytical sensitivity is obtained. Should the source emission line become wider than the absorption line, only the portion that significantly overlaps the absorption line will be absorbed, resulting in unabsorbed source radiation and less absorption for a given sample concentration.

This condition can usually be observed by running calibration curves (absorbance *versus* concentration) at various lamp currents. At lower currents, the linear portions of the curves usually all have the same slope. When the source linewidth begins to exceed the absorption linewidth at

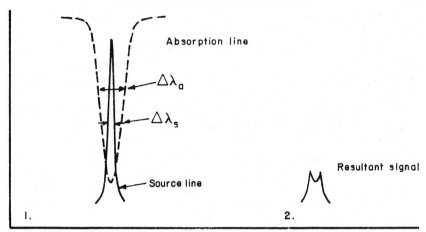

Fig. 6.14 Source emission and sample absorption line profiles in atomic absorption spectrometry.

higher currents, the slope begins to decrease (deviations from linearity).

Choosing the wavelength for a determination is usually a straightforward procedure. The most sensitive lines, and their relative sensitivities, have been tabulated by the various instrument manufacturers. Lacking this, one general approach is to select lines originating at (or very near) the ground state and having the highest transition probabilities (33). A thorough discussion of the selection of wavelengths for atomic absorption has been published by Parsons, Smith, and McElfresh (34).

b.   Continuum Sources

If one could use a continuum of radiation as the absorption source, an obvious advantage would be the elimination of perhaps the most important limitation of atomic absorption spectrometry, the need for a separate line source for each element to be determined. However, to achieve sensitivities comparable to those obtained with line sources, high spectral resolution is required. With a continuum source, the spectral region reaching the detector is determined by the spectral bandwidth of the monochromator. For conventional instruments, this spectral interval might be ~0.1 Å wide. Therefore the analyte atoms would absorb radiation only over the absorption linewidth, ~0.01 Å, and so considerable unabsorbed radiation would reach the detector and reduce the sensitivity. Conventional monochromators having a sufficiently narrow spectral bandwidth (~0.01 Å) would have rather low light throughput capabilities, a condition that would be aggravated by the relatively low intensity of a moderate power (total output) continuum over a ~0.01 Å spectral interval.

Actually, the situation is not quite as hopeless as it might at first seem. While *sensitivity* is important, the *detection limit* is equally important. If an extremely stable continuum source is employed, the detection limits obtained may well approach those obtained with line sources. One of the earlier investigations of continuum sources with conventional instrumentation is that of McGee and Winefordner (35), who obtained respectable detection limits for several elements. Later, Snelleman (36) described an AC scanning method to reduce the contribution of source fluctuations, allowing one to approach the theoretical detection limit. This approach was extended by Svoboda (37), who used a dual-frequency modulation system which eliminated both the source noise and scattering (nonspecific absorption) within the atomization cell.

While these techniques improve the detection limits obtained with continuum sources, the loss of sensitivity due to unabsorbed radiation remains unless the overall spectral bandwidth of the system is narrower than the absorption linewidth. More recently, two high-dispersion systems having both high resolution and high luminosity were investigated for atomic absorption and continuum sources. Veillon and Merchant (38) used a monochromator and a Fabry–Perot interferometer in series, the former serving as a prefilter (to isolate one free spectral range) for the latter. An overall spectral bandwidth of 0.013 Å was obtained, and essentially identical sensitivities with line and continuum sources were obtained. While this system works in principle, it is somewhat state-of-the-art-limited by the availability of broad-band, efficient reflective coatings (several sets of interferometer flats would be needed to cover the entire UV region).

Another interesting approach has been taken by Keliher and Wohlers (39). They employed a high-resolution echelle spectrometer and obtained sensitivities with a continuum source that were nearly as good as with line sources. Their measurements were confined to the 318.5 to 766.5 nm region, because of the rapid decline in source intensity below about 300 nm and the rather low luminosity for their echelle spectrometer.

Despite the encouraging results obtained thus far with continuum sources for atomic absorption, little real interest in this technique has developed. Perhaps this is due in part to the ready availability of stable hollow cathode lamp sources, or to the lack of a need for a highly versatile instrument, or to the use of emission techniques when high versatility (and/or simultaneous multielement determination) is needed. Continuum sources have found greater application and success in atomic fluorescence spectrometry, where they do not have an inherent sensitivity limitation. This application is discussed in Section D.3.c.

4. ATOMIC ABSORPTION INSTRUMENTS

a. Choice of Instrument

Several years ago, it was not uncommon for users to assemble their own atomic absorption instrument from components: burner, monochromator, appropriate electronics, and so on. This practice is still widely used today but is limited almost entirely to research applications in which maximum versatility of the system is required. For most other applications, the user has a wide range of rather sophisticated commercial instruments to choose from. A great many of the advances in atomic absorption instrumentation have come from the developments of numerous very capable scientists associated with instrument companies. Much of the instrumentation available today is well designed, easy to operate, reliable, and offers few traps for the unwary. Prices and features range from relatively simple instruments for a few thousand dollars to quite sophisticated ones well over $10,000. To someone contemplating the need for an atomic absorption analytical capability, choosing the instrument to fit his real needs may be confusing at least. One book has appeared regarding this problem (40), which presents some general suggestions regarding the choice.

Optically, atomic absorption instruments have changed little in the last decade. Some of the more recent developments are worthy of note. Modern instrumentation has become more sophisticated electronically, with integration features, digital readout, computer-compatible outputs, and even built-in micro processors. Two other major developments have been provisions for background correction and the nonflame atomization systems mentioned earlier. The background correction feature of many instruments uses a continuum source *and* a line source to correct for any nonspecific absorption (scatter, molecular absorption, etc.) in the sample readings. This technique takes advantage of the line-versus-continuum sensitivity difference mentioned earlier, but here a spectral bandwidth much larger than the absorption linewidth is used so line absorption from the continuum is negligible while scatter and other nonspecific absorption is not. This feature was adapted by many instrument manufacturers from the work of Koirtyohann and Pickett (41).

b. Operating Conditions

The choice of the proper instrumental conditions for an analysis are quite important. As mentioned earlier, optimum or near-optimum conditions for various instruments and determinations have been worked out by the various instrument manufacturers. Flame conditions and lamp current effects have been mentioned above. Electronic parameters vary too greatly from model to model to be profitably discussed here. One aspect of operating conditions that should be mentioned is the choice of monochromator spectral bandwidth (i.e., slit width). Most hollow cathode lamp sources produce a relatively "clean" spectrum, consisting of the lines of the cathode element and the fill gas. In most cases, the desired line is fairly well isolated, with few other nearby lines and little or no background continuum. Should there be a *nonabsorbing* line near the desired line, it is important that the monochromator spectral bandwidth be adjusted so that this other line is excluded. An example of a nonabsorbing line might be one of the fill gas lines, an impurity in the cathode material, or lines from the desired element, which originate well above the ground state (hence are not absorbed, since the lower state is not populated to any significant extent in the atomization cell). The presence of the nonabsorbing line reduces the sensitivity, just as in the case of the unabsorbed radiation from continuum sources mentioned earlier. An example of this is shown in Figure 6.15 for a nickel hollow cathode lamp; the 2320 Å line is an absorbing resonance line, while the 2316 Å line is nonabsorbing, being a nickel ion line. Here one has to adjust the wavelength setting and spectral bandwidth so as to utilize only the 2320 Å line.

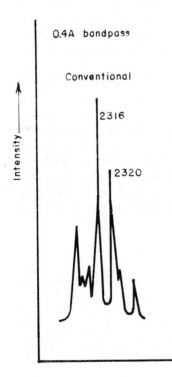

Fig. 6.15 Spectrum from a nickel hollow cathode lamp.

5. QUANTITATIVE ANALYSIS

a. General

Atomic absorption spectrometry has proved to be one of the most widely used and successful techniques for trace elemental analysis. It is quantitative, sensitive, specific, simple, rapid, and relatively inexpensive. It has reached the state of development now where it can be used to determine most of the metals in the periodic table in the microgram-per-milliliter to nanogram-per-milliliter range, and some at even lower concentrations. Recent developments, such as nonflame atomization systems, have significantly increased the analytical sensitivity of the method for many elements, making possible the investigation of many chemical systems previously difficult to study and making it a truly microanalytical method.

There are still some inherent limitations to the technique, which will require further investigation. It is still basically a "solution" method, and does not lend itself as well (at this point) to direct analysis of solids as, say, emission spectroscopy. For all practical purposes, a separate source is needed for each element to be determined. The technique is not well suited to simultaneous multielemental determinations, because each source must be on the optical axis; of the three techniques—emission, absorption, and fluorescence—absorption is the least well suited for this purpose. This is further complicated by the limited dynamic range of concentration determinable.

b. Standards

As in most quantitative techniques, atomic absorption instruments must be calibrated immediately before use with standards (refer to Chapter 2, Section A.3 and Appendix A). Because the greatest application of atomic absorption is for trace elemental determinations in aqueous solutions, standards are often made by dissolving the metal (or appropriate compound) in acid, diluting to volume (with water—see Chapter 3), and making further dilutions as necessary. If matrix effects are important, the standards must closely approximate the samples in composition or the standard addition method used. For very dilute concentrations, matrix effects are usually not important, but *stability* of the solutions is (refer to Chapter 2, Section A.3 and Appendix A). In the nanogram-per-milliliter region, it is not unusual to have standard and sample solutions change in concentration in a relatively short time, decreasing in concentration where adsorption occurs on the container walls and increasing in concentration as a result of contamination from the container or other sources.

c. Sample Preparation

The preparation of standards and samples for atomic absorption spectrometery is fairly straightforward (refer to Chapter 3). In the case of aqueous solutions, it is essentially a matter of dissolution and quantitative dilution. Because of the high analytical sensitivity and low detection limits for many elements, reagent purity, type, and cleanliness of containers, prevention of contamination, and so on, become very important. Some practical techniques for these procedures (and much other practical information associated with atomic absorption spectrometry) can be found in texts like the one edited by Dean and Rains (42) (also refer to Chapter 3).

d. Limits of Detection

Some representative detection limits for several elements by atomic absorption spectrometry are shown in Table 6.6. These data are taken from various sources (as indicated) for a variety of atomization systems. As pointed out earlier, differences of a fewfold are probably insignificant, while differences of an order of magnitude or more may be significant.

TABLE 6.6.   Some Detection Limits Reported for Some Elements by Atomic Absorption Spectrometry

Element	Detection limit (μg ml^{-1}) Flame[a]	nonflame[b]	nonflame[c]
Al	0.1 (N)	3×10^{-6}	1×10^{-6}
Sb	0.03 (A)	2×10^{-5}	5×10^{-6}
As	0.03[d]	—	8×10^{-6}
Ba	0.02 (N)	—	6×10^{-6}
Be	0.002 (N)	—	3×10^{-8}
Bi	0.05 (A)	1×10^{-5}	4×10^{-6}
B	2.5 (N)	—	2×10^{-4}
Cd	0.001 (A)	1×10^{-7}	8×10^{-8}
Ca	0.002 (A)	—	4×10^{-7}
Cs	0.05 (A)	—	4×10^{-7}
Cr	0.002 (A)	5×10^{-6}	2×10^{-6}
Co	0.002 (A)	4×10^{-6}	2×10^{-6}
Cu	0.004 (A)	1×10^{-6}	6×10^{-7}
Ga	0.05 (A)	—	1×10^{-6}
Ge	0.1 (N)	—	3×10^{-6}
Au	0.02 (N)	8×10^{-6}	1×10^{-6}
In	0.03 (A)	—	4×10^{-7}
Ir	1.0 (N)	—	—
Fe	0.004 (A)	3×10^{-6}	1×10^{-5}
Pb	0.01 (A)	6×10^{-6}	2×10^{-6}

Element	Detection Limit (μg ml^{-1}) Flame[a]	nonflame[b]	nonflame[c]
Li	0.001 (A)	—	3×10^{-6}
Mg	0.003 (A)	—	4×10^{-8}
Mn	0.0008 (A)	1×10^{-6}	2×10^{-7}
Hg	0.5 (A)	—	2×10^{-5}
Mo	0.03 (N)	—	3×10^{-6}
Ni	0.005 (A)	1×10^{-5}	9×10^{-6}
Pd	0.01 (A)	—	4×10^{-6}
Pt	0.05 (A)	—	1×10^{-5}
K	0.003 (A)	—	4×10^{-5}
Rh	0.02 (A)	—	8×10^{-6}
Rb	0.005 (A)	—	1×10^{-6}
Se	0.1[d]	—	9×10^{-6}
Si	0.1 (N)	8×10^{-6}	5×10^{-8}
Ag	0.001 (A)	3×10^{-7}	1×10^{-7}
Na	0.0008 (A)	—	—
Sr	0.005 (A)	—	1×10^{-6}
Ta	3.0 (N)	—	—
Te	0.05 (A)	—	1×10^{-6}
Tl	0.02 (A)	1×10^{-5}	1×10^{-6}
Sn	0.05 (A)	—	2×10^{-6}
Ti	0.1 (N)	—	4×10^{-5}
W	3.0 (N)	—	—
V	0.02 (N)	—	3×10^{-6}
Zn	0.001 (A)	6×10^{-8}	3×10^{-8}

[a]Flame atomization (from reference 43). Fuel is C_2H_2; oxidant is either N_2O (N) or air (A), except as noted.
[b]Graphite tube atomization (from reference 44).
[c]Graphite tube atomization (from reference 45). These data were obtained in an enclosed furnace with argon pressures between 1 and 6 atm and various furnace internal volumes.
[d]Argon–hydrogen–entrained air flame.

Where possible, the limits are based on a signal-to-noise ratio $= 2$ criterion and the assumption that a volume of 1 ml is the minimum required for a determination. For example, if an absolute detection limit is given (e.g., nonflame atomizer) as 10^{-9} g, this is expressed as a concentrational detection limit of 0.001 μg/ml. One must bear in mind that most current nonflame atomizers cannot handle samples larger than, say, 0.1 ml, and that most flame atomization systems cannot handle samples (for a reliable reading) of much less than 1 ml. The 1 ml criterion used in Table 6.6 is thus more for the purpose of direct comparison than for the very lowest possible detection limits obtainable under the most stringent conditions by the most avid analyst. Note also that these are not necessarily the lowest detection limits ever reported anywhere (the "world's record," in the words of Alan Walsh) but are taken from selected sources and are hopefully representative. One thing is clear from these data: nonflame atomization systems result in significantly lower detection limits than are obtainable with flame atomization.

e. Interferences

The interferences (also refer to Chapter 2, Section A.3) that occur in atomic absorption spectrometry with flame atomization are essentially the same as those that occur in flame atomic emission, and occur to approximately the same extent. Chemical interferences that affect the atomization process affect absorption to exactly the same extent as emission. Spectral interferences are of a somewhat different nature; in emission, radiation from extraneous species and background are important, while in atomic absorption the spectral bandwidth is determined by the source emission linewidth, and so nonspecific absorption is important.

With furnace atomization, matrix effects *appear* to be more serious, because of the transient conditions during a determination. However, this has neither been proved nor disproved, since a thorough investigation of matrix effects has not yet been performed with these devices.

Applications

Here we need only repeat earlier statements concerning atomic emission; atomic absorption has been applied to the determination of most of the metals in the periodic table in a great variety of sample types. The tremendous analytical sensitivity obtainable with nonflame atomization systems has recently increased greatly the scope of application in environmental and biological areas (which, ultimately, are related). Of particular importance will be investigations into the role of trace metals in biological systems, such as metalloenzymes and metalloproteins, in various clinical conditions.

D. ATOMIC FLUORESCENCE SPECTROMETRY

1. BASIS OF METHOD

Atomic fluorescence spectrometry is essentially a combination of the absorption and emission processes, combining some of the advantages of each and possessing some unique advantages of its own. The process can be represented as:

$$M + h\nu \text{ (from external source)} \rightarrow M^* \rightarrow M + h\nu \qquad (8)$$

Radiation from an external source is absorbed, resulting in excitation of the atoms, and a fraction of these excited atoms decay by emission which is detected and measured. It is an absorption-reradiation process, hence the name "fluorescence." Unlike atomic absorption, in which the source is placed on the optical axis, in atomic fluorescence the source is usually placed at a right angle to the atomizer-monochromator optical axis. This arrangement gives atomic fluorescence two of its greatest potential advantages over atomic absorption: (a) since it is a fluorescence technique, the sensitivity is directly proportional to the source intensity; and (b) source linewidth considerations are no longer as important, since the instrument does not "see" the source, only the reemitted radiation. Consequently, it is not necessary to use narrow line sources, and there is no inherent sensitivity loss with continuum sources. Only the source *intensity* over the absorption linewidth is important.

An additional potential advantage of the atomic fluorescence optical arrangement is that several line sources (or a continuum source) can be utilized for multielement determinations, taking advantage of the extremely simple fluorescence spectrum.

The principal limitations of atomic fluorescence are, at present, associated with the excitation source. As in any fluorescence technique, scattering by particulate matter in the atomization cell can be a problem. In the usual case of resonance fluorescence with line sources, scattering is difficult to correct for but, with continuum sources and wavelength scanning, the correction is straightforward. In practice, scattering is not usually a serious problem and depends to a great extent on the nature of the sample and the sample introduction-atomization systems. Suitably intense line sources are not yet commercially available for every element. Continuum sources, while they may have a high overall spectral output flux, are frequently of relatively low intensity over the $\sim$0.01 Å interval where the atoms absorb. In addition, most continuum sources decrease rapidly in intensity in the UV region, where most elements have their strongest resonance lines.

2. ATOMIZERS

Both flame and furnace devices are used in atomic fluorescence spectrometry. The requirements of an atomization system are basically the same as for atomic absorption, namely, efficient conversion of the sample into ground-state, gaseous atoms. Two additional desirable features are low background emission from the atomization cell (because one is basically making an atomic emission measurement), and low concentration of quenching molecules (with nonlaser sources) to maximize the fluorescence quantum efficiency of the process and thus maximize the fluorescence signal. However, the requirements here are not as stringent as in the case of atomic emission spectrometry, because the source radiation is usually modulated (and thus so is the fluorescence), and the emission arising within the atomization cell is discriminated against electronically.

a. Flame

Flame atomization systems have been widely used in atomic fluorescence, just as in atomic emission and absorption. In atomic absorption slot-type burners (air–C_2H_2 and N_2O–C_2H_2) have proved to be the most popular, while in atomic fluorescence the slot configuration is not geometrically well suited to the excitation-emission requirements. Instead, round burner head configurations have been the most popular.

While flame background emission is usually discriminated against electronically in the measurement system, this radiation still impinges on the detector, and the resulting shot noise can result in poorer detection limits. Thus it is desirable to utilize an atomization system having a low background emission intensity. An important advance in this respect has been the use of *separated* flames in atomic fluorescence spectrometry (46). The laminar, premixed flame may be separated by an inert gas sheath (nitrogen or argon) into the primary reaction zone (blue interconal zone) and the secondary combustion zone (outerconal zone), the region between the two zones having very low background emission. Several burner designs for separated flames have appeared; the one of Larkins (47) is illustrated in Figure 6.16, and a portion of the background emission spectrum is shown in Figure 6.17.

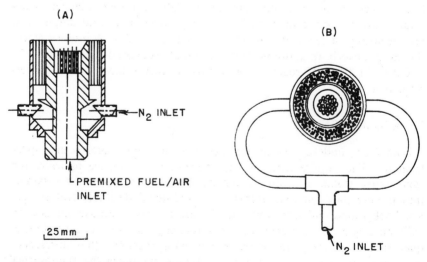

Fig. 6.16   Separated-flame burner [from Larkins (47)].

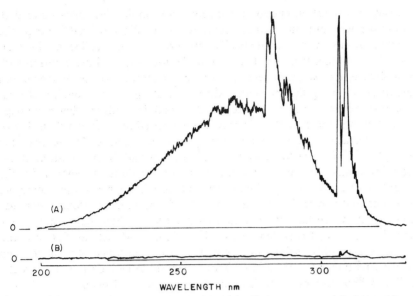

(A)

0 —

(B)

0 —

200   250   300

WAVELENGTH nm

Fig. 6.17   Portion of the air–C_2H_2 flame background emission spectrum of the burner shown in Figure 6.16. The lower trace is with flame separation, and the upper trace is that of the unseparated flame.

b. Furnaces

The same advantages of furnace atomization systems realized in atomic absorption are also realized in atomic fluorescence spectrometry. An additional advantage over atomic absorption is also realized in atomic fluorescence if an atomic gas like argon is used in the atomization system. This advantage stems from the decreased quenching of the fluorescence, that is, an increased quantum efficiency of the fluorescence process over that observed with molecular gases (e.g., nitrogen). In looking again at the fluorescence process,

$$M + h\nu \text{ (external source)} \rightarrow M^* \rightarrow M + h\nu \qquad (8)$$

the quantum efficiency can be defined as the ratio of the number of photons emitted to the number absorbed. The gas composition has a marked effect on this ratio. In atomic absorption, little or no effect by the gas is possible, because one measures only the fraction of photons absorbed. However, in atomic fluorescence, the fraction of the atoms, once excited, that decay by a radiative process depends on the gas composition. It was pointed out earlier in the discussion of emission that only a fraction of the atoms excited decays by radiative means as a result of transfer of excitation energy in collisions with molecular species and that, at atmospheric pressure and temperatures in the 2000 to 3000 K range, the excited atoms would (statistically) undergo several collisions in their $\sim 10^{-8}$ s lifetime. If an excited atom undergoes a quenching collision with a molecule, its excitation energy can be transferred to the molecule and distributed among the molecule's various electronic, vibrational, and rotational states. However, if the excited atom collides with another *atom* (e.g., argon), and if the excitation energy of the excited atom is less than that of the second atom, the excitation energy is not transferred. In this case, the energy is insufficient to excite the second atom and, being an atom, it has no vibrational-rotational states into which to distribute this energy. So, the quantum efficiency of the fluorescence process is improved (and therefore the sensitivity) with furnace atomization if one uses an atomic inert gas like argon, which has an excitation energy greater than that of any of the metals. This effect was demonstrated by Veillon, Mansfield, Parsons, and Winefordner (48), using an argon–hydrogen–entrained air flame, and sensitivity increases of up to 10-fold were obtained despite the greatly reduced flame temperature (compared to oxygen–hydrogen).

Several types of furnace atomization systems have been used in atomic fluorescence spectrometry. The graphite tube furnace configuration popular in atomic absorption does not lend itself to atomic fluorescence measurements, because of the geometry of the optical arrangement in fluorescence. Some versions that have proved popular for atomic fluorescence are similar to the West type shown in Figure 6.11 (cup configuration), the tantalum ribbon type, the platinum loop type (49), and the platinum tube furnace (50). The first three are used with discrete, small-volume samples. The last is used with continuous sample introduction but is limited to operating temperatures of about 1600°C. Several furnace atomization systems for atomic fluorescence have been evaluated by Murphy, Clyburn, and Veillon (51), including a novel graphite tube furnace system for use with continuous sample introduction, which has no background emission in the UV region. An improved version of the latter furnace system has been described by Clyburn, Bartschmid, and Veillon (52), and it was successfully applied to the determination of the metal stoichiometry in the DNA dependent RNA polymerase enzyme from *E. Coli* (52a).

3. SOURCES OF EXCITATION

The single most important part of an atomic fluorescence system is the excitation source. The sensitivity of an analysis by atomic fluorescence spectrometry depends directly on the source intensity. Four general types of sources have been investigated to a greater or lesser extent for use in atomic fluorescence. These can be classified: (1) low-intensity (relatively) line sources, such as hollow cathode lamps; (2) high-intensity line sources, such as electrodeless discharge lamps, metal vapor lamps, and some types of hollow cathode lamps; (3) continuum sources, all of which are effectively low intensity sources; and (4) extremely high-intensity sources, such as pulsed, tunable dye lasers.

a. Low-Intensity Line Sources

Conventional hollow cathode lamps, as widely used in atomic absorption spectrometry, are generally not suitable in atomic fluorescence because of their relatively low intensity. Detection limits with these sources are no better than those obtained in atomic absorption, or with the far more versatile continuum source–atomic fluorescence combination (52). In general, they are unsatisfactory sources for this purpose when operated in the usual manner (but see Section 3.b.(3))

b. High-Intensity (Nonlaser) Line Sources

These are by far the most widely used excitation sources in atomic fluorescence, and the bulk of the effort in the search for suitable sources has concentrated on this general type. Included in this category are metal vapor lamps, special types of hollow cathode lamps, specially operated (e.g., pulsed) conventional hollow cathode lamps, and electrodeless discharge lamps.

(1) Metal Vapor Lamps

Metal vapor lamps, such as the Osram and Philips type, are essentially sealed arc discharge devices. They can be made for several low-melting elements that have a significant vapor pressure within the temperature operating limits of quartz, such as mercury, cadmium, zinc, iridium, thallium, gallium, and the alkalis. Some of the earliest work in analytical atomic fluorescence spectrometry employed these sources (53). These sources provide intense line spectra, but care must be taken in their operating conditions to minimize line reversal.

(2) Specially Constructed Hollow Cathode Lamps

Specially constructed hollow cathode lamps have been developed (54) to produce higher-intensity resonance line spectra for use in atomic absorption, and these have also been used in atomic fluorescence (e.g., reference 55). A schematic diagram of a high-intensity hollow cathode lamp is shown in Figure 6.18. These operate like a conventional hollow cathode lamp, with an auxiliary discharge across the region above the cathode opening. The hollow cathode discharge serves mainly to generate an atomic vapor cloud of the cathode material, while the auxiliary discharge serves primarily to excite the atoms in this cloud, which greatly enhances the intensity of the resonance lines from the source.

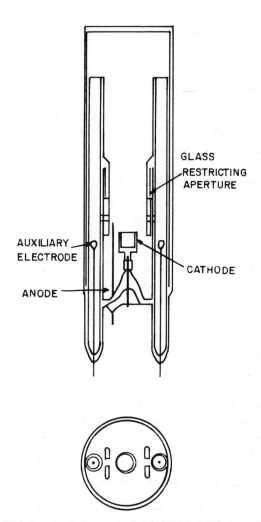

Fig. 6.18   High-intensity hollow cathode lamp [from Sullivan and Walsh (54)].

(3) Pulsed Hollow Cathode Lamps

Pulsed hollow cathode lamps are another means of achieving high spectral intensity. By pulsing a hollow cathode lamp at very high currents, maintaining the pulse width and repetition rate so that the average power is low, very high line intensities are obtained. Using detector or amplifier gating techniques so that the system measures only the fluorescence during the pulse, significant improvements in the signal-to-noise ratio (hence detection limit) can be achieved. This technique has been applied in the development of an atomic fluorescence instrument for simultaneous multielement analysis (56).

(4) Electrodeless Discharge Lamps

The high-intensity line source that has received the most attention in atomic fluorescence is the electrodeless discharge lamp. These are simple, easily constructed devices capable of producing intense, narrow line spectra of a great many elements. An electrodeless discharge lamp usually consists of a small, sealed quartz bulb containing the desired metal (if appreciably volatile) or volatile metal compound (e.g., chloride, iodide) and an inert gas (e.g., argon) at a pressure of about 1 torr. The lamp is placed in the microwave field produced via an antenna or resonant cavity fed by a suitable power supply. Popular power supplies have consisted of medical diathermy units (2450 MHz, 0 to 100 W Rf power) or an equivalent. A low-pressure plasma discharge is initiated in the lamp with a Tesla coil and is self-sustaining. The metal compound evaporates and dissociates, and the metal spectrum is produced in the discharge. Construction details have been described by Mansfield et al. (57), among others.

Until recently, a tremendous amount of work went into the investigation of these sources. While the emission lines from these devices are very intense (as well as narrow), there was considerable art and luck in the making of a really good (intense) electrodeless discharge lamp. Actually, the problem was more one of stability and reproducibility. The solution to this problem was finally found by Browner et al. (58), which proved to be a simple matter of temperature control of the lamp. They utilized a heated air bath system to control the lamp temperature, thus separating the vaporization and excitation processes and greatly increasing the stability of the source. This simple step is perhaps one of the most significant advances to date in atomic fluorescence spectrometry. It should also extend the use of these simple, inexpensive sources into other areas, such as atomic absorption spectrometry.

c. Continuum Sources

The feasibility of using a continuum source of excitation for atomic fluorescence was first demonstrated by Veillon, Mansfield, Parsons, and Winefordner (48), and has received periodic attention since that time. In principle, this approach offers significant advantages over emission and absorption. Only a single source is needed, the atomization system need not also excite the analyte atoms, the simplicity of the fluorescence spectrum eases the dispersion requirements of the monochromator, and it is readily adaptable to simultaneous multielement analysis. In practice, the main limitation (in achieving extremely low detection limits) stems from the relatively low intensity of a continuum over a spectral region as narrow as an atom absorption linewidth. This is further aggravated by the rapid intensity falloff of most continua in the UV region, where many elements have their most sensitive resonance lines. However, a recent study (52), using a single 150 W xenon continuum, nonflame atomization, and photon counting techniques has shown detection limits superior to flame emission and comparable to conventional atomic absorption without the expense of multiple line sources and with greater dynamic range.

d. Laser Excitation

Several recent publications (59–65) have appeared on the use of tunable dye lasers as excitation sources for atomic fluorescence spectrometry. These extremely intense sources have considerable promise for atomic fluorescence spectrometry. The present limitations of cost and wavelength coverage will no doubt be overcome in time, and the analytical use of atomic fluorescence employing these excitation sources should be greatly extended.

4. ATOMIC FLUORESCENCE INSTRUMENTATION

The instrumentation used in atomic fluorescence spectrometry is basically similar to that used in atomic emission or absorption. In fact, many instrumental setups can be used for all three types of measurement with a little modification and rearrangement of optical components.

a. Choice of Spectrometer

Because of the simplicity of the atomic fluorescence spectrum, as compared to that of emission sources and hollow cathode lamps, atomic fluorescence requires very little in the way of dispersion in the monochromator. Consequently, very simple, low-cost monochromators can be used, their light throughput usually being a more important consideration than their dispersion.

If one employs a line source that emits only the spectrum of the desired element (and other *nonabsorbing* lines, like fill gas, etc.), only the atoms (source element) in the absorption cell absorb and, consequently, fluoresce. This being the case, is a monchromator needed at all? The answer is a qualified *no*, and nondispersive atomic fluorescence has in fact been investigated (see references 66 and 67) and has been shown to be a sensitive analytical method employing extremely simple instrumentation. An example of the simplicity of the instrumentation is illustrated in Figure 6.19, which shows the instrumental arrangement used by Vickers and Vaught (67). With equally simple instrumentation Larkins (47), and Larkins and Willis (68), investigated nondispersive atomic fluorescence with air–C_2H_2 and N_2O–C_2H_2 flames, respectively.

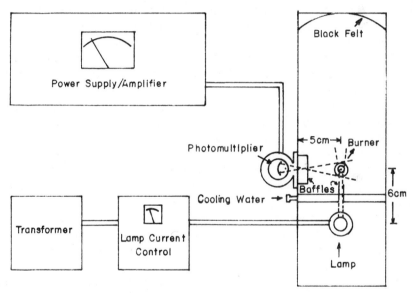

Fig. 6.19 Instrumental arrangement for nondispersive atomic fluorescence spectrometry [from Vickers and Vaught (67)].

b. Choice of Conditions

Optimization of the parameters affecting the atomization–sample introduction systems has essentially the same requirements as in other atomic spectroscopy techniques. Because of the simplicity of the fluorescence spectrum, the optimization of the dispersive-electronic systems also has simple requirements. For many of the low-background atomization cells used in fluorescence, the monochromator slit widths are usually quite wide to increase the sample radiation reaching the detector. Detector voltages are also quite high under conditions in which a maximum signal-to-noise ratio is achieved. Since one frequently tries to measure very weak signals superimposed on a weak background, the electronic system is usually of high gain, and various signal processing techniques are used to achieve the best signal-to-noise ratios, such as lock-in amplification, signal averaging, and long time constants.

c. Quantitative Analysis

Just as in atomic emission and absorption, standards must be used to calibrate the instrument, and the same factors that are important in samples and standards for atomic absorption are equally important in atomic fluorescence (also see Chapter 2, Section A.3, and Appendix A). Interferences occur to about the same extent; fluorescence has an advantage over absorption in that spectral interferences are essentially nonexistent, while scattering (high-solids samples) is usually more of a problem, particularly when line sources are employed. Chemical interferences affect all three methods equally.

In atomic absorption, the sensitivity is essentially defined as the slope of the linear portion of the absorbance-versus-concentration analytical curve (nonlogarithmic plot). Frequently, it is specified as a concentration that is, the concentration resulting in the absorption of 1% of the source line radiation (i.e., an absorbance of 0.004); this definition of sensitivity is confusing, and IUPAC does not recommend its use. In atomic fluorescence, sensitivity is again defined as the slope of the analytical curve plotted on linear coordinates; of course, all atomic methods are predicted (69) to have unity slopes at low concentrations when log-log analytical curves are prepared. The detection limit is defined in the same manner as in the other two techniques, that is, that concentration where the signal-to-noise ratio is 2. Some detection limits reported in the literature are given in Table 6.7. As before, these are not necessarily all of them or the lowest reported, and are based on a signal-to-noise ratio of 2 and a sample volume of 1 ml.

d. Applications

Atomic fluorescence spectrometry has been used in about 40 applications thus far, covering a wide variety of samples and problems. These include metals in oils, soil, air, rocks, alloys, water, food, and biological samples (serum, etc.). One recent application (52a) describes a sensitive

TABLE 6.7. Some Atomic Fluorescence Detection Limits Reported in the Literature

Element	Detection Limit ($\mu g\ ml^{-1}$)	System[a]	Reference[b]
Ag	0.0001	F, L, D	71
	0.0007	N, C, D	52
Al	0.1	F, L, D	71
As	0.1	F, L, D	71
Au	0.005	F, L, D	71
Be	0.01	F, L, D	71
Bi	0.005	F, L, D	71
	0.01	N, C, D	52
Ca	0.001	F, L, D	72
Cd	1×10^{-6}	F, L, D	71
	0.001	N, C, D	52
Co	0.005	F, L, N	47
Cr	0.05	F, L, D	71
	0.04	N, C, D	52
Cu	3×10^{-5}	N, L, D	51
	0.002	N, C, D	52
Fe	0.003	F, L, N	47
	0.02	N, C, D	52
Ga	0.01	F, L, D	71
Ge	0.1	F, L, D	71
Hg	0.0002	F, L, D	71
In	0.1	F, L, D	71
	0.01	N, C, D	52
Mg	0.0002	F, L, N	47
	0.003	N, C, D	52
Mn	0.006	F, L, D	71
	0.005	N, C, D	52
Mo	0.5	F, L, D	71

TABLE 6.7. *(Continued)*

Element	Detection Limit (μg ml^{-1})	System[a]	Reference[b]
Ni	0.003	F, L, D	71
	0.03	N, C, D	52
Pb	0.01	F, L, D	71
	0.01	N, C, D	52
Pd	0.04	F, L, D	71
Pt	0.3	F, L, N	47
Rh	3.	F, L, D	71
Sb	0.05	F, L, D	71
Si	0.6	F, L, D	71
Se	0.04	F, L, D	71
Sn	0.05	F, L, D	71
	0.02	N, C, D	52
Sr	0.03	F, L, D	71
Te	0.005	F, L, D	71
Tl	0.008	F, L, D	71
V	0.07	F, L, D	71
Zn	3×10^{-7}	N, L, D	51
	0.005	N, C, D	52

[a]First letter: F, flame atomizer; N, nonflame atomizer. Second letter: L, line source; C, continuum source. Third letter: D, dispersive; N, nondispersive.
[b]Reference 72 tabulated these values, with original references given in the tabulation.

atomic fluorescence system for use in the investigation of biological systems, and examples of its use in the investigation of metalloenzymes are given. Many of these applications are listed in recent (70) and earlier reviews.

REFERENCES

1. E. E. Pickett and S. R. Kortyohann, *Spectrochim. Acta*, **23B**, 235 (1968).
2. R. Mavrodineanu and H. Boiteux, *Flame Spectroscopy*, John Wiley, New York, 1965.

3. J. D. Winefordner and T. J. Vickers, *Anal. Chem.*, **36**, 1939 (1964).

4. P. W. J. M. Bourmans, *Theory of Spectrochemical Excitation*, Plenum Press, New York, 1966, p. 80.

5. M. Pinta, *Detection and Determination of Trace Elements*, Ann Arbor Science Publishers, Ann Arbor, 1966.

6. M. Slavin, *Emission Spectrochemical Analysis*, Wiley-Interscience, New York, 1971.

7. J. P. Walters and H. V. Malmstadt, *Anal. Chem.*, **37**, 1477 (1965).

8. M. Margoshes and B. F. Scribner, *Spectrochim. Acta*, **15**, 138 (1959).

9. M. Margoshes and B. F. Scribner, *J. Res. Nat. Bur. Stand.*, **67A**, 561 (1963).

10. L. E. Owen, *Appl. Spectrosc.*, **15**, 150 (1961).

11. S. E. Valente and W. G. Schrenk, *Appl. Spectrosc.*, **24**, 197 (1970).

12. P. Merchant, Jr., and C. Veillon, *Anal. Chim. Acta*, **70**, 17 (1974).

13. W. G. Elliott, *Am. Lab.*, August 1971, p. 45.

14. S. Greenfield, I. L. Jones, and C. T. Berry, *Analyst*, **89**, 713 (1964).

15. R. H. Wendt and V. A. Fassel, *Anal. Chem.*, **37**, 920 (1965).

16. G. W. Dickinson and V. A. Fassel, *Anal. Chem.*, **41**, 1021 (1969).

17. R. H. Scott, V. A. Fassel, R. N. Kniseley, and D. E. Nixon, *Anal. Chem.*, **46**, 75 (1974).

18. P. W. J. M. Boumans and F. J. deBoer, *Spectrochim. Acta*, **27B**, 391 (1972).

19. G. H. Morrison and Y. Talmi, *Anal. Chem.*, **42**, 809 (1970).

20. J. L. Jones, R. L. Dahlquist, and R. E. Hoyt, *Appl. Spectrosc.*, **25**, 628 (1971).

21. R. A. Sawyer, *Experimental Spectroscopy*, 3rd ed., Dover, New York, 1963.

22. M. Margoshes, *Spectrochim. Acta*, **25B**, 113 (1970).

23. P. W. J. M. Boumans and G. Brouwer, *Spectrochim. Acta*, **27B**, 247 (1972).

24. K. W. Busch, N. G. Howell, and G. H. Morrison, *Anal. Chem.*, **46**, 575 (1974).

25. D. G. Mitchell, K. W. Jackson, and K. M. Aldous, *Anal. Chem.*, **45**(14), 1215A (1973).

26. R. M. Barnes, *Anal. Chem.*, **46**(5), 150R (1974).

27. A. Walsh, *Spectrochim. Acta*, **7**, 110 (1955).

28. J. B. Willis, *Nature*, **207**, 715 (1965).

29. G. D. Christian and F. J. Feldman, *Atomic Absorption Spectroscopy; Applications in Agriculture, Biology and Medicine*, Wiley-Interscience, New York, 1970.

30. B. M. Gatehouse and A. Walsh, *Spectrochim. Acta*, **16**, 602 (1960).

31. B. V. L'vov, *Spectrochim. Acta*, **17**, 761 (1961).

32. M. K. Murphy and C. Veillon, *Anal. Chim. Acta*, **69**, 295 (1974).

33. C. H. Corliss and W. R. Bozman, NBS Monograph 53, U.S. Govt. Printing Office, 1962. (See also Monograph 32 Supplement, 1967.)

34. M. L. Parsons, B. W. Smith, and P. M. McElfresh, *Appl. Spectrosc.*, **27**, 471 (1973).

35. W. W. McGee and J. D. Winefordner, *Anal. Chim. Acta*, **32**, 429 (1967).

36. W. Snelleman, *Spectrochim. Acta*, **23B**, 403 (1968).

37. V. Svoboda, *Anal. Chem.*, **40**, 1384 (1968).

38. C. Veillon and P. Merchant, Jr., *Appl. Spectrosc.*, **27**, 361 (1973).

39. P. N. Keliher and C. C. Wohlers, *Anal. Chem.*, **46**, 682 (1974).

40. C. Veillon, *Handbook of Commercial Scientific Instruments—Atomic Absorption*, Vol. 1, Marcel Dekker, New York, 1972.

41. S. R. Koirtyohann and E. E. Pickett, *Anal. Chem.*, **37**, 601 (1965).

42. J. A. Dean and T. C. Rains, Eds., *Flame Emission and Atomic Absorption Spectrometry—Components and Techniques*, Vol. 2, Marcel Dekker, New York, 1971.

43. G. D. Christian and F. J. Feldman, *Appl. Spectrosc.*, **25**, 660 (1971).

44. S. Slavin, W. B. Barnett, and H. L. Kahn, *At. Absorpt. Newslett.*, **11**(2), 37 (1972).

45. B. V. L'vov, *Atomic Absorption Spectrochemical Analysis*, Adam Hilger, London, 1970, p. 228.

46. R. S. Hobbs, G. F. Kirkbright, M. Sargent, and T. S. West, *Talanta*, **15**, 997 (1968).

47. P. L. Larkins, *Spectrochim. Acta*, **26B**, 477 (1971).

48. C. Veillon, J. M. Mansfield, M. L. Parsons, and J. D. Winefordner, *Anal. Chem.*, **38**, 204 (1966).

49. M. P. Bratzel, R. M. Dagnall, and J. D. Winefordner, *Anal. Chim. Acta*, **48**, 197 (1969).

50. M. S. Black, T. H. Glenn, M. P. Bratzel, and J. D. Winefordner, *Anal. Chem.*, **43**, 1769 (1971).

51. M. K. Murphy, S. A. Clyburn, and C. Veillon, *Anal. Chem.*, **45**, 1468 (1973).

52. S. A. Clyburn, B. R. Bartschmid, and C. Veillon, *Anal. Chem.*, **46**, 2201 (1974).

52a. S. A. Clyburn, G. F. Serio, B. R. Bartschmid, J. E. Evans, and C. Veillon, *Anal. Biochem.*, **63**, 231 (1975).

53. J. M. Mansfield, J. D. Winefordner, and C. Veillon, *Anal. Chem.*, **37**, 1049 (1965).

54. J. V. Sullivan and A. Walsh, *Spectrochim. Acta*, **21**, 721 (1965).

55. J. Matousek and V. Sychra, *Anal. Chem.*, **41**, 518 (1969).

56. D. G. Mitchell and A. Johansson, *Spectrochim. Acta*, **26B**, 677 (1971).

57. J. M. Mansfield, M. P. Bratzel, H. O. Norgordon, D. O. Knapp, K. E. Zacha, and J. D. Winefordner, *Spectrochim. Acta*, **23B**, 389 (1968).

58. R. F. Browner, B. M. Patel, T. H. Glenn, M. E. Rietta, and J. D. Winefordner, *Spectrosc. Lett.*, **5**, 311 (1972).

59. L. M. Fraser and J. D. Winefordner, *Anal. Chem.*, **43**, 1693 (1971).

60. M. B. Denton and H. V. Malmstadt, *Appl. Phys. Lett.*, **18**, 485 (1971).

61. L. M. Fraser and J. D. Winefordner, *Anal. Chem.*, **44**, 1444 (1972).

62. N. Omenetto, N. N. Hatch, L. M. Fraser, and J. D. Winefordner, *Spectrochim. Acta*, **28B**, 65 (1973).

63. N. Omenetto, N. N. Hatch, L. M. Fraser, and J. D. Winefordner, *Anal. Chem.*, **45**, 195 (1973).

64. N. Omenetto, P. Benetti, L. P. Hart, J. D. Winefordner, and C. Th. J. Alkemade, *Spectrochim. Acta*, **28B**, 289 (1973).

65. N. Omenetto, L. P. Hart, P. Benetti, and J. D. Winefordner, *Spectrochim. Acta*, **28B**, 301 (1973).

66. P. L. Larkins, R. M. Lowe, J. V. Sullivan, and A. Walsh, *Spectrochim. Acta*, **24B**, 187 (1969).

67. T. J. Vickers and R. M. Vaught, *Anal. Chem.*, **41**, 1476 (1969).

68. P. L. Larkins and J. B. Willis, *Spectrochim. Acta*, **26B**, 491 (1971).

69. J. D. Winefordner, M. L. Parsons, J. M. Mansfield, and W. J. McCarthy, *Spectrochim. Acta*, **23B**, 37 (1967).

70. J. D. Winefordner and T. J. Vickers, *Anal. Chem.*, **46**(5), 192R (1974).

71. J. D. Winefordner and R. C. Elser, *Anal. Chem.*, **43**(4), 25A (1971).

72. H. V. Malmstadt and E. Cordos, *Am. Lab.*, August 1972, p. 35.

OPTICAL MOLECULAR SPECTROSCOPIC METHODS

P. A. ST. JOHN

American Instrument Company
Division of Travenol Laboratories, Inc.
Silver Spring, Maryland

A. INTRODUCTION

Electronic absorption spectrometry* has long played a dominant role in the qualitative and quantitative analysis of both minor and trace levels of materials of all types in diverse media. The origins of the technique are clouded in antiquity. Modern usage of light absorption as a quantitative technique received great impetus with the mathematical statements of Beer, Lambert, and Bouguer, which related in a precise manner the attenuation of a light beam to the concentration of absorbers and to the thickness of the absorbing sample. These statements, coupled with the development of photoelectric detectors, resulted in a long period of growth of the technique. This growth has continued to the present day.

Recent developments of alternative methods for the trace analysis of elements have somewhat overshadowed the use of absorptiometric methods. However, the ready availability of a wide variety of absorption instrumentation and an enormous literature of application information practically precludes the obsolesence of the technique.

Trace analysis by light absorption as discussed in this chapter is limited to the absorption of light in the wavelength range 190 to 1000 nm. No arbitrary distinction is made between UV light absorption and visible light absorption or colorimetry.

A discussion of spectrophotometric instrumentation was included in Chapter 5. No further general discussion is included here.

B. REAGENTS FOR TRACE ANALYSIS

The basic concepts of the absorption of radiation by atoms and molecules were discussed in Chapter 4. Only those features of the absorption

---

*Electronic absorption spectrometry has been historically called *absorption spectrophotometry*.

process that bear on the selectivity and sensitivity of the absorptiometric technique as applied to the condensed phase are repeated here.

1. SELECTIVITY

The absorption of UV or visible light by molecules in the condensed phase involves the resonant absorption of photons with energies in the range 50 to 150 kcal mole^{-1}. This energy range corresponds to the energy required to excite electrons from ground-state orbitals to various excited-state orbitals. In principle, the absorption of light by molecules should occur in sharply defined wavelength bands corresponding in energy to the quantized energy gaps between the ground electronic state and the excited electronic states, much in the same fashion as atomic absorption lines are observed in atomic vapors. Molecular absorption spectra are complicated by the presence of vibrational and rotational energy levels superimposed on the primary electronic levels. As a result, while there may be only a few electronic transitions in a molecule that correspond to the energy available in the UV-visible range, these transitions are split up into many separate quantized energy levels depending on the contributions of vibrational and rotational energy. In the vapor phase, where molecule–molecule interactions are infrequent, molecular absorption spectra appear as complex arrays of sharp absorption bands corresponding to the multiplicity of discrete energy levels available for excitation. In the condensed phase, where solute–solvent interactions are frequent, the net effect is to broaden the many sharp absorption bands into broad, rather featureless, absorption regions. These broad absorption bands severely restrict the degree of selectivity that can be obtained by the selection of the absorption wavelength. It is possible in some cases of simple mixtures to select a wavelength in which one compound absorbs and the others do not. However, in the vast majority of cases, selectivity must come from the choice of reagent used, the reaction conditions employed, and the extensive use of physical and chemical separation techniques.

As a rule, most inorganic compounds do not strongly absorb in the UV or visible region of the spectrum, whereas many organic compounds are, corresponding to the availability of $\pi \rightarrow \pi^*$ and $n \rightarrow \pi^*$ transitions in many organic compounds. The trace analysis of elements must usually be approached by reacting organic reagents with the desired inorganic ion in such a way that a strongly absorbing compound is formed. Preferably, the reaction product should have a markedly different spectral absorption characteristic than the reagent itself.

For these reasons, solution absorptiometric analysis of traces of elements is primarily concerned with the formation of complex ions and chelate compounds with organic reagents.

2. SENSITIVITY

The tendency of a molecule to absorb light is described by the molar absorptivity ϵ_{max} (or molar extinction coefficient):

$$\epsilon_{max} = 9 \times 10^{19} P\sigma a \qquad (1)$$

where P is the probability of the absorption transition, and σa is the cross-sectional area of the absorber, that is, the photon capture cross section. For a typical organic molecule, the area σa is on the order of 10^{-15} cm^2. For the limiting case in which the probability of the absorption transition is unity, the maximum attainable value for the molar absorptivity is about 10^5. The relationship of ϵ_{max} to the level of solute concentration that can be measured by absorptiometry is discussed in Section C.1. The important point to be made here is that there is a distinct physical limit to the extent to which a molecule absorbs light. This sets a limit on the sensitivity obtainable with the absorptiometric technique. Under favorable circumstances, this limit is on the order of 1×10^{-8} mole liter^{-1} (1).

3. COMPLEX FORMATION

a. The Effect of pH

Complex formation occurs when a ligand molecule having one or more available lone pairs of electrons donates these electrons to a cation which incorporates them into its own empty available orbitals. Such donor-acceptor behavior classifies the ligand as a Lewis base and the cation as a Lewis acid. A large number of organic complexing agents are in fact Brönsted acids. The relative ability of these acids to form complexes with various cations depends on the relative strengths of the ligand–cation complexes and on the hydrogen ion concentration in solution. With complexing reagents of this type, control of solution pH frequently provides a measure of selectivity. Elements that form strong complexes react under acidic conditions; elements that form weak complexes are not able to compete with hydrogen ions and do not significantly react.

The formation of complexes between cations and organic ligands frequently results in the same type of spectral changes as result, when an equivalent number of protons is removed from the ligand. In other words, the formation of an absorbing species (or a spectrally different species) is a result of the *effective* removal of a proton from the ligand and not the consequence of the association of the cation with the specific ligand. Even though the cation and the hydrogen ion are essentially the same in the Lewis sense, the formation of a ligand–cation complex has more nearly the effect of deprotonation of the ligand. The net result of this type of behavior is a lack of specificity in the applicability of the ligand. The reaction of a quantity of a cationic species with the ligand has roughly the same effect as the reaction of an identical quantity of a strong base (103). Spectral differences and differences in absorptivity exist between complexes of a specific ligand and various cations, but the differences are seldom sufficient to provide an effective means of selectivity determining one species of cation in the presence of other species.

b.   Approaches to Specificity

(1)   Chelate Complexes

The term "chelate" is derived from the Greek word for "claw," and refers to complexes formed with ligands having two or more donor atoms which simultaneously complex with a single cation to form stable ring geometries. Many well-known chelate complexers form highly stable complexes with a large number of cations. An approach that has not been widely investigated is the design of chelate complexing agents specifically to accommodate or "fit" only cations of certain rather specific sizes. One such reagent, Calcichrome (cyclo-tris-7-(1-azo-8-hydroxynapthalene)3,6-disulfonic acid), reacts specifically with calcium in highly alkaline media (2, 3). Strontium and barium ions are too large to fit into the chelate cage involved. Other ions that might react are either precipitated at the high pH or are present as unreactive anionic complexes.

The chelate cage approach has been used more extensively in the formation of fluorescent complexes in which the formation of rigid ring structures has the twofold benefit of increased size specificity and increased fluorescence sensitivity (4, 5).

(2)   Ternary Complexes

The great majority of ligands form what are called binary complexes. This means that one cation M forms a complex with one *type* of ligand L, which may be of the form ML, ML_2, ML_3, and so on. The problems of specificity with such complexes have been previously discussed. The formation of more complicated complexes involving the association of two different ligands L and Y with a single cation requires that a far more specific set of requirements be fulfilled, hence the relative potential for specificity is far greater. This type of complex may be of the form MLY, ML_2Y, MLY_2, and so on.

Ternary complexes appear to form by several different mechanisms. In some cases, the first ligand only partially fills the locations on the coordination sphere of the cation. The remaining positions are filled by the second ligand which presumably is suited to fill a specific geometric requirement (6). Another mechanism appears to be the formation of a strong, primary ionic complex with one ligand. This complex then acts as a separate entity which complexes with yet another ligand in a manner which might be schematically represented as (ML)Y (7, 8).

Many of the ternary complexes so far investigated have very high molar absorptivities. The ternary complex of zinc with 1,10-phenanthroline and rose bangal extra has a reported molar absorptivity of 95,000 in ethyl acetate (8).

The status of research into the formation of ternary complexes has been reviewed by Koch and Ackermann (9).

(3) Ligand Donor Atom Identity

The rather poor specificity of ligands with oxygen or nitrogen donor atoms leads naturally to the search for other chemically more specific donors. A list of potential donors includes sulfur, selenium, arsenic, and phosphorus. Sulfur analogs of many hydroxy-substituted reagents have been used extensively. These reagents have an increased specificity toward metals that have an affinity for sulfur, such as, mercury, silver, copper, rhenium, ruthenium, and several others. Arsenic is also finding increased usage in specific complexing reagents, particularly for the determination of rare earths (10).

4. APPROACHES TO HIGHER SENSITIVITY

The theoretical sensitivity limitation, which occurs at $\epsilon \cong 100,000$, may be circumvented to a degree by employing techniques that chemically serve to create absorbing species having *effective* molar absorptivities in excess of the absorptivity barrier. Two of these methods, the amplification reaction and the catalytic reaction are mentioned briefly.

An example serves to illustrate the approach taken in the amplification reaction procedure. Phosphate can be determined by the familiar 12-molybdiphosphate reaction. One equivalent of phosphate reacts with 12 equivalents of molybdate. The 12-molybdiphosphate can be extracted as phosphomolybdic acid into butanol–chloroform. This extract is then treated with ammonia, which destroys the acid and re-forms 12 equivalents of molybdate. The molybdenum is then determined by a colorimetric reagent such as 2-amino-4-chlorobenzenethiol (11). This reagent has a molar absorptivity of 36,000, however, there are 12 molybdate ions for every original phosphate so, from the standpoint of phosphate determination, the absorptivity has been *effectively* amplified by a factor of 12 ($\epsilon = 432,000$).

The applicability of the amplification approach is limited by the availability of unique reactions such as the 12-molybdiphosphate reaction. Consequently, the amplification reaction has not found widespread application.

An active area of research, particularly in the USSR, is the area of catalytic or kinetic approaches. The basic approach is straightforward: A trace level of an element has a catalytic effect on a reaction which is in turn monitored by a colorimetric procedure. A long known, but typical, example is the acceleration of the arsenic(III)–cerium(IV) oxidation–reduction reaction. Osmium, ruthenium, and iodine strongly catalyze this reaction. It has been used to determine osmium at the 0.001 ppm level and iodine at the 0.01 ppm level (12). The subject of kinetic methods has been recently reviewed by Mark (13, 14) and several others (15–17).

C. ABSORPTION TECHNIQUES FOR QUANTITATIVE ANALYSIS

1. BEER'S LAW

When a beam of light* of radiant flux ϕ_0 traverses a sample, the beam is reduced in power by losses due to absorption by the sample, by reflection at interfaces where there is a change of refractive index, and by scattering processes due to particles in the sample, including Rayleigh and Raman scattering by the sample molecules themselves. If only the absorption process is considered, and if the light beam is monochromatic and no light is reemitted, the power of the transmitted beam will be given by the familiar relationship:

$$\phi = \phi_0 10^{-abc} \tag{2}$$

where ϕ_0 is the radiant power of the incident beam, ϕ is the power of the transmitted beam, a is the absorptivity (a constant which indicates the ability of the sample to absorb light under fixed conditions of solvent, temperature, wavelength, and other factors), b is the path length traversed, and c is the concentration of the absorbing substance. This expression, the Lambert–Beer or Bouguer–Beer law, is the mathematical basis of quantitative absorption analysis. Detailed derivations of the law are found in numerous sources and are not repeated here.

A more common form of equation 2 is:

$$A = \log \frac{\phi_0}{\phi} = abc \tag{3}$$

*Light refers to electromagnetic radiation.

where A is called the *absorbance* (optical density) and is the quantity that is directly proportional to the sample concentration c. The proportionality constant a is called the *absorptivity coefficient* (extinction coefficient) and has units that are the reciprocal of the units of b and c. When the path length is expressed in centimeters and the sample concentration is expressed in moles per liter, a is designated ϵ, the molar absorptivity coefficient (in $M^{-1}\ cm^{-1}$ or liter mole^{-1} cm^{-1}).

The power ratio ϕ/ϕ_0 is called the *transmittance* T and is related to the absorbance by:

$$A = \log \frac{1}{T} \tag{4}$$

Commercially available instrumentation is almost universally equipped to display both absorbance and transmittance (or percent transmittance, 100% T).

Beer's law is additive for multicomponent mixtures, *provided* there is no interaction between the absorbing species:

$$A_{total} = \epsilon_1 b c_1 + \epsilon_2 b c_2 + \cdots \epsilon_n b c_n \tag{5}$$

This equation is the basis of the quantitative analysis of multicomponent mixtures by absorptiometry.

Equation 2 is derived on the basis of an ideal set of conditions, most of which are never completely fulfilled in practice. The following section discusses some of the deviations from the law (*apparent* deviations) and some means by which they can be minimized.

2. DEVIATIONS FROM BEER'S LAW

Deviations from Beer's law may be due to both chemical and instrumental causes. In most cases, the deviations are caused by failure to fulfill the conditions defined in the derivation of the absorbance equation. In some cases, the deviation is due to failure of the actual absorbing species to relate directly to the apparent* analytical concentration as anticipated by the analyst.

a. Chemical Equilibria

The shifting of chemical equilibria is perhaps one of the most common causes of apparent deviations from Beer's law. Equilibria of all types may be involved: dissociation, aggregation, complex formation, polymerization,

---

*Apparent concentration is the initial or starting concentration.

and many others. If the discrete species involved in the equilibrium do not have equal absorptivities at the analytical wavelength, any disturbance that shifts the equilibrium results in a change in absorbance which is not correlated with the total analyte concentration. Such equilibrium-related deviations can be minimized by driving the equilibrium strongly in one direction. Suppressing the ionization of a weak acid by lowering the pH is one example of such an approach.

If two absorbing species are in equilibrium, there will be a wavelength at which each species has the same molar absorptivity. This wavelength is called the *isosbestic point*. Absorbance measurements at the isosbestic wavelength are proportional to the total concentration of both species, regardless of the relative proportions of the two forms. The presence of an isosbestic point is a necessary, but not sufficient, condition for the presence of a two-component equilibrium. A drawback in using an isosbestic point as the analytical wavelength is that the molar absorptivities at the isosbestic point are usually much lower than the peak absorptivity of either species. The potential sensitivity of the analysis is thus not as great as would be realized if an absorbance maximum were used.

b. Temperature

Changes in temperature may have an effect by inducing shifts in chemical equilibria. In addition, changes in temperature may directly affect the absorbing species itself and produce changes in the absorption spectrum and the molar absorptivity. The great bulk of compounds does not vary significantly in terms of spectra or absorptivity in the room temperature range. Large decreases in temperature, from room temperature to liquid nitrogen temperature for instance, result in increased fine structure in the absorption spectrum and consequent changes in absorptivity. In the room temperature range, the molar absorptivity may be expected to decrease a few tenths of a percent per degree increase in sample temperature. For most analytical applications, temperature is not a critical factor unless very high precision is required or wide temperature fluctuations are anticipated (e.g., comparison of measurements taken in and out of a cold room). Conversely, if reaction rate methods are being employed or temperature-sensitive chemical equilibria are involved, tight temperature control is mandatory.

c. Solvent Effects

Solvent-induced spectral shifts have been studied extensively in the molecular spectroscopy of compounds. On dissolution, most compounds exhibit a red shift relative to their vapor phase absorption spectrum. Other compounds, particularly carbonyl compounds involving $n \rightarrow \pi^*$ absorption transitions, exhibit blue shifts which increase as solvent polarity increases. Solvent effects involve complex interactions relating to solvent dielectric strength, hydrogen bonding, and solvent–solute interactions of many types. Adherence to Beer's law cannot be expected if solvent composition is allowed to vary in an analytical procedure.

d. Stray Light

Stray light in a monochromator is usually defined as light that passes through the monochromator but which is not composed of wavelengths within the intended spectral bandwidth of the instrument (18). In an absorption spectrometer, the stray light levels may include light leaks from the outside of the instrument and light from the source, which do not pass through the sample cell but reach the photodetector by chance optical paths.

The amount of stray light sets a limit on the maximum value the measured absorbance can reach. For example, if we assume that none of the stray light is absorbed by the sample, the displayed absorbance is given by:

$$A_{OBS} = \log\left(\frac{\phi_0 + \phi_s}{\phi + \phi_s}\right) \tag{6}$$

where ϕ_0 and ϕ are the incident and transmitted light beams as previously defined, and ϕ_s is the power of the stray light. In highly absorbing samples where ϕ may be small with respect to ϕ_s the expression becomes:

$$A_{OBS} = \log\left(\frac{\phi_0 + \phi_s}{\phi_s}\right) \tag{7}$$

If the spectrophotometer has a stray light level of 0.01%, the displayed absorbance will reach a limiting value of:

$$A_{OBS} = \log\left(\frac{1 + 0.0001}{0.0001}\right) \cong 4 \tag{8}$$

Figure 7A.1 shows the deviation of measured absorbance from true absorbance for several different stray light levels. Clearly, instruments used to measure highly absorbing samples must have very low stray light levels.

Stray light is often a problem in the UV region. The level of stray light frequently increases rapidly below 220 nm and may give rise to a spurious apparent absorption peak in the vicinity of 200 nm. The stray light level is

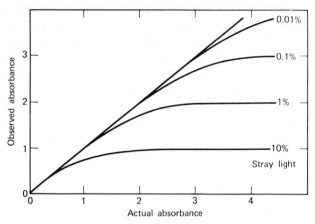

Fig. 7A.1 Observed absorbance versus actual absorbance for various stray light levels.

a function of the overall instrument design. A high-quality absorption spectrophotometer should have stray light levels of about 0.001% in the visible and near-UV regions and should be able to maintain this level to wavelengths as short as 220 nm.

e. Monochromator Spectral Bandwidth

One of the basic assumptions in the derivation of Beer's law is that the light traversing the sample must be monochromatic. If a light beam consisting of two discrete wavelengths passes through an absorber, two separate absorption expressions can be written for each wavelength measured separately:

$$A_1 = \log \frac{\phi_{01}}{\phi_1} = a_1 bc \tag{9}$$

and

$$A_2 = \log \frac{\phi_{02}}{\phi_2} = a_2 bc \tag{10}$$

If the absorbance is measured for both beams taken together, the expression is:

$$A_3 = \log \left(\frac{\phi_{01} + \phi_{02}}{\phi_1 + \phi_2} \right) \tag{11}$$

which is identical to equations 9 and 10 only if $a_1 = a_2$. The same reasoning can be extended to light beams with multiple wavelengths. If the spectral absorption band is broad, the condition that $a_1 = a_2$ is easily met, and the monochromator spectral band width, hence its slit width, can be quite large. If the absorption band is narrow, or if the measurement is being made on the side of an absorption band where the absorptivity is a strongly varying function of wavelength, the monochromator bandpass must be very narrow to fulfill the requirement of adequate monochromaticity. The use of narrow monochromator slits reduces the power of the incident light beam, which in turn may result in an elevated instrument noise level. Attempting to measure absorbance at wavelengths other than the absorption band maximum also places additional demands on the wavelength set-point stability of the monochromator, since even a slight shift in wavelength may result in a large change in the observed absorbance value.

f. Reflection

When light passes from a medium of refractive index n_1 to another medium of refractive index n_2, a fraction of the light is reflected. If the light beam is perpendicular to the media interface, the fraction of light reflected is:

$$f = \left(\frac{n_1 - n_2}{n_1 + n_2} \right)^2 \tag{12}$$

Reflection of light at the air–cuvet and cuvet–solution interfaces is not considered in the derivation of Beer's law. Apparent deviations in absorbance caused by reflection may take two forms: those related to variations in the amount of reflected light, sample to sample or sample to reference, and those related to multiple reflections within the sample cuvet. If the refractive index of the blank (reference) solution is different than that of the sample solution, there will be an error due to the difference in reflected light losses. Similarly, if the refractive index of the sample solution varies from sample to sample, non-solute-related variations in the displayed absorbance will result.

A fraction of light reaching the photodetector has undergone multiple reflections inside the sample cell, thus increasing the effective path length and giving a slight positive bias to the observed absorbance, which is not corrected by the blank. The error is not great, typically 1 or 2 ppt in the visible region for aqueous solutions, but may be a consideration in very high-precision measurements (19, 20).

g. Miscellaneous Effects

(1) Fluorescence

The reemission of light absorbed by the sample may result in spurious low absorbance readings, particularly if the sample absorbs light in the UV and emits it in the near-UV or violet region. Photodetector response frequently decreases sharply as wavelength decreases, and the detector may actually be manyfold more sensitive to the fluorescent wavelengths than to the primary absorption wavelength. The actual extent of fluorescence interference depends to a great extent on the optical configuration of the sample cell and detector area of the spectrophotometer. Fluorescence is emitted isotropically, that is, it is emitted in all directions regardless of the direction of the exciting light beam. If the cell is located a sufficient distance from the photodetector, say 10 cm, the percentage of the fluorescent light that strikes the photodetector will be small and the resultant absorbance error will be small. If the sample cell is closely coupled to the photodetector, an arrangement favored for use with turbid samples, the resulting error may be great. Some absorption spectrometers have two alternate sample cell positions, one is as far as possible from the photodetector for use with fluorescent samples, and the other is as close as possible to the detector for use with nonfluorescent or turbid samples.

(2) Turbidity

Attenuation of a light beam by particulate matter follows a law very similar to Beer's law and is the basis of the techniques of turbidimetry (attenuation of light beam) and nephelometry (measurement of the scattered light). The presence of particulate matter is detrimental, primarily because of the difficulty in reproducing the quantity of suspended particles from sample to sample and between sample and reference. Absorbance measurements in turbid samples are somewhat more successful if the sample cell is closely coupled to the photodetector. This is because a larger proportion of slightly scattered light is received by the photodetector under such circumstances than would be the case if the cell were located some distance from the detector. The technique of dual-wavelength spectrophotometry (21–23) is extremely useful for making accurate absorbance measurements in turbid samples.

(3) Photodecomposition

Photodecomposition is usually a minor consideration in absorptiometry, unless the sample is extremely photosensitive. The silver–thio–Michler's ketone complex is one such example (24). Spectrophotometer designs that place the sample between the source and the monochromator rather than between the monochromator and the detector are more prone to photo-chemical problems, because of the higher intensity of irradiating light in this position. Such designs are now almost totally limited to use in the IR region of the spectrum.

3.   PRECISION AND DYNAMIC RANGE

The precision and dynamic range of the absorptiometric technique may be limited by both chemical and instrumental factors. If the analytical procedure is chemically well behaved (highly reproducible sample to sample) and if specific trivial variations due to mechanical, optical, and electrical considerations in the instrumental system are eliminated, the limit of precision will be determined by the more fundamental limitations of inherent system noise level and data display resolution.

In a sense, resolution-limited data display may be considered a trivial design limitation, since the elimination of this limitation is simply a matter of improved instrumental configuration. However, there are so many data-display-limited instruments in use that the subject warrants considera-tion here.

It is desirable to measure the sample solution under conditions in which a given change in solute concentration produces the largest possible change in transmittance. This can be determined from Beer's law:

$$abc = -\log T \tag{13}$$

then

$$\frac{dc}{dT} = -\frac{0.434}{abT} \tag{14}$$

In an analytical procedure it is usually the relative change in concentration with a change in transmittance that is of interest:

$$\frac{dc/dT}{c} = \frac{0.434}{T \log T} \tag{15}$$

Figure 7A.2 illustrates the variation in $(dc/dT)/c$ with transmittance. The function goes through a minimum at $T = 0.368$ $(A = 0.434)$ and increases very rapidly at either end of the transmittance range. If a working range of $T = 0.1$ to $T = 0.75$ is assumed, an absorbance dynamic range of about 10 to 1 is indicated. If the instrument's data display system reads out in units of transmittance on an analog linear device such as a moving-needle meter or a servorecorder, it will be the analyst's ability to visually discern

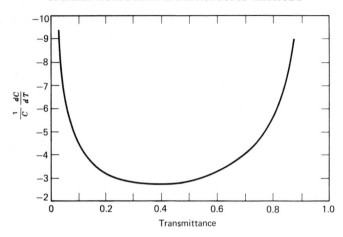

Fig. 7A.2   Relative concentrational error $(1/C)(dC/dT)$ as a function of transmittance.

changes in the needle or pen position that will limit the precision of the analysis and the dynamic range of the analysis to approximately 10 to 1.

Display-limited precision can be improved by various instrument modifications, such as replacing the analog meter with a high-resolution digital panel meter or providing means for electronic scale expansion. Many commercial instruments now incorporate such display improvements. In addition, many instruments now use electronic logarithmic ratio circuits which provide a direct display of absorbance. In some cases, such instruments have a display resolution of 0.0001 A. With these instruments, measurement precision is usually limited by photodetector and light source flicker noise. The usable dynamic range of such instruments may exceed 100 to 1.

Instruments that are not display-limited are usually limited by the inherent noise level of their photodetector. Instruments equipped with photomultiplier photodetectors are limited by shot noise which arises as a consequence of the random nature of light striking the photocathode of the detector. Signal-to-noise ratio calculations for such instruments (25, 26) indicate that minimal relative concentrational error occurs at about 0.86 A. The error level increases rapidly below about 0.1 A, but *does not* increase significantly in the high-absorbance range *above* 0.86 A. Thus it appears that instruments equipped with photomultiplier photodetectors and non-resolution-limited displays provide increased measurement precision primarily in the measurement of moderate to highly absorbing solutions.

The measurement of weakly absorbing solutions is likewise limited by system noise level. Under low-absorbance conditions, the photomultiplier detector has no clear-cut advantage over alternate detectors such as the photoconductive cell. The precision of measurements of weakly absorbing solutions is dependent on light source flicker noise, photodetector noise, and electronic noise filtering, as well as many optical and mechanical design considerations (27, 28).

Data readout limitations in simple instruments can be surmounted by the use of "chemical" scale expansion techniques. These techniques, which are known by the terms "precision spectrophotometry" and "differential spectrophotometry," are based on the use of standard solutions to set the limits of meter zero and meter full scale. These techniques are discussed in Section C.4.

4. PRECISION (DIFFERENTIAL) ABSORPTION SPECTROMETRY

The limitations of measurement precision inherent in the use of absorption spectrometers equipped with simple analog data display devices can be overcome quite easily by use of the technique of precision or differential spectrometry (29–31). Differential spectrometry consists of three separate types of photometric measurements which are classified according to the manner in which the photometer data display meter's zero and full-scale points are established. Each procedure is a form of meter scale expansion, with the degree of expansion dictated by the use of standard solutions.

In the conventional colorimetric procedure, meter zero corresponds to 0% T, which is adjusted with the source light shuttered, and meter full scale corresponds to 100% T, which is adjusted with a blank solution in the sample cuvet. In differential methods the meter end points are set by the use of solutions of known concentrations c_1 and c_2. Table 7A.1 gives the boundary conditions for each method as compared to the conventional method.

TABLE 7A.1. Boundary Conditions for Precision Spectrophotometry

Method	Zero-point Standard, 0% T	Full-Scale Standard, 100% T
Conventional	Darkness, $c_1 = \infty$	Blank, $c_2 = 0$
High-absorbance (transmittance ratio)	Darkness, $c_1 = \infty$	Solution of c_2, $c_2 < c$ unknown
Low-absorbance (trace analysis)	Solution of c_1, $c_1 > c$ unknown	Blank, $c_2 = 0$
Ultimate precision	Solution of c_1, $c_1 > c$ unknown	Solution of c_2, $c_2 < c$ unknown

Two instrumental requirements must be met if these techniques are to be applied. The instrument must have a means for electrically adjusting the meter zero point. This control must have sufficient range such that a relatively strong signal, for example, a meter full-scale reading, can be offset (bucked out) to read zero on the meter face. The instrument must also have a means for adjusting the meter full-scale point, either by a variable amplifier gain, a light level adjustment, or both.

a.   High-Absorbance (Transmittance Ratio) Measurements

The high-absorbance method is used to increase measurement precision when highly absorbing solutions are encountered. The meter full-scale deflection is set with a standard solution of concentration c_2 in the sample cuvet rather than a blank (100% T) solution. Meter zero (0% T) is adjusted with the source shuttered as it would be in the conventional procedure. The solution c_2 is chosen to be slightly more dilute than the anticipated unknown solution.

This procedure expands the low-transmission range of the display with the 0% T point left unchanged. The *indicated* absorbance value taken from the meter scale becomes the difference between the absorbance of the sample and the absorbance of the standard:

$$A = abc - abc_2 \qquad (16)$$

Figure 7A.3a indicates the section of the transmission scale expanded by this technique.

b.   Low-Absorbance (Trace Analysis) Measurements

The low-absorbance or trace analysis method involves the expansion of the opposite end of the transmission scale. The full-scale (100% T) point is set with a blank solution as in the conventional procedure. The meter zero point is set with a standard solution c_1 of slightly greater concentration than that of the anticipated unknown. The meter set points must be readjusted several times, since in this case the adjustments will interact. Figure 7A.3b indicates the portion of the transmission range expanded by this technique. The indicated absorbance in this case is *not* a linear function of solute concentration. An analytical curve is required for each set of experimental conditions.

c.   Ultimate Precision Measurements

This approach is a logical combination of the two previous procedures. The two meter end points are set with two standard solutions, such that $c_1 > c > c_2$. Figure 7A.3c illustrates the expansion procedure. As with the

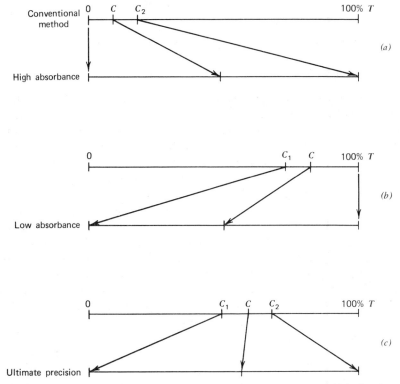

Fig. 7A.3 Scale expansion techniques in precision spectrophotometry: (*a*) high absorbance, (*b*) low absorbance, (*c*) ultimate precision.

low-absorbance technique, several reiterative adjustments of the zero control and the amplifier gain (or light level) are required, since the controls will interact. The indicated absorbance varies in a nonlinear fashion with concentration; a separate analytical curve is required.

d. Application of Differential Techniques

The use of differential techniques to improve reading precision in simple meter display systems is beneficial up to the point where scale expansion is so great that electrical noise, source and mechanical stability, cell positioning errors, and similar sources of imprecision become significant. As a rule of thumb (32), all the techniques will improve precision if no visible noise is evident when the conventional mode is being used. If noise is evident under these conditions, a moderate improvement in precision will result from use of the low-absorbance or ultimate precision technique, but a large improvement may result from the use of the high-absorbance technique.

D. APPLICATIONS

1. PRACTICAL NOTES AND PRECAUTIONS

a. The Blank

The magnitude and reproducibility of the blank nearly always sets the limit of useful analytical sensitivity in spectrophotometric methods of analysis. In most cases the dispersion of the blank values is the most important limitation and is usually traceable to the analytical methodology rather than to instrumental sources. Obviously, it is desirable to both control the variability of the blank and reduce its magnitude if the absorbance technique is to be utilized to its fullest potential. The poor inherent selectivity of the spectrophotometric technique frequently leads to lengthy sample preparation procedures involving various forms of sample manipulation such as extraction, precipitation, and distillation. The possibility of contamination of the sample by extraneous materials increases drastically with the number of manipulative steps in the procedure. Not only is it desirable to keep the number of steps to a minimum, but also to understand potential sources of contamination so that they can be dealt with in a logical and economical fashion.

Elevated or variable blanks are commonly encountered in trace analytical procedures. Depression of blank levels by contamination occurs less frequently. Usually a depressed blank occurs when an unsuspected constituent in the reagent reacts with the analyte of interest in such a way as to prevent further reaction with the prime reagent. Reagent-borne depression-type contamination can be detected by carrying standards through the complete analytical procedure. Sample-borne contamination of this type can be controlled by use of the technique of constant addition or internal standards.

b. The Laboratory Atmosphere

Contamination of samples and reagents by airborne particulates, aerosols, and volatile compounds constitutes a serious problem in trace analysis at the sub-parts-per-million level. Not only are airborne contaminants difficult to control, but they vary in level with atmospheric conditions: prevailing wind, humidity, time of day, and so forth. Analysis of laboratory air samples (33) showed major amounts of calcium, iron, aluminum, copper, silicon, nickel, potassium, magnesium, and managanese, as well as traces of many other elements. Correlation of atmospheric exposure and elevated levels of contaminants extends to many less common elements, including titanium and boron (34), cadmium (35), and arsenic (12).

Approaches to the control of contamination from airborne materials should be keyed to the level of sensitivity required. For analyses at moderate sensitivity levels attention to laboratory cleanliness, covering of containers, and minimized exposure to laboratory air may be all that is required. If analyses at the ultratrace level are required, but the study is limited in scope or number of samples handled, clean hoods, small plastic enclosures, or glove boxes may be used to advantage. Laminar-flow clean hoods are available which utilize high-efficiency particulate air (HEPA) filters (36) which remove the bulk of particles 0.5 μm in diameter or larger. For routine analysis at the ultratrace level, or where a very large number of samples must be processed, the use of a specifically designed clean room may be required (37). Such rooms are tightly sealed and supplied with clean air at a slight positive pressure via a large HEPA filter. Entrance to the room is usually through a double door "air lock" which minimizes an abrupt inflow of contaminated air on entry. Construction of such rooms is a specialized subject. Materials must be selected to minimize the use of metals of all types, paints with metallic pigments, and materials that might tend to generate dust with age (36–38).

c. Reagents

Distilled water as supplied by the central distribution systems commonly found in many large laboratories is seldom adequate for trace analytical purposes. A single distillation through a Pyrex still typically reduces the level of total heavy metals to below 0.1 ppm, with the concentration of any specific metal well below that figure. Ion exchange purification systems are available, which are capable of reducing the total heavy-metal level to the low parts-per-billion range. An ion exchange system coupled with a final distillation from a quartz still may be expected to produce water with metallic loading of less than 1 ppb of any metallic element. Purification of water at this level has been extensively reviewed (47, 48).

Ultrapure water is an extremely powerful solvent which has the ability to extract impurities from the most inert of containers. Such water has been shown to extract materials from carefully prepared Teflon, the best generally agreed on material for containers. For this reason, it is best to prepare high-purity water as it is required, rather than to attempt to store large quantities.

The common reagent acids, HCl, HNO_3, and H_2SO_4, typically contain heavy metals in the sub-parts-per-million range. While this is adequate for many procedures, considerable improvement can be obtained by simple single-stage distillation in a Pyrex still. Ultrapure reagents are commercially available from many sources. These reagents generally represent large improvements over the more commonly available materials. The problems of contamination in packaging, storage, and shipment are such that verification of reagent purity by the analyst is mandatory before the reagents are used in demanding applications.

Dramatic reductions in the levels of trace metal contamination in volatile reagents can be obtained by the technique of subboiling distillation (40–43). Simple distillation is limited in its ability to remove nonvolatile components, because of the formation of aerosols during bubble breakage in the boiling process and the creepage of undistilled liquid onto the condenser. The aerosols are entrained in the vapor phase and eventually contaminate the condenser. Subboiling distillation avoids this problem by heating only the surface of the liquid in the still pot with IR radiation. Only the surface layer of the liquid is strongly heated, and no bubbling or turbulence of the bulk liquid is generated. Creepage contamination is minimized by the use of cold finger–type condensers with very long surface paths between the still pot and the condenser. Quartz subboiling stills have been used for the preparation of H_2O, HCl, HNO_3, $HClO_4$, and H_2SO_4. All Teflon stills have been used to produce high-purity HF. Subboiling stills are very inefficient in the removal of volatile contaminants such as organic compounds and many anions.

High-purity solutions of volatile reagents can be effectively produced by titrating water directly with the gaseous reagent (44, 45). This technique has been used commonly with HCl, HBr, and NH_3. An alternate approach is so-called isopiestic distillation in which a vessel of pure water is placed along with a vessel of the desired reagent in a sealed container and allowed to equilibrate for some time. HCl, HBr, HF, acetic acid, and NH_3 solutions have been prepared by this method (46).

d. Standard Solutions

The stability of standard solutions at low concentrations is the subject of extensive discussion in virtually every text on trace analysis. The universal recommendation of maintaining concentrated stock solutions and diluting as required remains sound advice. Various other aspects of the preparation and storage of standard solutions were covered in Chapter 3 and are not repeated here. A word of caution is in order, however, on the selection of containers for the makeup and storage of solutions. Containment and storage of solutions presents a double problem; the solution may be contaminated by materials leached from the container, and materials in solution may be lost to the container via adsorption, ion exchange, absorption, or reaction. Not only must containers be as inert as possible to the solutions of interest and have very low levels of trace impurities, but they must be extremely well cleaned prior to use (34, 50, 51).

Containers made of polyethylene, polypropylene, Teflon, and many other types of plastics are finding increased usage in many laboratories. It must not be assumed that solute losses via adsorption and other surface interactions are limited to vitreous or metallic containers. Many examples of solute losses in plastic containers have been reported (12, 49). Such losses are more pronounced at alkaline or neutral pH, but they have been known to occur in low-pH solutions as well.

2. GENERAL APPLICATIONS

Table 7A.2 is a listing of sensitivities of absorptiometric techniques for various elements. This table is included as a general indicator of the relative sensitivites that may be expected in absorptiometry. The data in Table 7A.2 were compiled primarily from the texts (see Bibliography) by Cheng, Pinta, and Sandell. The sensitivity is quoted (in $\mu g \ cm^{-2}$) for an absorbance of $A = 0.001$ (Sandell's sensitivity).

**TABLE 7A.2. Comparison of Sandell's Sensitivity Index
for Various Metal Complexes**[a]

Metal	Reagent	Wavelength (nm)[b]	Extinction Coefficient ($\epsilon \times 10^3$)[c]	Sandell Sensitivity ($\mu g \ cm^{-2}$)[d]	Reference
Ag	Diethyldithio-carbamate	560	—	0.008	52
	TMK	520	88.5	0.0012	53
Al	Xylenol orange	540	10.3	0.0026	54
	Oxine	260	—	0.0035	55
		395	—	0.0025	
	Eriochrome cyanine R	530	—	0.0006	56
As	Molybdenum blue	850	—	0.0018	57
Au	TMK	545	150.0	0.0013	58
Ba	o-Cresolphthalein complexone	575	—	0.004	59
Be	4-(p-nitrophenylazo)orcinol	525	—	0.005	12
Bi	Xylenol orange	545	5.2	0.04	60

Metal	Reagent	Wavelength (nm)[b]	Extinction Coefficient ($\epsilon \times 10^3$)[c]	Sandell Sensitivity ($\mu g\ cm^{-2}$)[d]	Reference
Ca	*o*-Cresolphthalein complexone	575	—	0.0014	12
Cd	Dithizone	518	79.0	0.0014	12
Ce	H_2SO_4	320	—	0.025	12
Co	PAN	640	36.5	0.0016	61
Cr	Diphenylcarbazide	540	29.0	0.0018	12
Cu	2, 2′-Biquinoline	545	3.2	0.01	62
	Biscyclohexanone oxalyldihydrazone	595	16.0	0.004	63
Fe	Bathophenanthroline	533	22.35	0.0025	64
Ga	Oxine	392.5	6.4	0.011	65
	Rhodamine B	550	18.8	0.004	66
Ge	Phenylfluorone	510	72.7	0.0009	67
	Gallein	550	19.3	0.0035	68
Hf	Xylenol orange	530	15.7	0.011	69
	Methylthymol blue	570	18.7	0.010	69
	Quercetin	425	32.5	0.0055	70
Hg	TMK	550	9.6	0.002	58
In	PAN	545	24.5	0.005	71
Ir	$SnCl_2$–HBr	402	—	0.0039	12
Li	Thoron	486	—	0.0012	72
Mg	Thiazole yellow	535	—	0.006	73
	AHDCAHS	510	—	0.0005	74

Metal	Reagent	Wavelength (nm)[b]	Extinction Coefficient ($\epsilon \times 10^3$)[c]	Sandell Sensitivity ($\mu g\ cm^{-2}$)[d]	Reference
Mn	Tetrabase	475	—	0.0001	75
	Formaldoxime	450	—	0.005	76
Mo	Dithiol	675	—	0.005	77
Nb	Thiocyanate	383	—	0.0024	78
	Xylenol orange	535	16.0	0.0058	79
	PAR	550	38.7	—	80
Ni	DMG	445 or	14.0	0.0052	81
	PAN	560	61.0	0.001	82
Os	Tetraphenyl-arsonium chloride	346	—	0.02	12
Pb	Dithizone	520	—	0.0031	12
Pd	2-Nitroso-1-naphthol	370	20.3	0.005	83
Pt	*p*-Nitrosodimethyl-aniline	525	—	0.0029	84
Re	α-Furildioxime	532	24.0	0.008	85
Rh	2-Mercapto-4,5-dimethylthiazole	400	—	0.0067	86
Ru	*p*-Nitrosodimethyl-aniline	610	—	0.003	87
Sb	Rhodamine B	565	71.8	0.0017	88
Sc	Xylenol orange	535	27.5	0.002	89
	PAR	530	—	0.0044	90
Se	3,3'-Diaminobenzidine	340	28.9	0.0054	91
Sn	Pyrocatechol violet	555	62.0	0.0019	53

TABLE 7A.2. (*Continued*)

Metal	Reagent	Wavelength $(nm)^b$	Extinction Coefficient $(\epsilon \times 10^3)^c$	Sandell Sensitivity $(\mu g\ cm^{-2})^d$	Reference
Ta	Pyrogallol	325	—	0.04	92
Te	Bismuthiol II	335	28.0	0.005	93
	Gold chloride	570	—	0.001	94
Ti	Cupferron	350	6.2	0.008	95
	Xylenol orange	502	14.9	0.0033	96
Tl	Rhodamine B	560	104.5	0.0023	97
Th	Eriochrome black T	700	35.0	0.004	53
U	PAN	570	23.0	0.009	98
V	Xylenol orange	562	24.0	0.002	99
W	Thiocyanate	400	—	0.013	100
	Dithiol	630	—	0.009	101
Zn	PAN	560	27.8	0.0023	69
Zr	Xylenol orange	535	24.2	0.004	102
	Methylthymol blue	560	21.7	0.004	69

[a]AHDCAHS, sodium salt of 1-azo-2-hydroxy-3-(2,4-dimethylcarboxyanilido)-naphthalene-1-(2-hydroxybenzene-5-sulfonate); DMF, N,N-dimethylformamide; DMG, dimethylglyoxime; PAR, 4-(2-pyridylazo)resorcinol; PAN, 1-(2-pyridylazo)-2-naphthol, tetrabase, 4,4'-tetramethyldiaminodiphenylmethane; TMK, thio–Michler's ketone: 4,4'-bis (dimethylamino)thiobenzophenone.

[b]Wavelength corresponds to absorption maximum.

[c]Extinction coefficient is evaluated at λ_{max} (in 1 mole^{-1} cm^{-1}).

[d]Sandell's sensitivity is defined by: $cb = A/a = 10^{-3}/a$, where c is the concentration of metal (in $\mu g/cm^3$), and b is the absorption path length (in cm). For a 1 cm^2 cross-sectional area, that is, for a cell with a 1 cm^2 cross-sectional area containing sample, the sensitivity values are essentially absolute detection limits (in μg).

Few if any analytical techniques have a more extensive volume of literature available than the technique of absorptiometry. The relatively poor general selectivity of the technique requires that a rather specific set of conditions and separation steps be employed with each different type of matrix studied. The reviews of analytical chemistry, both fundamental and applied, that appear yearly in the journal *Analytical Chemistry* are an excellent source of information on recent developments. Reference texts, particularly those (see Bibliography) of Sandell, Boltz, and Pinta, provide valuable information on established methods and their applicability under a variety of conditions.

BIBLIOGRAPHY

R. P. Bauman, *Absorption Spectroscopy*, John Wiley, New York, 1962.

D. F. Boltz, Ed., *Colorimetric Determination of Nonmetals, Chemical Analysis*, Vol. 8, Interscience, New York, 1958.

G. Charlot, *Colorimetric Determination of Elements*, Elsevier, Amsterdam, 1964.

Cheng, K. L., *Advan. Anal. Chem. Instrum.*, **9**, 321 (1971).

Cheng, K. L., in *Trace Analysis: Physical Methods*, G. H. Morrison, Ed., Interscience, New York, 1965, pp. 161–192.

J. A. Howell and D. F. Boltz, "Spectrophotometry and Spectrofluorimetry," in *Modern Analytical Techniques for Metals and Alloys, Techniques of Metal Research*, Vol. III, Part 1, R. F. Bunshah, Ed., Interscience, New York, 1970, pp. 225–274.

M. G. Mellon, Ed., *Analytical Absorption Spectroscopy*, John Wiley, New York, 1950.

G. A. Parker and D. F. Boltz, "Colorimetry," in *Modern Methods of Geochemical Analysis*, R. E. Wainerdi and E. A. Ukon, Eds., Plenum Press, New York, 1971, pp. 97–126.

E. B. Sandell, *Colorimetric Determination of Traces of Metals, Chemical Analysis*, Vol. 3, 3rd ed., Interscience, New York, 1959.

M. Pinta, *Detection and Determination of Trace Elements*, Ann Arbor-Humphrey, Ann Arbor, 1970.

F. D. Snell and C. T. Snell, *Colorimetric Methods of Analysis Including Some Turbidimetric and Nephelometric Methods*, Vol. 4AAA, 3rd ed., Van Nostrand, Princeton, N. J., 1971. See also, *Colorimetric Methods of Analysis, Including Photometric Methods*, Vol. 4AA, Van Nostrand-Reinhold, New York, 1970.

R. E. Thiers, *Methods of Biochemical Analysis*, D. Glick, Ed., Vol. V, Interscience, New York, 1957.

T. S. West, *Chemical Spectrophotometry in Trace Characterization*, W. W. Meinke and B. F. Scribner, Eds., NBS Monograph 100, U. S. Govt. Printing Office, Washington, D. C., 1967.

J. H. Yoe and H. J. Koch, Jr., Eds., *Trace Analysis*, John Wiley, New York, 1957.

REFERENCES

1. E. A. Braude, *Determination of Organic Structures by Physical Methods*, Academic Press, New York, 1955.

2. R. W. Close and T. S. West, *Talanta*, **5**, 221 (1960).

3. M. Herrero-Lancina and T. S. West, *Anal. Chem.* **35**, 2131 (1963).

4. R. M. Dagnall, R. Smith, and T. S. West, *J. Chem. Soc.*, (**1966**), 1595.

5. D. C. Freeman and C. E. White, *J. Am. Chem. Soc.*, **78**, 2678 (1956).

6. R. Belcher, M. A. Leonard, and T. S. West, *Talanta*, **2**, 92 (1959).

7. R. M. Dagnall and T. S. West, *Talanta*, **11**, 1533 (1964).

8. B. W. Bailey, R. M. Dagnall, and T. S. West, *Talanta*, **13**, 753 (1966).

9. O. G. Koch and G. Ackermann, *Z. Chem.*, **12**, 410 (1972).

10. N. U. Perišić-Janjić, A. A. Muk, and V. D. Canić, *Anal. Chem.*, **45**, 798 (1973).

11. G. F. Kirkbright and J. H. Yoe, *Talanta*, **11**, 415 (1964).

12. E. B. Sandell, *Colorimetric Determination of Traces of Metals*, Interscience, New York, 1959.

13. H. B. Mark, Jr., *Talanta*, **20**, 257 (1973).

14. H. B. Mark, Jr., *Talanta*, **19**, 717 (1972).

15. P. R. Bonchev, *Talanta*, **19**, 675 (1972).

16. A. M. Gary and J. P. Schwing, *Bull. Soc. Chim. Fr.*, **9**, 3657 (1972).

17. H. V. Malmstadt, C. J. Delaney, and E. A. Cordos, *CRC Crit. Rev. Anal. Chem.*, **2**, 559 (1972).

18. F. J. J. Clarke in *Accuracy in Spectrophotometry and Luminescence Measurements*, R. Mavrodineanu, J. I. Shultz, and O. Menis, Eds., NBS Special Publication 378, U. S. Govt. Printing Office, Washington, D. C., 1973.

19. C. T. Chen and J. D. Winefordner, *Canadian J. Spectros.*, **19**, 120 (1974).

20. R. W. Burnett, *Anal. Chem.*, **45**, 383 (1973).

21. B. Chance, *Rev. Sci. Instrum.*, **22**, 634 (1951).

22. J. C. Cowles, *J. Opt. Soc. Am.*, **55**, 690 (1965).

23. T. J. Porro, *Anal. Chem.* **44**, 93A (1972).

24. K. L. Cheng, *Mikrochim. Acta*, **5**, 820 (1967).

25. B. G. Wybourne, *J. Opt. Soc. Am.*, **50**, 84 (1960).

26. J. D. Ingle, Jr., and S. R. Crouch, *Anal. Chem.*, **44**, 1375 (1972).

27. J. J. Cetorelli, W. J. McCarthy, and J. D. Winefordner, *Anal. Chem.*, **45**, 98 (1968).

28. J. J. Cetorelli and J. D. Winefordner, *Talanta*, **14**, 705 (1967).

29. C. N. Reilley and C. M. Crawford, *Anal. Chem.*, **27**, 716 (1955).

30. C. V. Banks, J. L. Spooner, and J. W. O'Laughlin, *Anal. Chem.*, **28**, 1894 (1956).

31. C. V. Banks, J. L. Spooner, and J. W. O'Laughlin, *Anal. Chem.*, **30**, 458 (1958).

32. J. D. Ingle, Jr., *Anal. Chem.*, **45**, 861 (1973).

33. J. Ruzicka and J. Stary, *Substoichiometry in Radiochemical Analysis*, Pergamon Press, New York, 1968, pp. 54–58.

34. I. P. Alimarin, Ed., *Analysis of High-Purity Materials*, Israel Program for Scientific Translations, Jerusalem, 1968, pp. 1–31.

35. J. K. Taylor, Ed., NBS Technical Note 545, U. S. Govt. Printing Office, Washington, D. C., December 1970, p. 53.

36. H. Gilbert and J. H. Palmer, *High Efficiency Particulate Air Filter Units*, TID-7023, USAEC, Washington, D. C., August 1961.

37. *Federal Standards*, Circular 209a, GSA Business Service Center, Boston, Mass., 1966.

38. B. D. Stepin, I. G. Gorshteyn, G. Z. Blyum, G. M. Kurdyumov, and I. P. Ogloblina, *Methods of Producing Superpure Inorganic Substances*, Joint Publications Research Service JPRS 53256, U. S. Govt. Printing Office, Washington, D. C., 1971.

39. J. W. Mitchell, C. L. Luke, and W. R. Northover, *Anal. Chem.*, **45**, 1503 (1973).

40. E. C. Kuehner, R. Alvarez, P. J. Paulsen, and T. J. Murphy, *Anal. Chem.*, **44**, 2050 (1972).

41. J. M. Mattinson, *Anal. Chem.*, **44**, 1715 (1972).

42. *Highly Pure Water Generator*, Leaflet 51-5A, Quartz Products Corp., Plainfield, N. J.

43. K. D. Burrhus and S. R. Hart, *Anal. Chem.*, **44**, 432 (1972).

44. R. E. Thiers, in *Trace Analysis*, J. H. Yoe and H. J. Koch, Eds., John Wiley, New York, 1957, pp. 637–66.

45. R. E. Thiers, in *Methods of Biochemical Analysis*, D. Glick, Ed., Vol. 5, Interscience, New York, N. Y., 1957, pp. 274–309.

46. H. Irwing, J. J. Cox, *Analyst*, **83**, 526 (1958).

47. R. C. Hughes, P. C. Müran, and G. Gundersen, *Anal. Chem.*, **43**, 691 (1971).

48. V. C. Smith, in *Ultrapurity*, M. Zief and R. Speights, Eds., Marcel Dekker, New York, 1972, pp. 173–91.

49. J. W. Mitchell, *Anal. Chem.*, **45**, 492A (1973).

50. D. E. Robertson, in *Ultrapurity*, M. Zief and R. Speights, Eds., Marcel Dekker, New York, N. Y., pp. 208–250.

51. D. N. Hume, *Analysis of Water for Trace Metals*, Advances in Chemistry Series, Vol. 67, American Chemical Society, Washington, D. C., 1967, pp. 30–44.

52. V. Vasak and V. Sedivec, *Chem. Listy*, **46**, 341 (1952).

53. K. L. Cheng, in *Trace Analysis, Physical Methods*, G. H. Morrison, Ed., Interscience, New York, 1965.

54. M. Otomo, *Bull. Chem. Soc. Japan*, **36**, 809 (1963).

55. A. Claassen, I. Bastings, and J. Visser, *Anal. Chim. Acta*, **10**, 373 (1954).

56. A. I. Cherkesov and A. I. Busev, *Zh. Anal. Khim.*, **12**, 268 (1957).

57. C. J. Rodden, *J. Res. Natl. Bur. Stand.*, **24**, 7 (1940).

58. K. L. Cheng and P. F. Lott, in N. D. Cheronis, Ed., *Microchemical Techniques: Proceedings of the International Symposium on Microchemical Techniques, Pennsylvania State University, Aug. 14–18, 1961*, Interscience, New York, 1961, pp. 317–331.

59. F. H. Pollard and J. V. Martin, *Analyst*, **81**, 348 (1956).

60. K. L. Cheng, *Talanta*, **5**, 254 (1960).

61. K. L. Cheng and R. H. Bray, *Anal. Chem.*, **27**, 782 (1955).

62. K. L. Cheng and R. H. Bray, *Anal Chem.*, **25**, 655 (1953).

63. E. Jacobesen, F. J. Langmyhr, and A. R. Selmer-Olsen, *Anal. Chim. Acta*, **24**, 579 (1961).

64. E. Booth and T. W. Evett, *Analyst*, **83**, 80 (1958).

65. J. W. Collat and L. B. Rogers, *Anal. Chem.*, **27**, 961 (1955).

66. H. Onishi and E. B. Sandell, *Anal. Chim. Acta*, **13**, 159 (1955).

67. A. Hillebrant and J. Hoste, *Anal. Chim. Acta*, **18**, 569 (1958).

68. P. J. Sun, *Formosan Sci.*, **14**, 81 (1960).

69. K. L. Cheng, *Anal. Chim. Acta*, **28**, 41 (1963).

70. E. Cerrai and C. Testa, *Energia Nucl.* (Milan), **7**, 477 (1960).

71. S. Shibata, *Anal. Chim. Acta*, **23**, 434 (1960).

72. P. F. Thomason, *Anal. Chem.*, **28**, 1527 (1956).

73. S. Samson and S. Ijlstra, *Chem. Werkblad*, **50**, 213 (1954).

74. C. K. Mann and J. H. Yoe, *Anal. Chem.*, **28**, 202 (1956).

75. E. M. Gates and G. H. Ellis, *J. Biol. Chem.*, **168**, 537 (1947).

76. E. G. Bradfield, *Analyst*, **82**, 254 (1957).

77. L. J. Clark and J. H. Axiey, *Anal. Chem.*, **27**, 2000 (1955).

78. C. E. Crouthamel, B. E. Hjelte, and C. E. Johnson, *Anal. Chem.*, **27**, 507 (1955).

79. K. L. Cheng and B. L. Goydish, *Talanta*, **9**, 987 (1962).

80. R. Belcher, T. V. Ramakrishna, and T. S. West, *Talanta*, **10**, 1013 (1963).

81. K. L. Cheng and B. L. Goydish, *Microchem. J.*, **7**, 166 (1963).

82. S. Shibata, *Anal. Chim. Acta*, **23**, 367 (1960).

83. K. L. Cheng, *Anal. Chem.*, **26**, 1894 (1954).

84. J. J. Kirkland and J. H. Yoe, *Anal. Chem.* **26**, 1340 (1954).

85. S. A. Fisher and V. W. Moloche, *Anal. Chem.*, **24**, 1001 (1952).

86. D. E. Ryan, *Analyst*, **75**, 557 (1950).

87. J. E. Currah, A. Fischel, W. A. E. McBryde, and F. E. Beamish, *Anal. Chem.*, **24**, 1980 (1952).

88. S. H. Webster and L. T. Fairhall, *J. Ind. Hyg. Toxicol.*, **27**, 184 (1945).

89. S. S. Berman, G. R. Duval, and D. S. Russell, *Anal. Chem.*, **35**, 1392 (1963).

90. L. Sommer and M. Hnilickova, *Anal. Chim. Acta*, **27**, 241 (1962).

91. K. L. Cheng, *Anal. Chem.*, **28**, 1738 (1956).

92. J. I. Dinnin, *Anal. Chem.*, **25**, 1803 (1953).

93. K. L. Cheng, *Talanta*, **8**, 301 (1961).

94. H. W. Lakin and C. E. Thompson, *Science*, **141**, 42 (1963).

95. K. L. Cheng, *Anal. Chem.*, **30**, 941 (1958).

96. M. Omoto, *Bull. Chem. Soc. Jap.*, **36**, 1577 (1963).

97. H. Onishi, *Bull. Chem. Soc. Jap.*, **29**, 945 (1956).

98. K. L. Cheng, *Anal. Chem.*, **30**, 1027 (1958).

99. M. Omoto, *Bull. Chem. Soc. Jap.*, **36**, 137 (1963).

100. K. L. Cheng, G. W. Goward, and B. B. Wilson, WAPD CTA (GLA)-180 (Rev. 1), AEC Report, April 1957; *Chem. Abstr.*, **53**, 13886a (1959).

101. S. H. Allen and M. B. Hamilton, *Anal. Chim. Acta*, **7**, 483 (1952).

102. K. L. Cheng, *Talanta*, **2**, 61 (1959).

103. T. S. West, in *Trace Characterization—Chemical and Physical*, W. W. Meinke and B. F. Scribner, Eds., NBS Monograph 100, U. S. Govt. Printing Office, Washington, D. C., 1967.

FLUOROMETRIC METHODS FOR TRACES OF ELEMENTS

P. A. ST. JOHN

American Instrument Company
Division of Travenol Laboratories, Inc.
Silver Spring, Maryland

A. INTRODUCTION

Records of the observation of inorganic crystal luminescence in materials such as barite date back to the early fifteenth century. Perhaps the first recorded observation of solution luminescence was that of David Brewster who, in 1833, described the red fluorescence of chlorophyll in plant extracts. By the early 1900s, fluorescent dyes, such as disodium fluorescein, had found use in tracing subterranean water flow, but little substantive work was reported utilizing fluorescence or phosphorescence as qualitative or quantitative analytical techniques. The use of the luminescence phenomenon as a quantitative analytical tool has developed since approximately 1945. With the development of the fluorescence spectrometer with its photoelectric detector and high-intensity arc lamp, the field of luminescence analysis commenced a period of rapid growth.

A review of the literature from 1950 to 1964 (1) lists over 5000 titles pertaining to fluorescence. A review of fluorometric analysis (2) for the years 1971 to 1973 alone lists over 1000 references. It is expected that this growth rate will continue, particularly as applications are found in the areas of biochemistry and medicine.

In view of this growth as a technique and in view of the specific advantages fluorimetry and phosphorimetry possess over absorptiometry and other competitive techniques, it may appear surprising that luminescence techniques have not made greater inroads into the specific area of trace elemental analysis. Why this is so is open to interpretation. Certainly the development of atomic absorption spectrometry as a high-sensitivity analytical technique has exerted an effect. Perhaps the isolation of biochemists from chemists, which frequently occurs on the college campus, has channeled the development of luminescence techniques away from the

213

field of elemental analysis and into the areas of biochemistry and medicine.

In view of the extreme sensitivity and versatility of luminescence analytical procedures, it seems likely that this method of trace elemental analysis will grow apace as one of the powerful techniques for trace characterization and quantitation.

The discussion in this chapter is limited to solution luminescence. Usually this involves absorption and emission of light* in the wavelength range 200 to 850 nm. Atomic fluorescence and X-ray fluorescence are discussed in Chapters 6 and 9, respectively. Any discussion of the trace analysis of elements by luminescence techniques is almost entirely a discussion of the luminescence of metal chelate systems and molecular complexes. With only a few exceptions, metallic elements require complexation with an organic molecule to become fluorescent. Therefore what follows is essentially a review of molecular luminescence analysis as applied to trace elemental analysis.

B. THEORY OF MOLECULAR LUMINESCENCE SPECTROSCOPY

The terminology employed in molecular luminescence spectroscopy is confused to a degree by the use in everyday language of terms that have very specific meanings when used in the spectroscopic context. For example, "fluorescent" lights utilize inorganic crystal fluorescence, a very different type of phenomenon which is not discussed here; "phosphorescent" planktons and fireflies are examples of bioluminescence; "phosphorescent" chemicals are examples of chemiluminescence; and so forth.

The basic principles of absorption and emission of radiation have been covered in Chapter 4. This chapter briefly reviews this material as it applies specifically to solution molecular luminescence.

The terms "fluorescence" and "phosphorescence" are applied to the reemission of previously absorbed light. In appearance, both types of luminescence are similar. The emitted light is composed of wavelengths longer than those of the exciting light. Each type of emitted light has a characteristic wavelength distribution (spectrum) which is independent of the excitation light. This spectrum is dependent in varying degrees on the environment of the emitting molecule. Each type of luminescence has a distinct lifetime (decay time) during which the light continues to be emitted after termination of the exciting light. This lifetime too is dependent on the environment of the molecule.

*Light is used synonomously with radiation in this chapter.

In appearance, fluorescence and phosphorescence differ in that the wavelength of the fluorescence emission is always shorter than that of phosphorescence, and the lifetime of the fluorescence emission is always much shorter than that of phosphorescence. Fluorescence lifetimes usually fall in the range 10^{-9} to 10^{-6} s, and phosphorescence lifetimes usually fall in the range 10^{-3} to 10 s. The most striking difference is the conditions under which each type of luminescence is observed. Fluorescence is usually seen at moderate temperatures in liquid solution. Phosphorescence is seen in rigid media, usually at very low temperatures.

1. EXCITATION

Electronic excitation of a molecule involves the promotion of an electron from a ground-state orbital to an excited-state orbital with a corresponding increase in molecular energy, changes in molecular symmetry, and occasionally changes in multiplicity (net electron spin). The energy for such electronic excitation may come from many sources. In luminescence analysis, we are concerned with the excitation resulting from the resonant absorption of photons.

Figure 7B.1 is a schematic representation of the electronic energy levels of a hypothetical molecule (Jablonski diagram). Superimposed on each electronic level or state are vibrational levels which correspond to vibrationally excited states of each specific electronic configuration.

At room temperature in liquid solution, most organic molecules exist primarily at the ground vibrational level. Consequently nearly all electronic transitions originate from the ground vibrational level. Absorption of light by the molecule results in the direct excitation of an electron to an excited-state orbital. Transitions may occur between the ground electronic state and any of the excited singlet electronic states. Radiationless transitions between excited *vibrational* levels in a given electronic state occur very rapidly when compared to the time required for an electronic transition. Therefore any excess vibrational energy is quickly lost as the molecule returns to the ground vibrational level of the particular electronic state involved. Similarly, radiationless transitions between higher-energy excited singlet electronic states and the first excited state occur very rapidly. For this reason, nearly all electronic processes can be explained by considering only the ground state and the first singlet and triplet excited states.

Transitions between electronic states of different multiplicities—different net electron spin—are referred to as *spin-forbidden transitions*. This simply means that the probability of a change in net electron spin during an electronic transition is extremely small, whether the transition is radiative or nonradiative. This is why direct transitions from the ground

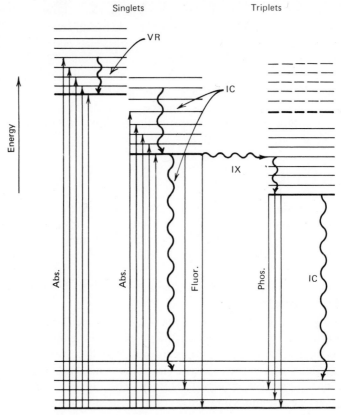

Fig. 7B.1 Electronic energy level diagram (Jablonski diagram) for a hypothetical molecule showing absorption (Abs.), vibrational relaxation (VR), fluorescence (Fluor.), internal coversion (IX), and phosphorescence (Phos.).

singlet to the first excited triplet are virtually unobservable. In the Jablonski diagram, each excited singlet state is shown with a companion triplet state at a lower energy level. This is an extension of Hund's rule from atomic spectroscopy, which states that the electronic configuration with the greatest number of unpaired electrons is the most stable configuration. This observation can be qualitatively explained by the fact that electrons with like spins tend to be further separated in space (in separate orbitals) and experience less coulombic interaction.

2. DEEXCITATION

There are five basic intramolecular processes by which excitation energy can be dissipated: vibrational relaxation, internal conversion, fluorescence, intersystem crossing, and phosphorescence. Figure 7B.1 is a simplified energy level diagram which shows the transition involved in each of these processes.

a. Vibration Relaxation

A molecule at an excited vibrational level may lose this excess vibrational energy either by the emission of IR light or by the transfer of energy to its surroundings. In the gas phase where molecule–molecule collisions are infrequent, infrared emission is observed. In the liquid or solid phase, such collisions are so frequent that all excess vibrational energy is lost to the surrounding media in 10^{-13} to 10^{-11} s.

When a molecule absorbs light of sufficient energy, it is frequently excited to one of the higher vibrational levels of an excited electronic state. Vibrational relaxation is so efficient that this excess vibrational energy is quickly lost, and it maybe assumed that any subsequent emission of light originates from the ground vibrational level of the excited electronic state involved.

b. Internal Conversion

The collective nonradiational processes by which an electronically excited molecule can return to its ground state are referred to as *internal conversion*. These processes comprise the dominant routes of excited-state deactivation. This is why many more materials absorb light than reemit it. Internal conversion proceeds at a rate that is strongly dependent on the relative energy difference between the excited states involved.

The first excited singlet state and the higher energy excited singlet state above it are usually closely spaced in terms of energy, such that the higher vibrational levels of a lower electronic state may overlap the ground vibrational level of a higher electronic state. Under these conditions, internal conversion proceeds at a very rapid rate, quickly deactivating the molecule to its lowest *excited* singlet state. The energy difference between the first excited singlet and the ground state is usually sufficiently large that this vibrational overlap is small or essentially nonexistent, and internal conversion proceeds at a much slower rate. Because of these rate differences, it can be assumed that, regardless of the excited state initially populated by absorption, the molecule will undergo rapid deactivation to the lowest vibrational level of the lowest excited singlet state before fluorescence or any other subsequent deactivation process can occur.

c. Fluorescence

Fluorescence is the direct radiational deactivation of an excited singlet electronic state. Photon emission occurs from the ground vibrational level of the excited electronic state to any of the vibrational levels of the ground electronic state.

The mean lifetime of an excited singlet state is typically 10^{-9} to 10^{-7} s. Consequently, fluorescence decay times are of the same order of magnitude. The fraction of excited molecules that emit fluorescence (the fluorescence quantum efficiency) may vary from essentially unity, as in disodium fluorescein or rhodamine B, to very nearly zero for the bulk of organic compounds. The photon emitted as fluorescence will be of lower energy than the absorbed light by an amount equal to the energy lost by vibrational relaxation in the excited state and the excess vibrational energy remaining in the ground state after the transition has occurred.

This is why the fluorescent light is generally of longer wavelength than the absorbed light. This should be contrasted with atomic fluorescence in which the fluorescence may be the same wavelength as the exciting light (atomic resonance fluorescence).

d.   Intersystem Crossing

An additional process for the depopulation of the excited singlet state is the spin-forbidden conversion from the excited singlet to its companion, but lower-energy, triplet state. This process, called *intersystem crossing*, is a vibrationally coupled process of the same general type as internal conversion and involves the highly unlikely reversal of electron spin. In spite of the spin-forbidden nature of this transition, it proceeds at a rate roughly comparable to fluorescence deactivation. This is a result of the small energy gap between the singlet and triplet levels, which favors a vibrationally coupled crossover. The rate of crossover can be increased by introducing an atom of high atomic number into the molecular electronic domain either as a component of the molecule itself or as part of the solvent (the *heavy atom effect*). Introduction of a paramagnetic species into the electronic domain also increases the crossover rate.

Following intersystem crossover, vibrational relaxation occurs, resulting in population of the ground vibrational level of the excited triplet state.

e.   Phosphorescence

Phosphorescence is the direct radiative deactivation of the excited triplet state. The spin-forbidden nature of the triplet–singlet transition and the smaller energy separation between the triplet level and the ground singlet level combine to make nonradiative internal conversion processes the dominant modes of deactivation of the triplet state. The radiative lifetime of the triplet state is so long (10^{-3} to 10^2 s) that in general it is necessary to immobilize the triplet molecule in a rigid medium at a low temperature in order to observe phosphorescence. In analytical applications utilizing phosphorescence, this is usually accomplished by freezing samples at liquid nitrogen temperature. Cooling and immobilization minimizes collisional deactivation and vibrationally coupled deactivation. Under these conditions, phosphorescence emission becomes a favorable mode of deactivation for some molecules.

Phosphorescence emission is shifted to wavelengths longer than those of fluorescence emission by an energy indicative of the difference in energy between the first excited singlet and the excited triplet levels.

The presence of a heavy atom or a paramagnetic species can increase the probability of phosphorescence emission. In the case of the heavy atom, the higher the atomic number the greater the spatial delocalization of its "electron cloud" and the more likely it is that the electron undergoing the transition will spend a small amount of time associated with the atom either through charge transfer or electron exchange. This serves to increase the overall singlet nature of the complete system and results in a higher probability of spin reversal. As the probability of the triplet–singlet transition increases, whether radiative or nonradiative, the lifetime of the excited state decreases. Halogen substitution in a phosphorescent compound illustrates this effect; for example, the phosphorescence decay times of α-fluoro-, α-chloro-, α-bromo-, and α-iodonaphthalene in a solvent of diethyl ether, isopentane, and ethanol ($5:5:2$ v/v) at 77 K are 1.5, 0.3, 0.18, and 0.0025 s, respectively.

The effect of a paramagnetic species can be qualitatively explained by considering the electron spin states that may arise when the paramagnetic material forms a weak association or a charge transfer complex with a triplet molecule. If the paramagnetic species is a doublet, the spin state of the complex will be either a doublet or a quartet. Since the ground-state complex is a doublet, there is an allowed, that is, not spin-forbidden, transition between one component of the perturbed triplet and the ground state.

3. QUANTUM EFFICIENCY

The quantum efficiency of a luminescence process can be defined as the number of emission transitions per unit time divided by the number of absorption transitions per unit time. By identifying the competing processes involved in the deactivation of the excited state, rate equations can be written, rate constants assigned to each process, and a complete rate equation written for the complete process of interest. Such equations are of little direct use to the analyst, because of the difficulty in experimentally evaluating the rate constants. The equations can be simplified for several limiting cases which can be used to illustrate the effects of temperature and of luminescence quenchers on the intensity of fluorescence and phosphorescence.

a. Fluorescence at High Temperatures

At room temperature and above the fluorescence quantum efficiency is given approximately by:

$$Y_F = \frac{K_F}{K_F \sum {}_i K_i + K_Q(Q)} \tag{1}$$

where K_F is the rate constant for the fluorescence process (in s^{-1}), $\sum {}_i K_i$ is the summation of all competing rate constants excluding K_F and K_Q (in s^{-1}), and K_Q is the rate constant for a quenching process that may occur when a quenching agent of concentration (Q) (in mole liter^{-1}) collides with the excited solute molecule (in s^{-1} liter^{-1} mole^{-1}).

Both the vibrational deactivation processes included in $\sum {}_i K_i$ and the collisional process K_Q increase with temperature. Thus we expect and generally observe a decrease in fluorescence quantum efficiency and therefore in fluorescence intensity as temperature increases. Also, it is clear that dilution of the quencher (and solute) minimizes the effect of collisional quenching.

b. Phosphorescence at High Temperatures

A similar expression can be written for the phosphorescence quantum efficiency:

$$Y_P = \frac{K_P}{K_P + \sum {}_i K_i + K_Q(Q)} \tag{2}$$

where the rate constants are similarly defined but do not necessarily represent the same deactivation or quenching processes.

The radiative lifetime of the triplet state is long, resulting in a small value for K_P. Also, the energy gap between the triplet state and the ground singlet is small, favoring efficient vibrational deactivation and collisional quenching. Therefore at room temperature and higher and in solutions of low viscosity $\sum {}_i K_i + K_Q(Q) \gg K_P$ and $Y_P \to 0$, as is experimentally observed.

c. Luminescence at Low Temperatures

At very low temperatures (≈ 77 K), solution viscosity is high, and both collisional quenching and vibrational deactivation are minimized. Under these conditions, equations 1 and 2 become:

$$Y_F = \frac{K_F}{K_F + K_{IX}}$$

(3)

$$Y_P = \frac{K_{IX}}{K_F + K_{IX}}$$

(4)

where K_{IX} is the rate constant for the intersystem crossing process from the first excited singlet to the excited triplet level (in s^{-1}). Under these conditions, $Y_F + Y_P = 1$, and the dominant modes of deactivation are radiational. The relative magnitude of Y_F and Y_P depends on the molecular structure.

In the intermediate temperature range between room temperature and very low temperature, the relative intensities of fluorescence and phosphorescence vary according to the relative efficiency of the various competing deactivation processes. For a typical molecule that is strongly phosphorescent at low temperatures, we may observe weak fluorescence at room temperature. As the sample is cooled, the fluorescence intensity may increase up to the point where solution temperature and viscosity begin to favor phosphorescence emission. As the solution is cooled further, Y_F becomes constant and Y_P increases until $Y_P + Y_F \approx 1$.

d. Delayed Luminescence

The discussion in the preceding paragraphs represents an extreme simplification of the processes involved in the deactivation of excited molecular electronic states. Several mechanistically more complex modes of light emission have been observed. For example, E-type delayed fluorescence (as seen in eosin) exhibits the fluorescence spectrum of eosin and the decay time of eosin phosphorescence.

Introductory discussions of this general topic can be found in references 3 and 4.

The analytical application of these delayed luminescence processes remains to be developed. In most cases, the solute concentration range required is far in excess of the range required for trace analysis. Frequently, the conditions required for the observation of delayed luminescence emission are quite specific. The luminescence emission is observed only over a narrow range of concentration, temperature, or solution viscosity. Usually, the emission intensity is weak when compared to the emission intensity because of the primary processes of fluorescence and phosphorescence.

4.   LUMINESCENCE INTENSITY*

According to the definition of quantum efficiency and the Beer–Lambert law of absorption, the total emission intensity is given by

$$I_L = Y_L \left[I_0 - I_0 \exp\left(- \epsilon b c \right) \right] \qquad (5)$$

where I_L is the total isotropic luminescence intensity (in relative intensity units), Y_L is the quantum efficiency (dimensionless), I_0 is the intensity of exciting radiation (in relative intensity units), ϵ is the molar absorptivity. coefficient (in $1 \ mol^{-1} \ m^{-1}$), b is the sample thickness (in m) and c is the concentration of the luminescent solute (in $mol^{-1} \ l^{-1}$).

If the fraction of the incident light absorbed is small, approximately 5% or less, equation 5 reduces to

$$I_L = 2.303 \, Y_L I_0 \epsilon b c \qquad (6)$$

This equation is the mathematical basis of quantitative luminescence analysis. Luminescence intensity is directly proportional to solute concentration. The single limitation is that the fraction of light absorbed from the excitation light beam must be kept small, a condition easily met in virtually all analytical applications. This linear relationship of emission intensity versus sample concentration frequently extends over concentration ranges of 10^3 to 10^4. Note also that the emission intensity is directly proportional to the excitation intensity.

By appropriate choice of light source, sensitivity can be increased simply by increasing the excitation intensity.

If the fraction of the exciting light absorbed becomes large, that is, 100%, equation 5 simplifies to:

$$I_L = Y_L I_0 \qquad (7)$$

This is called the *quantum counter effect*. The emission intensity depends only on the quantum efficiency and the excitation intensity and is independent of solute concentration. This effect has been used as a means of monitoring the constancy of the excitation intensity I_0.

C.   LUMINESCENCE AND MOLECULAR STRUCTURE

This section deals briefly with the considerations of structure and luminescence that are of importance in the general application of

*For a more thorough discussion of the quantitative relationship refer to Chapter 4.

luminescence reagents for trace analysis. Further introductions to the broader field of molecular luminescence spectroscopy can be found in the texts by Winefordner (5), Hercules (3), Parker (4), and Zander (6).

1. MOLECULAR STRUCTURE

Electronic excitation in strongly luminescent molecules nearly always results from $\pi \rightarrow \pi^*$ absorption transitions. For this reason, generally only those molecules having cyclic conjugated bond systems luminesce. Of the aliphatic and alicyclic compounds, only biacetyl and a very few other compounds exhibit luminescence. Given a series of aromatic compounds, those that are the most planar, rigid, and sterically uncrowded are the most fluorescent.

Frequently, weakly fluorescent or nonfluorescent *aromatic* compounds are strongly phosphorescent. Usually this signals the involvement of $n \rightarrow \pi^*$ absorption transitions. Such n, π^* excited states are observed to have smaller energy gaps between the excited singlet and triplet levels and have longer excited-state lifetimes. These conditions thus favor population of the triplet state and lead to phosphorescence. Carbonyl-substituted aromatic compounds frequently exhibit this behavior. This "either-or" relationship of fluorescence and phosphorescence emission in aromatic compounds can often be used to advantage when selecting the most sensitive luminescence technique.

Unsubstituted aromatic hydrocarbons are usually fluorescent. As the number of fused rings comprising the molecule increases, the emission wavelength increases, indicating a decrease in the energy of the first excited singlet state. For example, benzene fluoresces in the UV, anthracene in the blue region, and pentacene in the red region.

Substitution of a single functional group in an aromatic structure produces relatively predictable changes in the emission intensity. Substitution of more than one functional group leads to more complex interactions which are difficult to predict. In Table 7B.1 the effects of various *single* substituients introduced into an aromatic ring system are summarized.

Aromatic carbonyl compounds, ketones, aldehydes, and carboxylic acids, are not as predictable as most of the compounds listed in Table 7B.1. This is because of the involvement of the n, π^* excited singlet state which has a longer lifetime and lower energy relative to the π, π^* system. As a result, these compounds are sometimes phosphorescent but not fluorescent, for example, acetophenone, anthraquinone, and benzophenone. Substitution of additional functional groups which have the capability of hydrogen bonding, —O.H, —NH$_2$, and so on, has the effect of nullifying the activity of the oxygen nonbonding electrons. Thus hydroxy- or amino-substituted carbonyls are frequently fluorescent, even though the parent compound is not. Also, in some cases, carbonyls may be fluorescent in highly polar hydrogen-bonding solvents, whereas they are not fluorescent in nonpolar solvents.

TABLE 7B.1. Summary of Relative Effect of Functional Groups Added to Aromatic Structure

Functional Group	Fluorescence Intensity	Phosphorescence Intensity
Alkyl	Variable	Increase
CN	Increase	—
CO_2H	Large decrease	Large increase
F ⎫	Decreasing	Increasing
Cl ⎬	Decreasing	Increasing
Br ⎪	Decreasing	Increasing
I ⎭	Decreasing	Increasing
NH_2, NHR, NR_2	Increase	Increase
NO_2, NO	Quenching	Variable
OH, OCH_3, OC_2H_5	Increase	Increase
SH	Decrease	—
SO_3H	No effect	—

Nitrogen heterocyclic compounds tend to be nonluminescent (occasionally they are phosphorescent). Oxygen and sulfur heterocycles are usually nonluminescent, unless one or more aromatic ring is fused to the heterocycle, such as in coumarin and dibenzothiophene.

Organic dyes are too complex structurally to generalize as to the relationship between structure and luminescence. Obviously, since dyes are highly colored, they absorb in the visible region, and their luminescence is shifted even further into the red or IR region of the spectrum. Because of the low excitation and emission energies involved, it is not unusual for dye molecules containing normally quenching functional groups to be luminescent. The most important single factor in anticipating dye luminescence is the generally observed requirement for *rigid, planar structures*. Many dyes are complex nonrigid, noncoplanar structures. These conditions all tend to militate against fluorescence or phosphorescence. If such a dye is capable of complexing an ion in solution, the resultant chelate or complex may be a rigid planar structure and thus become luminescent. This in fact is observed in many of the most successful reagents.

2. REQUIREMENTS FOR THE ANALYSIS OF ELEMENTS

Several generalizations can be made regarding the availability of reagents for the determination of cations. As a rule, cations that tend to form noncolored complexes or compounds tend to form fluorescent complexes. Aluminum, beryllium, calcium, magnesium, and zinc all form such complexes with many reagents. Conversely, cations that form colored complexes, for example, iron, copper, cobalt, chromium, and nickel, seldom form fluorescent complexes. The low-lying energy levels in these ions offer relatively direct paths for nonradiative deactivation of the excited singlet levels of the cation–fluorophore complex. It is not uncommon for a nonfluorescent complex to be counterbalanced by being phosphorescent, because of the heavy-atom effect of the cation or the effect of a paramagnetic cation. Both influences increase the rate of intersystem crossing and the resulting likelihood of phosphorescence emission. Copper is a good example of this type of behavior. Copper forms several phosphorescent complexes but very few fluorescent complexes. This is an example of the complementary nature of fluorescence and phosphorescence that applies to both organic and inorganic materials. The heavy-atom effect can also be observed within a periodic group of atoms that form fluorescent complexes. As the atomic number of the central ion increases, the fluorescence intensity decreases owing to the competing process of intersystem crossing which populates the triplet state.

Relatively few fluorimetric techniques have been developed for anions and nonmetals. Most of the methods that have been developed involve indirect procedures, such as the determination of fluoride by the quenching of the aluminum–$2,2',4'$-trihydroxyazobenzene-5-sulfonic acid complex. Anions that have been determined by indirect fluorimetry include cyanide, fluoride, oxalate, phosphate, sulfate, and sulfide.

A reagent for trace elemental analysis ideally should be nonfluorescent or nonphosphorescent. On reaction with the specific material of interest, it should become intensely and quantitatively luminescent. The materials most likely to fulfill these requirements are molecules with nonrigid, nonplanar aromatic structures in their *unreacted* state. These molecules may also contain carbonyl functional groups which lead to nonfluorescent or weakly fluorescent n, π^* excited singlet states. The structures of these molecules should be such that they are capable of complexing or reacting in some fashion so as to form a rigid, planar structure and/or effectively change the nature of the n, π^* excited state to the more intensely luminescent π, π^* excited state. Many reagents fulfill this simple set of requirements. Structurally, these requirements are frequently met by ortho-substituted aromatic moieties bridged by conjugated bond systems such as Schiff's base or azo linkages.

3. LUMINESCENCE REAGENTS

Luminescent complexes have unique potential as a means for sensitive and selective trace analysis of inorganic cations and anions. This is because of the rather specific requirements that must be met if the complex is to be luminescent. The relative scarcity of fluorimetric reagents as compared to colorimetric reagents is indicative of the potential for selective analytical procedures. It is true that this scarcity can also be considered a limitation to the application of the technique. However, it should be pointed out that this is an area still in the early stages of development, and research is increasing rapidly, particularly in regard to rare-earth chelates which are of interest in laser and semiconductor development. The number of reagents available has grown rapidly, and at this point a lack of reagents is no longer a serious disadvantage of the luminescence technique.

Luminescence techniques are comparable in sensitivity and cost to the techniques of atomic emission, atomic absorption, and atomic fluorescence, without the concomitant loss of sample, a consideration often extremely important when dealing with toxic, radioactive, or rare materials. Molecular luminescence techniques are considerably less expensive and require less manipulative skill than the techniques of neutron activation, X-ray spectrometry, and mass spectrometry. Of course, each technique has its specific advantages and disadvantages, but it is clear that luminescence analytical procedures belong in the repertory of the analyst engaged in the analysis of traces of elements.

a. Schiff's Bases

The parent compound of this class of reagents is salicylidene-*o*-aminophenol. Other variants of the Schiff's base linkage include *N*-salicylidene-2-amino-3-hydroxyfluorene, bissalicylidene ethylenediamine, and *N*,*N*'-bissalicylidene-2,3-diaminobenzofuran (Figure 7B.2*A*). These reagents have been used extensively for aluminum, gallium, magnesium, and other elements that form noncolored complexes.

b. Azo Compounds

The parent compound of this class of reagents is *o*,*o*'-dihydroxyazobenzene (Figure 7B.2*B*), and includes 2,4,2'-trihydroxyazobenzene-5'-sulfonic acid (alizarin garnet R), 2,2'-dihydroxyazonaphthalene-4'-sulfonic acid, sodium salt (pontachrome blue black R or superchrome blue), 2,2'-dihydroxy-1,1'-naphthalene-5-sulfonic acid, sodium salt (pontachrome violet SW), 6-(5-chloro-2-hydroxy-3-sulfophenylazo)-5-hydroxy-1-naphthalenesulfonic acid, 1-(8-quinolinol-7-azo)-2-naphthol-4-

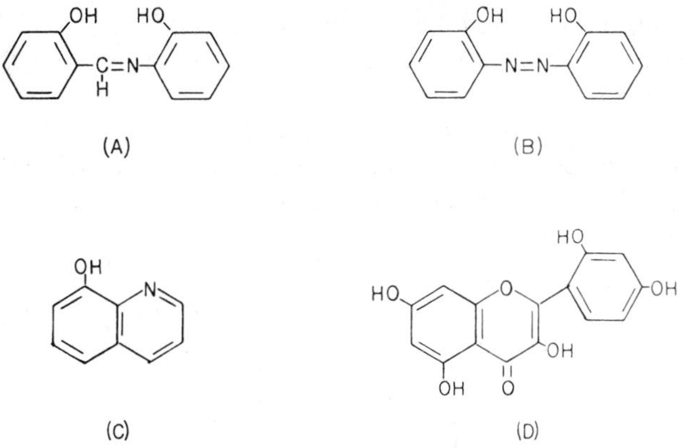

Fig. 7B.2 Luminescence reagents. (*A*) Schiff's base linkage: salicylidene-o-aminophenol. (*B*) Azo linkage: *o,o'*-dihydroxyazobenzene. (*C*) 8-Quinolinol (oxine). (*D*) 2',3,4',5,7-pentahydroxyflavone (morin).

sulfonic acid, 2-hydroxy-3-sulfo-5-chlorophenyl azobarbituric acid, 1-(2-hydroxy-3-sulfo-5-chlorophenylazo)-2'-hydroxynaphthalene, and 2,2',4-trihydroxy-5'-chloroazobenzene-3'-sulfonic acid.

These azo compounds form fluorescent complexes primarily with aluminum, gallium, and magnesium, and with other elements that form noncolored complexes.

c. 8-Quinolinol (Oxine)

Oxine is a general nonspecific reagent which has been shown to form fluorescent chelates with 25 metals including silver, beryllium, lithium, sodium, and tin(II)(7). Most of the transition metals do not form *fluorescent* complexes with this compound. Fluorometric procedures with oxine parallel the procedures used with the same material in absorptiometry, except that the oxine complex is generally extracted into chloroform prior to the measurement of fluorescence. Considerable care is required to avoid interferences which may result from the nonspecificity of this reagent (Figure 7B.2*C*).

d. Flavonols (Hydroxy Flavones)

These compounds are derivatives of 2-phenyl-1,4-benzopyrone. They occur widely in nature as plant pigments related structurally to anthocyanin pigments. Morin (Figure 7B.2*D*) was reported as a fluorescent reagent for aluminum as early as 1867. Compounds in this group that have been used as fluormetric reagents include flavonol, 3-hydroxy-flavone, morin, quercetin, and naringenin. Katyal (8) has reviewed the use of this class of reagents.

Flavonol in acidic solutions is used for tin(IV), zirconium(IV), thallium-(IV), and tungsten(VI). Morin is used for aluminum in acidic solutions and beryllium in alkaline solutions. Boron also forms a fluorescent complex with morin.

Flavonol complexes are strongly temperature-dependent and solvent-sensitive. The ligands have very high molar absorptivities and care must be used to avoid quenching by excess reagent.

e. Benzoin

Benzoin (Figure 7B.3A) is the cyanide-catalyzed, self-condensation product of benzaldehyde. It is a very specific and highly sensitive reagent for boron, germanium, and silicon. It has also been used for the qualitative determination of zinc, antimony, and copper. Silver, cobalt, and nickel react to a lesser degree and may interfere.

Benzoin is easily air-oxidized in alkaline solution. Deoxygenation of solutions with nitrogen gas is usually required to prevent reagent loss. Benzoin is also readily photodecomposed. Excitation at 405 nm and only brief exposure to the exciting light minimize this problem.

f. β-Diketones

β-Diketones, such as dibenzoylmethane (Figure 7B.3B) and benzoyl-acetone, are unique reagents for transition and rare-earth metals. While most ligands both absorb *and* reemit light, these ligands absorb light but transfer the excitation energy to the central cation of the complex where it is emitted with the characteristic atomic emission spectrum of that cation. This type of absorption-emission behavior arises where the cation has excited f- or d-electronic levels lying close to the ligands' excited triplet levels. Thus d^*/d emission is seen in some transition metal complexes, and f^*/f emission is seen in many rare-earth complexes (9, 10).

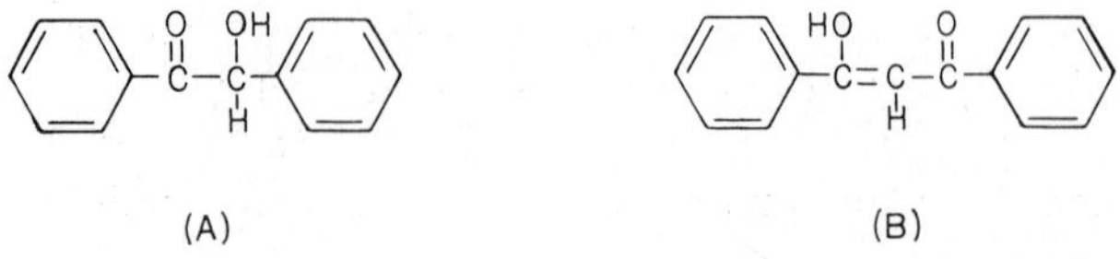

(A)   (B)

Fig. 7B.3 Luminescence reagents. (*A*) Ethine linkage: benzoin. (*B*) *p*-Diketone linkage: dibenzoylmethane.

g. Calcein

An alternate approach to the formulation of fluorogenic reagents for metal ions is the addition of strong chelating groups to fluorescent dyes. An example is the condensation of iminodiacetic acid and formaldehyde with fluorescein to form a complex mixture of fluorescent chelators known collectively as *calcein* (11). One of the active chelators in this mixture has been identified as 2,4-bis[N,N'-di(carboxymethyl)aminomethyl] fluorescein (12). This compound, and the related compound fluran (fluorescein complexone), have been extensively used to determine calcium and magnesium in a variety of biological materials (13).

h. Miscellaneous Reagents

A considerable variety of other reagents has been used for the fluorometric and phosphorimetric analysis of elements. These compounds and many others are listed in Table 7B.4.

A few of the more commonly used compound groups are mentioned here.

Salicylaldehyde, salicylic acid, and their substitution products have been used to determine various noncolored metal ions.

The intensely fluorescent dyes rhodamine B and rhodmine 6G, (also known as rhodamine 6Zh) have been used for gallium(III), gold(III), and thallium(III).

The hydroxyanthraquinones have been used to determine beryllium and thorium, and also react with boron in very acidic solution.

A reagent of note for copper is 1,1,3-tricyano-2-amino-1-propene, which is sensitive to about 20 ng ml^{-1}.

Iridium can be determined with 2,2',2"-terpyridene to about 2 ppm.

Zinc can be determined at the nanogram level with either 2,2'-methylenebibenzothiazol, picolinealdehyde-2-quinoylhydrazone, or *p*-tosylaminoquinoline. The latter is also useful for cadmium.

Oxygen, ozone, the halides, and water have all been determined by a variety of direct or indirect procedures.

In Table 7B.3, a listing of relative limits of detection for the elements as determined by various selected reagents is given.

D. INSTRUMENTATION

1. TYPES OF LUMINESCENCE INSTRUMENTS

A generalized luminescence instrument may be considered to be made up of six basic components: a source of light, an excitation monochromator for the selection of specific bands of radiation from the light source, a sample cell, an emission monochromator, a photodetector, and a data readout device. In addition to these basic components, several special-purpose components may be used, such as a repetitive shutter mechanism which is required for the observation of phosphorescence. Figure 7B.4 is a block diagram of the generalized luminescence instrument.

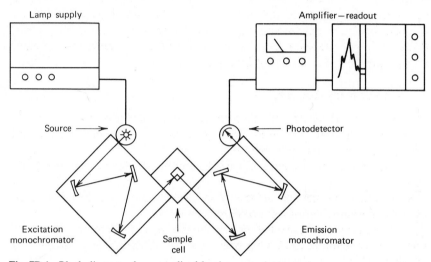

Fig. 7B.4   Block diagram of a generalized luminescence instrument.

The six basic components of the generalized luminescence instrument can be assembled in many varying degrees of complexity depending on the requirements of the user. For the purpose of discussion, fluorescence instruments can be further catagorized as filter fluorimeters, spectrophotometer accessories, spectrofluorimeters (fluorescence spectrometers), and compensating spectrofluorimeters.

The nomenclature requires some explanation. The prefix "spectro" implies that at least one dispersive monochromator is used in the instrument, that is, one grating or prism monochromator. The presence of the word "photo" in a name usually implies that photoelectric detection is being used. This terminology may be considered obsolete, since virtually all instruments now use such detection. Finally, *fluorimeter* and *fluorometer* are identical and are used interchangeably frequently even within the same publication.

Filter fluorimeters are characterized by their simplicity, high sensitivity, low cost, and lack of selectivity. The use of optical filters as very simple monochromators results in very high excitation light levels and efficient light detection, but as a consequence sacrifices the selectivity that can be obtained by more precise selection of excitation and emission wavelengths. The small physical size of filters results in compact portable instruments. Filter fluorimeters are frequently used for routine repetitive analyses where the lack of versatility is not a major drawback and where the high sensitivity may be a distinct advantage.

The fluorescence microscope can be considered a type of filter fluorimeter. This instrument usually consists of a standard dark-field microscope with a filtered mercury arc lamp as a source of exciting light. The microscope slide and cover slip function as the sample cell, the microscope optics serve to transfer and image the sample, and the eye serves as monochromator and photodetector.

Fluorescence accessories for spectrophotometers generally consist of an external light source, an excitation filter, and a special sample cell. An external light source is necessary because the UV lamps used in most absorption spectrometers (hydrogen arc lamps) are not intense enough for use in fluorimetry. This type of attachment enables the user to record fluorescence *emission* spectra and adds a degree of selectivity to the analysis as compared to a filter fluorimeter. More complex attachments include a prism or grating monochromator instead of an excitation filter. A spectrophotometer fitted with such an attachment performs most of the functions of a spectrofluorimeter. Caution should be exercised in selecting such a system, however, because the components are unlikely to be optimized for fluorescence usage, particularly in the area of photodetector type, amplifier linearity and dynamic range, and photodetector sensitivity.

Spectrofluorimeters may be recognized by their grating or prism monochromators. Some, but not all, have provision to scan each monochromator automatically to generate either excitation or emission spectra. Spectrofluorimeters are single-beam devices which do not compensate for variations in instrumental parameters such as wavelength distribution of the light source, the transmission of the optical system, and the response of the photodetector. As a result, the measured spectra are distorted by instrumental response. Intensity measurements are never calibrated in absolute terms but are expressed simply as relative intensity units which are reasonably constant for any given instrument under fixed conditions of slit widths, photodetector voltage, amplifier gain settings, and so forth.

Lamp stability is the limiting factor in the long-term stability of the spectrofluorimeter. For this reason, the level of precision attainable in fluorescence measurements is seldom what one might expect from, for example, a double-beam absorption spectrophotometer. In practical analytical usage, however, the level of precision is adequate for virtually all applications.

Most commercial instruments offer a complete complement of accessories such as flow cells, microcells, sample changers, and chromatogram scanners.

Compensating spectrofluorimeters provide automatic means for correcting the instrumental artifacts associated with the single-beam spectrofluorimeter. These instruments utilize a reference photodetector whose output signal is directly proportional to the flux of radiation falling on it irrespective of wavelength (within the wavelength region of interest). This detector is used to monitor the excitation lamp output and, by utilizing feedback techniques, to correct the lamp output for variations in intensity as a function of wavelength and as a function of time. The reason such reference detectors are not used to measure the fluorescence of the sample intensity itself is simply that they are very insensitive and useful only in high-intensity applications. Correction for photodetector response characteristics and emission monochromator transmission characteristics is usually accomplished electrically by adjusting the photodetector amplifier gain as a function of wavelength. The amplifier gain is adjusted by reflecting the regulated light from the excitation monochromator into the emission monochromator and making gain corrections point by point over the desired wavelength range. Compensating fluorimeters may have from 10 to 30 individual "trimmer" potentiometers to fully adjust the photodetector response.

The difficulties associated with fluorimeter compensation are formidable. The instruments are optically and electrically complex and as a result are usually at least twice as expensive as an uncorrected spectrofluorimeter. An excellent discussion of the optical and electrical considerations going into the design of a compensated spectrofluorimeter is given in reference (14).

2. INSTRUMENT COMPONENTS

a. Sources of Excitation Light

The primary factors to consider when selecting a light source for fluorescence instrumentation are lamp intensity, wavelength distribution of emitted light, and stability. Scanning spectrofluorimeters require light sources that emit light continuously over a wide spectral range. Filter fluorimeters or spectrofluorimeters, used specifically for analytical measurements at fixed wavelengths, may utilize atomic spectral line sources. Various combination sources, such as phosphor-coated fluorescent lamps, may be used in some cases. These various lamps are discussed in the following sections. A more detailed discussion of light sources for luminescence spectrometry can be found in reference (5).

(1) High-Pressure Arc Lamps

High-pressure, DC xenon arc lamps are used in nearly all commercial spectrofluorimeters because they emit an intense and relatively stable continuum of radiation which extends from the UV to the IR regions of the spectrum. The spectral output of the xenon arc lamp approximates that of a "black body radiator" with a temperature of about 7000 K and an emissivity of 0.06. This means that the arc emits according to the Planck distribution law, as does the ordinary tungsten lamp, with a wavelength distribution dependent primarily on arc temperature. Several xenon atomic emission lines appear in the visible region, and the light output drops sharply in the UV region, thus strongly distorting single-beam excitation spectra. Figure 7B.5 shows the relative intensity of the xenon arc and the tungsten lamp as a function of wavelength.

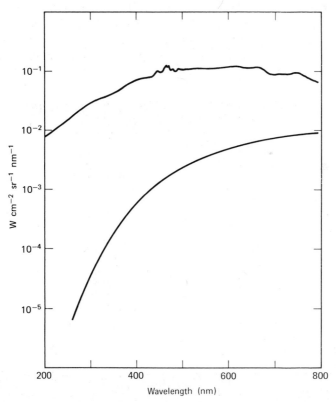

Fig. 7B.5 Intensity of the 75 W xenon arc lamp (top curve) and the 75 W tungsten lamp (bottom curve).

High-pressure mercury arcs and mixed mercury–xenon arcs find some applications where added sensitivity is required from a spectrofluorimeter. The mercury arc output consists of strong mercury emission lines superimposed on a weak continuum, whereas the mercury–xenon arc consists of mercury lines superimposed on the strong xenon continuum. Neither lamp has any utility if *excitation* spectra must be recorded.

(2) Low-Pressure Mercury Lamps

Low-pressure mercury vapor lamps are most frequently used in filter fluorimeters. These lamps are true arc lamps, however, the fill gas pressure is quite low, in the neighborhood of 10 torr. The resulting arc discharge is spatially very diffuse and much less intense than that of a high-pressure arc. The "effective temperature" for these lamps is much lower than that for high-pressure arc lamps, and consequently the stability of low-pressure lamps is generally better. These lamps may be phosphor-coated to emit a more nearly continuous spectrum (as in the familiar household fluorescent lamp), or may have a clear bulb of UV-transmitting material to permit use of the individual mercury lines. Table 7B.2 lists the useful emission lines from a typical low-pressure mercury filter fluorimeter lamp. Interference filters can be used to select individual mercury lines for excitation, or bandpass filters may be used to select for several lines.

TABLE 7B.2.   Useful Emission Lines Emitted from Typical
Low-Pressure Mercury Arc Lamp

Wavelength (nm)		Intensity
253.65		Very strong
312.57	313 nominal	Moderate
313.17		
365.02		
365.44	365 nominal	Moderate
366.33		
404.66		Moderate
407.78		Weak
435.84		Strong
546.07		Strong
576.96		Moderate
579.07		Moderate

(3) Other Lamps

The multiple requirements of intensity, stability, and UV emission severely restrict the choice of light sources. Tungsten lamps can be made extremely stable but cannot be used at wavelengths shorter than about 375 nm (Figure 7B.5). Furthermore, when used in filter fluorimeters, the intense visible light output from the tungsten lamp causes the scattered light level, that is, the blank level, to be unacceptable; this is because the filters used do not completely reject light outside their bandpass, and the higher the intensity of the source outside the desired wavelength region the more light that will traverse both the excitation *and* the emission filter to appear as unwanted stray light at the photodetector.

Metal vapor lamps similar to the mercury lamp can be constructed with cadmium, zinc, gallium, indium, thallium, and other volatile metals. These produce very intense line spectra of the fill elements. The cadmium lamp (primarily 228.8 nm radiation) and the zinc lamp (213.9 nm) find some use in special-purpose instruments for applications requiring excitation light shorter than the mercury 254 nm line. These lamps are currently being manufactured by Osram, Phillips, and others.

Hydrogen, deuterium, and other gas-filled low-pressure lamps such as those frequently found in double-beam absorption spectrophotometers have too low fluxes to be used effectively in fluorescence analysis.

Pulsed lamps (flash tubes) filled with xenon, hydrogen, and a variety of other gases have found limited use in some special-purpose instruments. These lamps emit continuum and/or line radiation in short but very intense pulses. Such lamps are used in instruments designed to measure luminescence decay times.

b. Monochromators*

The excitation monochromator serves to illuminate the sample with a band of wavelengths selected from the light source. The emission monochromator similarly serves to select a band of wavelengths from the light emitted by the sample. Monochromators must fulfill several requirements: they must prevent reflected or scattered light from reaching the photodetector; they must provide sufficient resolution such that the desired spectral information can be obtained from the excitation and emission spectra; they must provide for sufficient selectivity to be utilized in quantitative analytical applications; and perhaps most importantly they must transmit radiation characteristic of the analyte so that it can be detected and measured at useful concentrations levels.

(1) Filters

Filters function as very simple, optically efficient monochromators. Obviously little or no spectral information can be obtained with a filter instrument, and the selectivity afforded by accurate selection of excitation

*Also refer to Chapter 5.

and emission wavelengths is sharply reduced. In many applications, these faults are more than offset by the high sensitivity resulting from the light transmission characteristics of the filter.

Excitation filters are generally bandpass types which transmit a rather broad band of wavelengths. Figure 7B.6 illustrates the spectral characteristics of a Corning 7-51 "black glass" UV transmitting filter. Note that this filter transmits from about 315 to 410 nm, a range that includes several of the mercury lines listed in Table 7B.2. Filters are available in many sizes, 2×2 in.^{2} square being the most common. The thinness of filters allows the lamp and photodetector to be very closely coupled to the sample cell.

Emission filters are usually of the sharp-cutoff type which pass long wavelengths and attenuate shorter wavelengths. Groups of these filters are

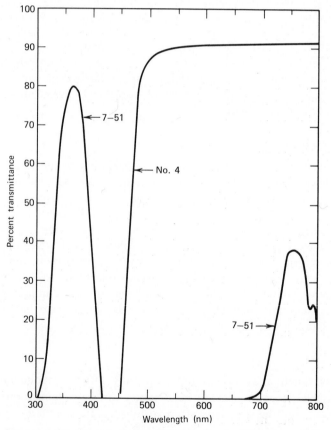

Fig. 7B.6 Spectral transmission characteristics of the Corning 7-51 filter and the Wratten No. 4 filter.

available with cutoff points extending from the near UV to the far-red portion of the spectrum. Figure 7B.6 also includes a transmission curve of the Wratten (Kodak) No. 4 gelatin filter, a typical yellow sharp-cutoff emission filter.

Most filters used in filter fluorimetry are of colored glass, plastic film, or gelatin construction. A great variety of filters is available from such sources as Corning Glass Works, Eastman Kodak (Wratten filters), and Schott Optical Company. Sill (15) has given a summary of the transmission spectra of a large number of filters. A point to remember in the use of color filters are that bandpass filters often have more than one transmission band or "window" (Figure 7B.6). Photodetectors often have a *low-level* response over an unexpectedly wide wavelength range. This coupled with an unexpected filter band may result in a high stray light level. Sharp-cutoff filters, particularly glass filters, are frequently fluorescent themselves. Every effort should be made to avoid having the excitation light strike the emission filter directly. If a combination of filters is used, the least fluorescent filter should face the sample cell.

Another type of filter called the *interference filter* finds some usage. This filter, made by vacuum deposition of a thin, semitransparent film on a glass substrate, offers a compromise between sensitivity and selectivity. Transmission bandwidths of 20 nm with a peak transmission of 35% at the band center can be obtained. This is comparable to a grating monochromator with large entrance and exit slits. Usually in fluorimetry such filters are used to isolate single mercury lines. The advantages of interference filters are nullified to a great extent by their high cost, lack of versatility, and low light transmittance. A further problem related to thin film construction is that the off-band transmission, hence the stray light level, is usually quite high when compared to color filters. This is attributable to unavoidable pin holes and defects in the film coatings.

(2) Gratings

Gratings disperse light by the diffraction of light by many equally spaced grooves or slits ruled on a reflecting or transparent material. Diffraction occurs as a consequence of the wave nature of light and the precisely ordered grooves of the grating. Figure 7B.7 shows a cross-sectional view of a typical grating. The most frequently used reflection-type gratings consist of 100 to 1500 grooves per millimeter precisely spaced on a highly reflective material like aluminum.

The diffraction law states that a given wavelength is dispersed at many different angles, each differing by a positive or negative integer value. Thus a second-order line ($m = 2$) of 250 nm is diffracted at the same angle as a first-order line of 500 nm, and so forth. This aspect of grating behavior can

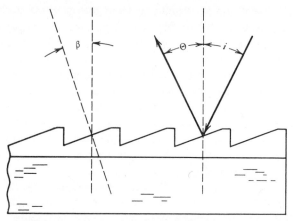

Fig. 7B.7 Cross-sectional view of an idealized diffraction grating.

be troublesome in fluorescence instruments where the higher-order diffracted light is not used or desired. The intensity profile of a grating is somewhat dependent on the groove contour. Proper choice of the *blaze angle β* results in most of the light being dispersed in the first order for a specified range of wavelengths (Figure 7B.7). A grating ruled in this manner is called an *echelette grating*. Excitation monochromators are usually blazed for 250 or 300 nm, and emission monochromators are blazed for 450 or 500 nm. Filters may also be used to block out higher-order diffracted light. Some compensating spectrofluorimeters have such filters built in to eliminate higher-order light automatically.

Grating monochromators are currently being used in virtually all commercial spectrofluorimeters. Advances in the technology of grating production have resulted in inexpensive reflection-type gratings capable of high resolution and dispersion, high numerical aperature (light gathering ability) and high reflectivity. Also, optical design considerations make the Czerny–Turner grating monochromator design popular with instrument designers (Figure 7B.4).

(3) Prisms

Dispersion of radiation by a prism is based on the angular separation of different wavelengths as a result of differences in the refractive index of the prism material at different wavelengths. Prism materials must have a refractive index that varies significantly over the wavelength region of interest; they must be reasonably transparent in their useful region; and they must be shaped into a form that causes the angle of the incident light to differ from that of the dispersed light in a useful manner.

For any given material, the angular dispersion of a prism *increases* as the wavelength *decreases*. With a quartz prism, this results in good dispersion in the UV region but rather poor dispersion in the visible region. Prism monochromators have not been used extensively in luminescence instrumentation, primarily because of the expense associated with construction of large-aperture prism instruments.

(4)   Choice of a Monochromator

The choice of a spectrofluorimeter should be based in large part on the choice of monochromators used and on their performance. Essential characteristics which should be examined are the following: the angular dispersion R_d, the resolving power R, the spectral bandwidth $\Delta\lambda_s$, and the numerical aperture or f number. Stray light levels should also be examined.

For good spectral selectivity and ability to resolve spectral fine structure the monochromator should be able to resolve two lines 1.0 nm apart. This means that R_d should be 5.0 nm mm^{-1} or *less*; R should be *greater than* 5000; and the *minimum* spectral bandwidth should be less than 1.0 nm. For good sensitivity, the monochromator should have a reasonably large aperture, for example, approximately $f/5$.

The requirements of selectivity and sensitivity of the desired analysis should be considered carefully. Trace analysis requires large-aperture, high-sensitivity instruments with only modest resolution and stray light requirements. Conversely, instruments used in the study of spectra require medium to high resolution and very low stray light.

In selecting an instrument, some of the practical items to check are given below. The optical baseplate should be heavy and thick for rigidity and stability. The optical layout should not have light wasting mirrors to "fold" the optical paths to fit into a pretty cabinet. The lamp should be cooled to avoid heat being conducted to the optics or detector. All optical units (monochromators) should be completely sealed, including windows over the entrance and exit slits, or normal laboratory fumes will quickly attack the mirrors and grating, decreasing their reflectivity particularly in the UV region.

c. Sample Cells

(1)   Sample Cell Geometry

There are three basic arrangements for illuminating and viewing the sample in luminescence spectrometry—the right-angle method, the frontal method, and the straight-through, or transmission method (Figure 7B.8). The method of illumination determines the amount of stray or scattered

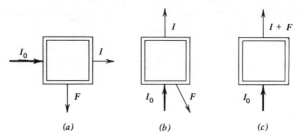

Fig. 7B.8 Sample cell geometries. (a) Right-angle mode, (b) frontal mode, (c) transmission mode.

light that reaches the photodetector and determines the maximum concentration of solute that may be observed before the inner-filter effect disrupts the linear relationship between luminescence intensity and solute concentration.* Of course, the source cell–detector configuration also determines the effective path lengths for absorption and emission, hence strongly influences sensitivity.

The right-angle geometry (Fig. 7B.8a) is used almost exclusively in commercial instruments. It is the only geometry that can be recommended for trace analytical applications. The 90° geometry is efficient, because none of the sample cuvet surfaces directly illuminated by the excitation beam are viewed by the emission monochromator, hence no cuvet fluorescence or light reflected from these surfaces enters the emission monochromator. Scattered light can originate only from the bulk of the solution itself.

The frontal method of cell illumination is used primarily for solutions that are highly absorbing, or for semiopaque materials or solids. The disadvantage of this approach is that the emission monochromator views directly illuminated cell surfaces with their own residual fluorescence and unavoidable reflected light.

The straight-through or transmission mode of illumination and observation is seldom used. The transmitted excitation beam I is directed straight into the emission monochromator, leading to high stray light levels even in the most efficient monochromators. Furthermore, this suffers most from the inner-filter effect, and multiple illuminated cell surfaces are viewed by the emission monochromator.

(2) Phosphorescence Sample Cells

Scattered light is not a problem in viewing phosphorescence, because the emitted light is viewed in the absence of the exciting light by use of a

*Also see Chapter 4.

repetitive shutter mechanism called a *phosphoroscope*. This is possible because the phosphorescence emission continues for a considerable length of time after termination of excitation, usually 10^{-3} to 10 s, as compared to 10^{-9} to 10^{-5} s for fluorescence emission.

The problems of inner filtering (prefilter effect) and cell luminescence remain and are essentially the same as discussed for fluorescence. Any of the three viewing modes can be used. The right-angle method is most often used, primarily because most instruments are designed basically for fluorescence assay and only convert to phosphorescence instruments by use of accessories.

Useful phosphorescence signals are obtained only by immobilizing the sample in a rigid matrix at a very low temperature. Usually this is accomplished by immersing the sample in liquid nitrogen while it is being excited and observed.

A typical phosphorescence attachment consists of a rotary shutter mechanism which is placed in the sample compartment in such a way that the sample cuvet is alternately exposed to exciting light and viewed by the emission monochromator. This cycle is repeated many times a second. A quartz Dewar flask with an unsilvered base is inserted inside the rotating shutter mechanism. A slender narrow-bore sample tube (1 to 2 mm i.d.) is inserted directly into the liquid nitrogen so that the tube is centered in the optical path. The excitation beam passes through the shutter, through the transparent double walls of the Dewar, and through the liquid nitrogen, and strikes the sample cell. Liquid nitrogen is transparent in the UV and visible region and does not block the excitation and emission beams.

Solvents that solidify into clear glasses at low temperature are required (16). If a solvent forms a cracked glass or microcrystalline "snow," multiple reflections by the emitted light will result in very poor sample-to-sample reproducibility.

Solvents that form cracked glasses or "snow," such as all aqueous solvents, can be used if the sample tube is spun rapidly around its long axis. This spinning, if done rapidly, permits the photodetector-amplifier system to effectively average the fluctuating intensity signal caused by the sample irregularities and results in a stable and reproducible signal (17).

d. Photodetectors

Spectrofluorimeters, almost without exception, utilize photomultiplier tubes for light detection and quantitation. Such tubes, which more properly should be called multiplier phototubes, consist of a single light-sensitive element called the *photocathode* and a series of electron-sensitive elements called *dynodes*. A feeble electron current is generated by light striking the photocathode. This current is accelerated toward each sucessive dynode, displacing additional electrons and yielding a net multiplication of total current flow to a level that can be detected easily by external electronic circuitry. There are numerous other types of photodetectors, however, none combine all the desirable features of the photomultiplier for use in spectrofluorimetry.

For radiation in the visible region of the spectrum the required work function to displace an electron from a photographic surface is quite small. The materials that fulfill this requirement are typically the alkali metals, and various alloys and interstitial compounds such as Cs–Sb, Cs–Bi, Ag–O–Cs, and O–Rb–O–Ag. In recent years, newer materials, such as gallium arsenide and gallium arsenide–phosphide have been used. A good indicator of photomultiplier sensitivity is the radiant sensitivity γ which is the number of amperes produced at the anode (i.e., the output) for each watt of radiant flux striking the cathode. Photomultipliers useful for fluorimetry have sensitivity factors of 10^4 to 10^5 A W^{-1}. A graph of γ versus wavelength clearly defines the useful range of the photomultiplier type. A typical response curve for the RCA 1P21 photomultiplier is given in Figure 7B.9. An important point to note is the very sharp drop in response at either end of the tube's useful range. The exact wavelength at which this drop-off occurs is variable, even within a specific group of tubes of the same type.

All photomultipliers register a small residual current flow even when they are in total darkness. This *dark current* is related to tube construction details such as cathode and dynode materials, residual gas pressure, surface contamination, and background radiation. It varies greatly from tube to tube (as does radiant sensitivity). The purchase of selected tubes with guaranteed gain and dark current specifications is mandatory if the tubes are to be used in demanding applications.

The current flowing through the photomultiplier is subject to random fluctuations known as *shot noise*,* which is due to the random nature of electron emission from the cathode and dynodes. Shot noise is the dominant noise in spectrofluorimetry. It can be minimized by selecting tubes with low dark current and by reducing system frequency response bandwidth Δf, that is, by increasing the electrical filtering in the detector-readout system (often referred to as *damping*).

As a general rule, tubes sensitive to red light tend to have higher dark current. This is because photocathode materials with work functions low enough to be efficient with low-energy red light are also more susceptible to thermionic emission. In general, it is also true that tubes with wide wavelength response, particularly in the red region, tend to have high dark

*Also refer to Chapter 2.

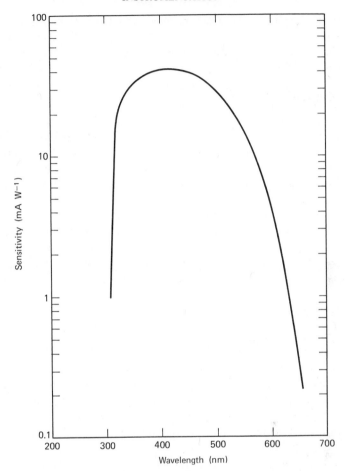

Fig. 7B.9 Spectral response of the 1P21 photomultiplier tube.

current and consequently higher noise levels. The best signal-to-noise ratio is obtained with a tube that responds only in the wavelength region of interest. Wide spectral range tubes are seldom used to their fullest extent in quantitative analytical procedures and may actually degrade the precision and long-term reproducibility of measurements.

A multitude of photomultiplier types is available. A recent catalog by one manufacturer alone lists over 135 different tubes. Of these, only a handful of tubes is suitable for applications in spectrofluorimetry. Most spectrofluorimeters use side window, cage dynode photomultipliers such as the RCA 1P21 (Figure 7B.9). This tube has very high gain, good stability, low dark current, and a useful spectral response from about 310 to 670 nm.

For response in the UV region the similar type 1P28, with a range of 210 to 630 nm, is a good choice. For measurements outside these wavelength intervals, it is best to consult the manufacturer of the spectrofluorimeter for specific advice.

e. Data Readout Devices

The basic considerations of electronic measurement were covered in Chapter 2. This section reviews the basic requirements for data readout devices for filter fluorimeters and spectrofluorimeters. Compensating spectrofluorimeters are electrically complex and beyond the scope of this discussion. The principles of dynamic range, blank subtraction, noise level, and so on, still apply, however.

The filter fluorimeter or spectrofluorimeter data readout device generally provides a source of high voltage for the photomultiplier tube, a means of amplifying the low-level current output of the photomultiplier to a useful level, and a means for subtracting blank or background signals as a convenience in the quantitative use of the instrument.

(1) Amplifiers

The output of a photomultiplier tube is a current which is directly proportional to the radiation flux falling on the photocathode. Since it is emission intensity that is proportional to sample concentration in fluorimetric measurements, it is desirable to amplify the low-level signal from the photomultiplier in a linear fashion, that is, in the form of a linear equation $y = ax + b$. This is in contrast with absorptiometry in which the concentration-dependent quantity, the absorbance, is the logarithm of the ratio of two intensities.

Photomultiplier output signals in fluorimetry usually vary from 10^{-9} to about 10^{-5} A. Thus for some fixed output level, such as a meter full-scale reading, the amplifier must have a gain variable by 10^4 to 1. For a conventional meter movement of 100 μA full scale an amplifier with a current gain of 10 on its least sensitive range and a current gain of 100,000 on its most sensitive range is required. Such a wide range is seldom actually required. Typically, a gain variability of 10^3 to 1 is adequate.

When the signal is displayed on a meter or graphically on a recorder, it is desirable for the signal to always be at least one-third of full scale to minimize visual error in reading the value. The variable gain control, usually called the *sensitivity* or *meter multiplier*, should be divided into increments of 1, 3, 10, 30, and so on, or 1, 2, 5, 10, 10, 50, and so on. The range-to-range accuracy should be better than 1%. In addition to the switch-selectable range settings, there should be a gain vernier or "fine sensitivity" adjustment over a range of perhaps 2 or 3 to 1. This permits the overall system sensitivity to be accurately reproduced day to day by the use of standard solutions or reference luminophors.

A means for subtracting background luminescence signals and dark current should be included in the amplifier. This blank subtracting control should be capable of subtracting (often called *bucking out* or *nulling out*) any signal up to perhaps one-third of full scale on the least sensitive range.

(2) High-Voltage Power Supplies

The sensitivity of a photomultiplier tube varies exponentially as a function of the applied voltage. Consequently, a well-regulated power supply is required if photodetector sensitivity is to remain constant. For filter fluorimetry, a fixed, regulated high voltage in the neighborhood of -750 V DC is typical. An adjustable but well-regulated supply offers a convenient means for varying the photodetector sensitivity over a wide range. An adjustable high-voltage supply should be used only to set the amplifier-readout system at a convenient sensitivity region. Tube gain and applied voltage are related in a nonlinear fashion, and it is virtually impossible to correlate gain and voltage accurately in a quantitative way. The most effective way to utilize the wide dynamic range fluorescence intensity versus concentration characteristic is to vary the gain of a precision amplifier. Photodetector systems whose range-to-range sensitivity cannot be correlated in a precise numerical fashion are of little use for quantitative fluorometric assays.

(3) Data Display

Commercial photodetector systems almost universally use standard moving-needle meters as primary data display devices. This may appear out of place when digital display devices are being used in devices ranging from computer displays to wristwatches, however, the simple meter has some advantages which recommend its use. It is highly reliable and much less expensive than a comparable digital display, and it can be used to display signals with relatively high noise levels superimposed on them. This is because the eye can easily average the position of a moving needle to arrive at a best-choice value, whereas the eye cannot easily average the flipping digits of a digital display without considerable bias. Disadvantages of the moving-needle meter include poor readability and poor visibility. There is little doubt that digital display devices will soon replace the older type of meter possibly for reasons not strictly related to function.

In addition to a meter display, a detection system should have outputs available for use with strip chart or *xy*-recorders. Recorder sensitivity varies widely from type to type, so an adjustable output voltage is a "plus" feature. Both meter display and recorder output signals are usually well filtered to reduce noise.

There are some applications, such as rapid kinetics, stopped flow, and phosphorescence decay measurements, that require detection devices that can accurately amplify rapidly changing signals. This may be accomplished by having a switch available to select the level of filtration desired or by having a separate output independent of the meter and recorder outputs that is not highly filtered.

Spectrofluorimeters frequently utilize xy-recorders for the display of excitation and emission spectra. Potentiometers are attached to the monochromator drives to generate a voltage-versus-wavelength signal. When the resulting voltage is fed to the x-axis for an xy-recorder, the recorder accurately tracks the monochromator position, regardless of the speed with which the monochromator is scanned. This type of display can also be generated with a strip chart recorder and a constant-speed wavelength drive on the monochromator, however, it is considerably less convenient to correlate strip chart graduations with wavelength.

Digital printers, paper tape punches, and computer interface devices are not offered, to any great extent, by the manufacturers of luminescence instrumentation. This is simply because the diversity of requirements as to end use, computer type, desired format, system speed, and so forth, make the offering of such components economically unattractive. Computer manufacturers, particularly those who manufacture small desk-top types are quite willing to offer assistance in interfacing instruments, such as fluorimeters. There are also many firms that specialize in the design of hardware specifically for the task of interfacing instruments with computers of all sizes. The fluorimeter manufacturer should be willing to act as technical liaison with such firms if the computerization of an instrument is desired.

E. EXPERIMENTAL CONSIDERATIONS

1. MEASUREMENT OF MOLECULAR PARAMETERS

a. Excitation and Emission Spectra

Excitation and emission spectra are obtained by the use of spectrofluorimeters or compensating spectrofluorimeters equipped with scanning monochromators. Some spectrofluorimeters are available that are equipped with manually adjustable monochromators only. These instruments are generally used only for quantitative analyses at fixed wavelengths, although spectra can be obtained by point-to-point measurements.

An excitation spectrum is obtained by manually locating the excitation and emission bands of the sample by trial-and-error adjustment of the monochromators. For solutions of single pure compounds, the wavelength of emission is independent of the wavelength of excitation, and therefore this trial-and-error procedure can be done very quickly. To obtain the best signal-to-noise ratio in the recorded spectrum, the wavelengths of maximum excitation and emission intensity are usually chosen. The photometer* is adjusted to give a full-scale or nearly full-scale reading. The emission monochromator is left on the emission band maximum, and the excitation monochromator is moved to the desired starting wavelength as dictated by the design of the specific spectrofluorimeter being used. The monochromator is then scanned through the wavelength region of interest to generate the excitation spectrum. Most instruments have a precision potentiometer coupled to each monochromator drive. The output of each potentiometer is a voltage that is linearly related to wavelength. This voltage is applied to the x-axis of an xy-recorder to generate an accurate recording of intensity versus wavelength regardless of rate of scan.

Excitation spectra recorded on single-beam, noncompensated spectrofluorimeters bear little resemblence to true absorption spectra recorded on double-beam spectrophotometers. This is a result of the distorting effects of instrumental response, particularly in the UV region. As a rule, absorption (excitation) peaks are shifted and skewed toward longer wavelengths, and peaks in the UV region shorter than about 250 nm appear weak (Figure 7B.10). Corrected excitation spectra are identical to absorption spectra in all but a very few instances. This fact serves as a means of verification of system performance for compensating instruments and as a criterion of purity of fluorescent compounds.

Manual correction of recorded excitation spectra has been described by Parker and Rees and others (18). These procedures are tedious and very inaccurate at best, particularly in the UV region below 250 nm, where lamp output is weak and poorly defined. Uncorrected excitation spectra are useful in qualitative analysis, *provided* comparisons are made only between similar instruments under very similar conditions. Ideally, standard spectra should be run on the same instrument being used to investigate unknown compounds.

Excitation spectra of very long-lived fluorophores or of phosphorescent compounds require some additional care in recording. The scan speed of the excitation monochromator must be sufficiently slow to allow for the growth and decay of the sample luminescence as the excitation wavelength shifts on and off the absorption bands.

---

*Photometer is used interchangeably with *detection system*, *amplifier*, and *electrometer*.

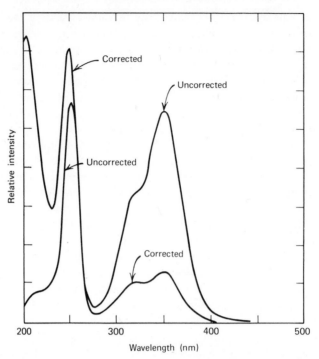

Fig. 7B.10   Corrected and uncorrected excitation spectra: quinine bisulfate (0.1 ppm) in 0.1 N H$_2$SO$_4$. (Aminco-Bowman spectrophotofluorometer with corrected spectra attachment.)

Emission spectra are recorded in essentially the same fashion as excitation spectra. The excitation monochromator is adjusted to a peak excitation band, and the emission monochromator is scanned through the emission region of the spectrum. As with excitation spectra, emission spectra are distorted by instrumental responses unless a compensating spectrofluorimeter is used.

Taken together, excitation and emission spectra form useful "fingerprints" for the qualitative analysis of unknown compounds, uncorrected spectra only for comparison with spectra taken on similar instruments, and corrected spectra for comparison with data taken on any quality compensating spectrofluorimeter or absorption spectrophotometer.

Monochromator slit widths must be chosen with care in the recording of either excitation or emission spectra. When excitation spectra are being recorded, the excitation monochromator slits must be sufficiently narrow to resolve spectral fine structure. Emission slits can be as wide as necessary to obtain a sufficient signal-to-noise ratio. Conversely, when emission spectra are being recorded, the emission monochromator slits must be narrow, and the excitation monochromator slits as wide as required. In either case, of course, the slits should not be so wide as to allow overlap of the monochromator bandpasses at any time during the scan.

The electronic response time* of the photodetector and data readout system must also be considered as a potential source of spectral distortion. A rule of thumb is to maintain the scan speed (in nm s^{-1}) at a smaller value than $\delta\lambda/t_r$, where $\delta\lambda$ is the half-width of the most narrow spectral band to be observed (in nm), and t_r is the overall system response time (in s).

b. Quantum Efficiencies

The quantum efficiency for any luminescence process is defined as the number of quanta of light emitted by that process divided by the total number of quanta absorbed. The quantum efficiency is a characteristic molecular parameter for a species under fixed conditions of solvent type, temperature, and other environmental factors. To be useful for trace analytical purposes, fluorescent compounds usually must have quantum efficiencies of greater than 0.01. Experimentally quantum efficiencies are usually determined as *relative* quantum efficiencies by comparison of the area under the *corrected* emission spectrum of the unknown compound with the area under the corrected emission spectrum of a compound with a known quantum efficiency:

$$\frac{\text{Area}_{\text{unk}}}{\text{Area}_{\text{std}}} = \left(\frac{Y_{\text{unk}}}{Y_{\text{std}}}\right)\frac{A_{\text{unk}}}{A_{\text{std}}} \tag{8}$$

where A_{unk} and A_{std} are the absorbances of the two solutions (4). This procedure requires that the emission spectra be corrected in terms of relative quanta per unit frequency interval. Some commercial instruments are corrected in terms of constant energy per unit wavelength interval. Conversion from one form of presentation is straightforward but laborious. More recently, compensated instruments have offered either mode of display (14).

c. Excited-State Lifetimes

The lifetimes of the fluorescent excited states (τ_F) of most fluorescent organic compounds range from about 1 to 100 ns. This means that, when the exciting light is terminated, the fluorescence emission signal drops to a value of $1/e(=0.368)$ times the original intensity in 1 to 100 ns. Most pure organic compounds exhibit first-order exponential decays, and therefore a

*Also refer to Chapter 2.

plot of the logarithm of signal versus time to decay yields a straight line. Any deviation from linearity of such a plot indicates either impurities in the sample or the operation of more complex energy transfer processes in the sample.

The instrumentation required to detect and measure such rapidly changing intensities is electronically complex and varies in approach. Techniques that have been applied include phase shift and stroboscopic techniques, and time-correlated single-photon counting (19). Optically, the instruments are typically simple filter fluorimeters with effective numerical apertures of $f/2$ or better.

Under fixed environmental conditions, fluorescence lifetime (decay time) is a characteristic molecular parameter of each fluorescent material. The potential exists for the use of decay times in the qualitative identification of unknown materials, but the difficulty and expense in obtaining such data have all but precluded its use.

The lifetimes of phosphorescent excited states (τ_p) may range from microseconds to tens of seconds. Phosphorescence decay times are thus considerably easier to measure than fluorescence decay times. For compounds such as metal chelate systems or halogenated aromatic hydrocarbons with lifetimes in the microsecond to millisecond range, techniques similar to those used in fluorescence are applied. For compounds with moderately slow decay times in the 1 to 100 ms range, a spectrofluorimeter equipped with a fast repetitive shutter (the so-called phosphoroscope) can be used with the photodetector signal viewed on an oscilloscope. Compounds with long decay times can be studied by terminating the exciting light with a mechanical shutter and recording the slowly decaying light on a time base recorder or direct-coupled oscilloscope (6).

As with fluorescence, phosphorescence decays generally follow first-order kinetics; however, there are considerable deviations from this rule because of the long triplet-state lifetimes.

2.   INSTRUMENT CALIBRATION

a.   Wavelength

The use of recorded excitation and emission spectra for qualitative identification of luminescent compounds requires that the monochromators be accurately calibrated. Low-pressure mercury vapor discharge lamps are generally used for this purpose. The very sharp line emission spectra of these lamps provide accurate wavelength reference points for the adjustment of monochromators.

Calibration procedures are dictated by the design of each specific instrument, and the manufacturer's instructions deserve study. Usually the emission monochromator is adjusted first, by placing the mercury lamp in the position of the sample cuvet. Start with the smallest slit widths possible and increase them carefully until the desired mercury line is located. Table 7B.2 gives a listing of useful lines in the low-pressure mercury arc. Considerable care must be taken not to overload the photodetector. The relatively weak visible emission from these lamps is deceptive. The UV emission lines are *very* intense. Eye protection designed to block UV radiation should be used. Common plastic eyeglasses are *not* adequate.

Excitation monochromators can be conveniently calibrated by scattering light from the normal excitation source through the sample cuvet and into the emission monochromator. A colloidal suspension of silica (Ludox), a magnesium carbonate block, or similar material can be used to scatter light into the emission monochromator. The excitation monochromator is then adjusted to match the wavelength calibration of the emission monochromator point by point as desired. This comparison-type procedure is mandatory if the excitation monochromator collimating mirror is focused on the arc lamp arc column itself rather than on an actual mechanical entrance slit. In this case, the arc itself becomes the effective entrance slit, and any motion of the arc affects wavelength calibration. Replacement or adjustment of the lamp necessitates recalibration. Arc wander in such designs limits the calibration accuracy to about ± 1 nm.

b. Intensity

To maintain constant sensitivity, single-beam spectrofluorimeters and filter fluorimeters rely completely on the stability of their lamps and photodetector systems. Excitation lamps are by far the least stable components in fluorimeters, making some means of sensitivity adjustment a necessity. High-pressure arc lamps rarely "strike" or ignite to the same position on the electrodes. Considerable changes in spectral output occur as the lamps warm up to operating temperature and pressure. Also, as the lamps age, tungsten from the electrodes is deposited on the bulb's inner surface, progressively blackening the bulb and decreasing output. Low-pressure arc lamps and phosphor-coated arc lamps (fluorescent lamps) are considerably more stable than high-pressure arc lamps; however, these lamps also exhibit temperature effects and long-term aging effects.

For these reasons, a means of adjustment of overall system sensitivity is mandatory. Generally, sensitivity is adjusted by varying amplifier gain or photodetector voltage.

The most frequently used standard for the adjustment of instrument sensitivity is a solution of quinine bisulfate in 0.1 N H_2SO_4. Many other materials have been suggested as standard materials, but quinine continues to be popular because of its very high sensitivity, reasonable stability, and relative freedom from oxygen quenching. Other materials may serve equally well as secondary intensity standards. Care should be taken in chosing materials that are stable, have a small temperature coefficient of emission, and are relatively free from oxygen quenching. Fluorescent glasses have been studied as durable permanent standards for the comparison of fluorescent intensities. The principal drawback with these materials is their high temperature coefficient.

3. LUMINESCENCE QUENCHING

Luminescence quenching refers to any effect, chemical or physical, that tends to decrease luminescence intensity. Some investigators restrict the use of the word "quenching" to factors influencing the excited electronic states only (4), however, common usage reflects the broader definition which includes ground-state effects also.

a. Sample Concentration Quenching

The linear dependence of emission intensity on sample concentration is based on the condition that the excitation intensity is constant throughout the sample solution. Practically speaking, this means that the excitation light beam must not be attenuated by the sample by more than about 5% at most. This restriction applies to both the exciting light and the emitted light, be it fluorescence or phosphorescence. This form of quenching is called the *inner-filter effect* (prefilter) and is manifested by a decrease in the slope of the plot of signal versus sample concentration. The inner-filter effect can result in erroneous readings in quantitative analysis, as well as severe distortion in excitation and emission spectra. The cure for this type of quenching, fortunately, is simply dilution of the sample to the point where the excitation beam intensity does not significantly vary throughout the sample volume. Inner-filter-type quenching can equally, and more deceptively, be caused by the presence of one or more foreign species in the sample cuvet, whose absorption bands overlap either the absorption or emission bands of the sample. Of course, this effect can be caused by an absorbing solvent as well as trace materials.

The extent to which inner-filtering affects the luminescence signal is dependent on the type of sample viewing geometry utilized. Front surface illumination is least affected and is thus best suited for solids, turbid solutions, and concentrated solutions. The 90° geometry is best for general-purpose applications, having the widest range of linearity and sensitivity. The straight-through geometry is poorest from both the standpoints of quenching and stray light level.

In addition to the bulk attenuation of the excitation beam by sample absorption there are other less dominant concentration-dependent effects.

As the concentration of analyte increases, analyte-analyte encounters correspondingly become more frequent, and analyte dimers or higher aggregates may form. Such aggregates are separate chemical entities and have their own characteristic absorption bands and emission spectra. Pyrene exhibits an interesting variation of this behavior. Pyrene concentrations in excess of approximately 10^{-4} M form no ground-state dimers, however, it appears that ground-state molecules dimerize with an excited-state molecule to form electronically excited dimers with their own characteristic blue emission spectrum (20). This combination of a ground-state molecule and an excited singlet molecule has been termed an *excimer* (21).

b. Oxygen Quenching

Oxygen is a compound-specific quencher of fluorescence in liquid solutions (22). The magnitude of the quenching is dependent on the viscosity of the solution (i.e., a diffusional process), the oxygen concentration in solution (about 10^{-3} M in air-saturated ethanol), the lifetime of the excited singlet state, and other more poorly understood factors (23). The fluorescence of anthracene in ethanol, for example, is quenched by about 50% by saturation with pure oxygen at atmospheric pressure. Quinine bisulfate, however, shows very little sensitivity to oxygen quenching.

The theoretical involvement of triplet-state molecules as quenchers was discussed in Section 7B.**B**.2.E. In liquid solution at room temperature, the combined effects of oxygen quenching, long triplet-state lifetime, and collisional quenching cause very rapid quenching of phosphorescence. Low temperatures and rigid media are usually necessary for the observation of phosphorescence. Under these conditions, oxygen has not proved to be an important factor in analytical applications.

From a practical standpoint, the deoxygenation of samples during analysis has not been applied to any major extent. However, it is an important effect which potentially could strongly influence the precision of luminescence measurements if oxygen concentrations are allowed to vary from sample to sample or during the time course of an analysis.

c. Temperature and Viscosity

The radiative modes of deactivation of electronic excited states—fluorescence and phosphorescence—compete with nonradiative modes of deactivation as means of depopulating the excited state. Nonradiative modes of deactivation are primarily collisional modes, that is, they are proportional to the frequency of encounters of molecules in solution. The frequency of encounter of molecules in solution is directly proportional to temperature and inversely proportional to viscosity; cooling and increasing the viscosity of a luminescent sample enhance the luminescence efficiency to the point where other radiative processes become dominant.

Fluorescence intensities generally increase a few percent per degree Celsius decrease in temperature in the region near room temperature. As with oxygen quenching, the effect is highly compound dependent.

The high-intensity arc lamps used for fluorescence excitation can result in a considerable temperature rise in the sample cell. Thermostatted cell holders can minimize variations due to sample temperature coefficients. Also, making the luminescence measurement as quickly as possible helps keep sample temperature constant.

The first excited triplet electronic state and the ground electronic state are usually separated by relatively small energy differences. Therefore, collisional and vibrational deactivation are very efficient modes of deactivation. This, plus the efficiency of oxygen quenching in the triplet excited state, explain why phosphorescence is rarely observed at room temperature, and *then only in rigid media*. The observation and the analytical uses of phosphorescence depend primarily on the fact that nonradiative processes can be minimized by both cooling the sample and increasing the solution viscosity, usually by freezing the sample into a clear, rigid glass at liquid nitrogen temperatures.

4. SOLVENTS

The selection of a solvent for a luminescence analysis procedure follows essentially the same guidelines as those applied in absorptiometry. Several obvious criteria apply; the solvent must by transparent in the spectral region of interest; it must not luminescence in this region; it must not quench the desired luminescence process; and it must dissolve the sample, at least at the concentration range under study.

The primary difference between solvent selection for luminescence procedures and for absorptiometry is that the relatively higher sensitivity levels frequently attained with luminescence techniques require a correspondingly higher level of solvent purity.

Solvent purification procedures should be tailored to each specific problem of trace analysis under consideration. The preparation of highly purified solvents represents an enormous investment of time and money. Judicious use of such solvents is usually just common sense.

Commercial solvents are available in several high-purity forms. Those intended for use in gas chromatography, "pesticide" grades, and those intended specifically for fluorimetry are generally very good. The "spectral" grade solvents offered by most suppliers do not necessarily have low fluorescence backgrounds (and frequently do not). *Caveat emptor* is the watchword in solvent selection.

Water purification by distillation is the time-honored procedure for quality water preparation. Distilled water delivered from laboratory to laboratory via a pipe distribution system has justifiably fallen into disrepute as a solvent adequate for any task other than the rinsing of routine glassware. Contamination by metals in the pipe system, leaching of organic materials from plastic tubing, and growth of microorganisms in stagnant portions of the system are a few of the reasons why such water is of poor quality. Single-stage distillation from glass stills directly at the laboratory can produce remarkably good water. Triple distillation adds additional improvement but must be done with extreme care or the benefits may be nullified.

Laboratory-scale ion exchange deionizing systems are capable of producing very low levels of dissolved solids. Activated carbon filters coupled with such systems can reduce total organic compounds to the sub-parts-per-million range. The principal drawback to such systems is the maintenance required to keep them in peak operating condition. Experience has shown that small deionizing systems are frequently neglected and, in some cases, have actually degraded the quality of water passing through them.

Several commercial deionizing-type water conditioning systems are available on a contract basis. Complete systems, consisting of prefilter, carbon filter, and deionizing bed, have been shown to yield very low-conductivity water and low fluorescence background luminescence. Maintenance can be arranged on a routine basis as part of the contract price, thus avoiding the problems of neglect.

Regardless of the type of water system used, a final filtration through a fluorocarbon membrane filter of 200 to 500 nm pore diameter produces water that is essentially particle-free and sterile, effectively reducing the problems of particulate scatter and microbial contamination.

The fluorescence and phosphorescence background luminescence of organic solvents can be considerably reduced by *carefully* distilling them through a glass helix packed column. A well-equilibrated column and slow distillate takeoff is required for good yields. Simple single-stage distillations do not, as a rule, produce adequate enough results to justify the labor involved.

Organic solvents that tend to form peroxides; dioxane, ethers, and Cellosolves should be used with caution in qualitative analytical applications. The spectra of analytes made up in these solvents tend to change with time, apparently because of the formation of peroxide complexes.

The variety of solvents available for use in phosphorimetry is dependent on the type of sample cell employed. Recent developments in the use of the rotating sample cell (17) permit the use of aqueous solvents and other solvents that form polycrystalline solids or cracked glasses at low temperature, as well as solvents that form clear rigid glasses. The older type of stationary sample tube requires the use of solvents that form clear glasses when rapidly cooled. This limits the choice of solvents to a few single-component systems. A variety of multicomponent solvents has been devised (16). The most popular of these has been EPA, a $5:5:2$ v/v mixture of diethyl ether, isopentane, and ethanol. Ethanol itself is an excellent solvent for phosphorimetry and may be combined with many inorganic and organic acids, bases, and salts, as long as the water content does not exceed approximately 5% v/v. A good discussion of various other solvents for phosphorimetry can be found in reference 6.

5. SCATTERED LIGHT

It is a basic premise in luminescence analysis that the light reaching the photodetector is the result of a specific radiative deactivation process in the sample molecule of interest. Any light reaching the photodetector by any other process is undesirable. Various sources of undesirable "stray" light have been discussed in Section 7B.**D**; for example, light leaks into the instrument, unwanted reflections from optical elements, fluorescence of optical materials, and overlapping orders in grating monochromators. Three major processes remain that affect the measurement of sample fluorescence: Rayleigh scatter, Raman scatter, and particulate scatter. These processes are primarily of concern in fluorescence measurements. Phosphorimetry is essentially free of these problems, with the exception of the attenuation of emitted light by particulate scatter.

a. Rayleigh Scatter

Nonabsorbing particles smaller in size than approximately a wavelength of light scatter an incident beam of that light. The process involves no loss of energy from the scattered photon and thus involves no change in wavelength of the scattered light. Rayleigh scattering is an unavoidable property of the solvent irrespective of its purity. The scattering particles are the solvent molecules themselves. The intensity of the scattered light is strongly wavelength-dependent, varying inversely as the fourth power of the wavelength of the incident light beam, that is, λ^{-4}. Consequently, UV radiation is much more strongly scattered than visible radiation.

In fluorescence measurements, the height and width of the Rayleigh scattering peak are functions of the monochromator slit widths and excitation source intensity. Since continuum sources are almost invariably used in spectrofluorimeters, the intensity of the scattered light, that is, the peak height, is strongly dependent on the intensity of the lamp itself, the slit widths of the excitation monochromator, and the wavelength setting of the monochromator. The width of the scatter peak is primarily a function of the overlap of the bandpasses of the excitation and emission monochromators.

When a fluorescence emission band is located very close to the excitation band, that is, when the excitation and emission monochromator bandpasses must be adjusted very close to each other, there exists an optimum slit width combination that yields maximum fluorescence and minimum scatter. This combination of slit widths must be determined by trial and error. The fluorescence spectrum is strongly distorted under these conditions; however, quantitative data can nonetheless be obtained.

b. Raman Scatter

A beam of light passing through a molecular medium may lose energy to that medium through the nonresonant excitation of vibrational* modes of the molecular bonds. This process is similar to the resonant absorption of IR radiation by molecules but involves different modes of vibration and is a very much weaker phenomenon. As a result of the loss of energy from the light beam to the molecule, the Raman-scattered light is composed of wavelengths longer than that of the exciting light itself (anti-Stokes Raman spectra are of negligible importance).

The conditions under which Raman bands are observed in fluorimetry, relatively broad-band excitation and low emission monochromator resolution, result in a single Raman scatter peak located close to the Rayleigh peak but at longer wavelengths. This Raman peak is an unresolved aggregate of wavelengths due to the many different vibrational modes excited. The Raman band is always separated from the main Rayleigh scatter peaks by a constant *energy* interval. Raman peaks therefore can be recognized by comparing the inverse of the wavelength separation between the two scatter peaks at several different excitation wavelengths. Figure 7B.11 illustrates the Raman bands from cyclohexane, showing the characteristic Raman shifts in wavelength separation and peak heights as excitation wavelength is varied.

Raman scatter is a very weak phenomenon and generally is observed only at the limits of fluorimeter sensitivity where it contributes to the overall background or "blank" light levels.

c. Particulate Scatter

Particles of dust, lint, precipitates, and so on, in the sample solution result in light from the excitation beam being reflected directly into the emission monochromator. This reflected light, along with Rayleigh scatter, comprises the main scatter peak in fluorimetry.

*Rotational Raman is of no separate interest because it appears as simply scatter.

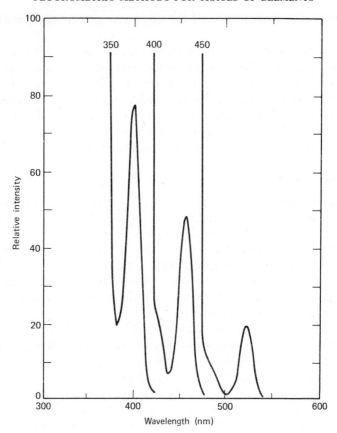

Fig. 7B.11   Raman bands of cyclohexane excited at 350, 400, and 450 nm. (Aminco-Bowman spectrophotofluorometer.)

Particulate scatter is of special concern in filter fluorimetry where excitation sources are effectively much more intense than in spectro-fluorimeters, and where the off-band rejection of unwanted light by the filters is seldom as good as that found in monochromators. In addition to problems of stray light, turbidity in the sample can lead to attenuation of both the excitation beam and the emitted luminescence.

Particulate contamination can be minimized by careful attention to cleanliness, filtration, and centrifugation. Membrane filters are available that can ensure virtually complete removal of particles as small as 200 nm in diameter. Colloidal particles are very efficient light scatterers. Minute chips of glass from ground-glass stoppers and joints have been shown to lead to anomalous results in very high-sensitivity analyses. High-speed centrifugation can minimize this particular problem, but the use of Teflon stoppers in volumetric flasks and similar glassware is probably a better long-term solution.

F. APPLICATIONS IN TRACE ANALYSIS

1. ANALYTICAL PROCEDURES

a. Characteristics of Quantitative Luminescence Analysis

Quantitative luminescence analysis shares many features with absorptiometry. The basic aspects of the techniques by which quantitative information is obtained by either method were discussed in Chapters 2 and 3. Luminescence analysis also possesses several unique features which make it an extremely useful tool for trace analytical applications.

The single most characteristic feature of the luminescence technique is the direct proportionality between analyte concentration and analyte luminescence signal. Here the analytically important quantity is the emission signal itself. A change in analyte concentration results in a linearly related change in emission signal, a proportionality that frequently exceeds four decades of concentration. This should be contrasted with absorptiometry in which the analytical quantity of interest is the logarithm of the ratio of two signals (intensities). With absorptiometry, low concentrations of absorbing materials produce a small percentage change between two large signals. This leads to a rapid loss of measurement precision at low absorbance values.

Luminescence techniques frequently offer a greater degree of selectivity over absorption techniques. Since absorption is the necessary prerequisite to luminescence, and since luminescence is only one of several modes of dissipation of the absorbed energy, it follows that many more types of molecules absorb light than reemit it. The trade-off is of course that luminescence techniques are not as broadly general in applicability as absorptiometric techniques.

In addition to the natural selectivity of the luminescence technique, one can further increase the selectivity of the method by adjusting several other experimental parameters. The analyst can choose the wavelength of excitation (excitation resolution), the wavelength at which the emission is observed (spectral resolution), and the time at which the emission is observed (time resolution). Time resolution is seldom applied to fluorescence techniques, because of the extremely rapid decay of fluorescence after termination of the exciting light. Time resolution is routinely applied in phosphorimetry in which the phosphorescence emission may continue for many seconds after the termination of excitation.

Both fluorescence and phosphorescence procedures are subject to interference by materials that absorb light in their excitation wavelength region. They are also subject to interference by materials that emit and/or absorb light in the desired emission wavelength region. These processes, as well as the compound-specific modes of quenching have been discussed in Section 7B.E.3.

b. Quantitative Determination of Concentration

The various procedures for quantitative analysis detailed in Chapter 2, Section E, all can be utilized in luminescence. Of the methods outlined, the analytical calibration curve method (also called *standard curve* or *working curve*) is the most frequently used method. The analytical curve is constructed of a series of standards of various concentrations, which are carried through the complete analytical procedure. The emission signal of each sample is recorded, corrected for blank luminescence, and recorded in the form of a signal-versus-concentration graph. Unknown samples are carried through the identical procedure, and their concentration determined by interpolation from the analytical curve. For screening applications, where wide variations in sample concentrations are encountered, the data are recorded on log-log graph paper. Where sample concentration falls into a more narrow range of variation, the use of linear coordinates results in higher precision.

Luminescence analytical curves are usually linear over a concentration range of 10^4 or more. Negative* curvature frequently occurs in the 10^{-3} M concentration range where the inner-filter effect, molecular aggregation, and concentration quenching processes become significant. As concentration is increased, the curvature increases until the emission signal is independent of sample concentration (the quantum counter effect). A few compounds show a *decrease* in emission signal at very high concentrations due to the combined effects of quenching processes. This leads to a double value for a single signal reading. Dilution of the sample avoids this problem, but points out the hazard of extrapolating the analytical curve beyond the established data points at its high-concentration end. The actual shape of the analytical curve in the high-concentration region is compound specific. Some materials exhibit a linear response right up to the limit of solubility.

Variations in instrumental sensitivity must be routinely considered when

---

*Curvatures toward the abscissa, that is, dS_L/dc, become less as c increases, where S_L is the luminescence signal.

using single-beam fluorimeters or spectrofluorimeters. The combined effects of lamp aging, phototube fatigue, and thermal effects result in changes in sensitivity.* A solution of a reference standard material should be measured at the time the working curve is determined, the intensity value recorded, and the instrument sensitivity† adjusted to duplicate this standard value each subsequent time the instrument is used. The reference standard may be a material such as quinine bisulfate in 0.1 N H_2SO_4, or it may be a known concentration of the analyte itself. Fluorescent glass blocks are also available for this purpose. Instrument sensitivity should be varied only by adjusting amplifier gain or phototube high voltage, preferably the former.

The use of the methods of internal standards‡ or constant addition is highly recommended, particularly where the sample compound is to be determined in a complex matrix.

Various precautions concerning the contamination of samples and the potential loss of analyte during sample preparation procedures were discussed in Chapter 3. These precautions apply equally whether the quantitative procedure utilizes absorptiometry or fluorimetry. In fluorimetry, additional attention should be given to the problem of maintaining constant low blank levels. A large number of common materials normally present in the laboratory are stongly fluorescent in their own right and may find their way into samples, solvents, and reagents, resulting in high blank levels or, worse, in erratic blank values.

Of the variety of plastics normally encountered in the laboratory only polytetrafluoroethylene and polytrifluorochloroethylene can be used without concern as to potential contamination. Nearly every plastic contains plasticizers, modifiers, extenders, or lubricants that are potential sources of contamination. The relative risk of contamination depends on the time of exposure to the plastic and the relative polarity of the solvent. For example, polyethylene vials are frequently used with aqueous samples with no apparent problem. Use of the same vials with ethanolic solutions results in slowly increasing blank levels as materials are leached from the plastic.

Solvents should not be stored in plastic containers. The possibility of fluorescent materials being leached out of the container is too great. All bottle caps should have Teflon inserts. Paper or plastic inserts are inadequate. Aluminum foil is a good alternative, provided aluminum does

---

*Sensitivity is the slope of the analytical calibration curve of signal S_L in analyte concentration c.

†Instrumental sensitivity is the change in photodetector signal with the change in flux reaching the photodetectors.

‡Refer to Chapter 2.

not interfere in the analytical procedure. The flexible paraffin films used in quick temporary closures should not be used.

Dust and lint normally contain strongly fluorescent materials. Most laundry detergents now contain "optical brighteners" which are nothing more than intensely blue fluorescent dyes added to counteract the yellowing of white fibers. Lint from such fibers is a potential source of trouble. The use of such detergents as routine laboratory cleansers should be avoided.

Vinyl and rubber tubing both contain fluorescent modifiers. Vinyl in particular is heavily loaded with plasticizers which are very frequently fluorescent. Rubber stoppers and vial caps are frequent sources of contamination.

Smoke, tobacco or otherwise, is a potent source of fluorescent contaminants. Stopcock greases, particularly silicone high-vacuum grease, are fluorescent. Silicone greases are a particular nuisance because of their affinity for glassware.

c.   Limits of Detection*

The inherent capability of a compound to luminesce is a function of its molar extinction coefficient ϵ and its quantum efficiency Y for the luminescence process of interest. The product ϵY is a good indicator of analyte sensitivity. This method of expressing sensitivity has not been widely used, primarily because of the difficulty in evaluating the quantum efficiency. Most spectrofluorimeters in use are uncorrected single-beam instruments which do not lend themselves to the determination of quantum efficiencies.

From the standpoint of experimental utility, a more useful indicator of detection sensitivity is the experimentally determined limit of detection based on signal-to-noise ratios. Such ratios include instrumental influences, as well as molecular influences on the estimation of detection levels.

In any luminescence measurement, the level at which useful information can no longer be obtained from a signal is determined by the fluctuating non-signal components (noise) superimposed on the luminescence signal. A limit of detection can be defined as the concentration of solute that produces an arbitrary ratio of signal to noise. A signal-to-noise ratio of 2 has been used frequently in the past.

In the presence of a *constant* background (blank) signal, the limiting noise in luminescence measurements is primarily photomultiplier shot noise with a small contribution from lamp flicker. In practice, the background level is seldom constant from sample to sample. This in effect

*Also refer to Chapter 2.

introduces a very low-frequency noise into the measurement. If the fluctuation in the blank appreciably exceeds the instrumental system noise level, it is the blank that limits the overall limit of detection. This is frequently the case, and for this reason the publication of limits of detection based strictly on system noise level are generally very optimistic limits.

In molecular luminescence measurements, there does not, in general, exist an optimum slit width for excitation and emission monochromators. This is because of the broad absorption and emission bands of most compounds and because of the continuum light sources used for excitation. This is in sharp contrast to atomic spectroscopy in which an optimum set of conditions can almost always be determined. In practice, the greatest sensitivity and the best limits of detection in molecular luminescence measurements are obtained by using the widest slits available in the monochromators.

2. APPLICATIONS

Table 7B.3 is a listing of elements that have been determined by luminescence methods. The table also includes the *reported* limits of detection and a representative citation.

TABLE 7B.3. Elements, Reagents, and Representative Limits of Detection Obtained by Molecular Luminescence Spectrometry (Fluorimetry)

Element	Reagent	Limit-of-Detection ($\mu g\ ml^{-1}$)	Reference
Ag	Eosin plus 1,10-phenanthroline	0.004	24
	8-Hydroxyquinoline-5-sulfonic acid	0.01	25
	2,3-Naphthotriazole	0.02	26
Al	3-Hydroxy-2-naphthoic acid	0.0002	27
	Mordanted blue 9	0.0005	28
	Morin	0.0002	96
	Salicylidene-*o*-aminophenol	0.0003	30
	N-Salicylidene-2-amino-3-hydroxyfluorene	0.0008	31
Ag	Uranyl nitrate	50.0	32
Au	Rhodamine B	0.5	33
B	Acetylsalicylic acid	0.01	34
	Benzoin	0.01	35
	Dibenzoylmethane	0.0005	36
	Quinizarin	0.01	37

TABLE 7.B.3. (*Continued*)

Element	Reagent	Limit-of-Detection ($\mu g\ ml^{-1}$)	Reference
Be	1-Amino-hydroxyanthraquinone	0.2	29
	3-Hydroxyquinaldine	0.001	29
	3-Hydroxy-2-naphthoic acid	0.0002	38
	Morin	0.00004	35
Bi	Rhodamine B	0.5	39
C(CN$^-$)	*p*-Benzoquinone	0.2	40
	o-(*p*-Nitrobenzene sulfonyl)quinone monoxime	0.2	41
	Pd complex of sulfo-8-hydroxyquinoline	0.02	42
C(C$_2$O$_4{}^{2-}$)	Ce(IV)–As(III)reaction, Os catalyst	9.0	43
Ca	Calcein	0.01	44
	Dicarboxymethylaminomethyl-2, 6-dihydroxynaphthalene	0.01	45
	8-Quinolyhydrozaone of 8-hydroxyquinaldehyde	0.02	46
Cd	*p*-Tosylaminoquinoline	0.02	47
Ce	4-[(2, 4-Dihydroxyphenyl)azo]-3-hydroxy-naphthalene sulfonic acid	0.05	48
Cl	Luminol plus H$_2$O$_2$	0.05	49
Co	Salicylfluorene plus H$_2$O$_2$	0.0001	50
	1-(2-Pyridylazo)-2-naphthol	0.06	51
	Benzamido (*p*-Dimethyibenzylidene) acetic acid	0.0002	52
Cu	2, 9-Dimethyl-4, 7, diphenyl-1, 10-phenanthroline	0.01	53
	Salicylalazine	0.05	32
	Tetrachlorotetraiodofluorescein plus *o*-phenanthroline	0.001	29
	Thiamine	3.0	54
	1, 1, 3-Tricyano-2-Amino-1-propene	0.01	55
Dy	Calcium tungstate	5.0	32
	Sodium tungstate	0.01	70
Er	Calcium tungstate	10.0	32
Eu	Thenoyltrifluoroacetone	0.01	81

TABLE 7.B.3. (*Continued*)

Element	Reagent	Limit-of-Detection ($\mu g\ ml^{-1}$)	Reference
	Benzoyltrifluoroacetone plus trioctylphosphine oxide	0.07	31
	Calcium tungstate	0.5	32
	Sodium tungstate	0.005	69
F	Al complex of alizarin garnet R	0.001	35
	Al complex of eriochrome B	0.001	29
Fe	Stilbexone plus H_2O_2	0.001	82
	Luminol plus H_2O_2	0.0008	83
Ga	8-Hydroxyquinaldine	0.02	29
	8-Hydroxyquinoline	0.05	29
	Lumogallion	0.002	84
	2,2′-Pyridylbenzimidazole	0.07	79
	Salicylidene-*o*-aminophenol	0.007	29
	2′,2′,4′-Trihydroxy-5′-chlorobenzene-3′-sulfonic acid	0.001	80
Gd	Calcium tungstate	10.0	32
	Thenolyltrifluoroacetone	100.0	60
Ge	Benzoin	2.0	29
	2,2′,4′-Trihydroxy-3-arseno-5-chlorobenzene	0.004	85
Hf	Flavanol	0.1	42
	Quercetin	1.0	86
Hg	Rhodamine B	0.002	87
Ho	Thenoyltrifluoroacetone	~100.0	60
I(I⁻)	Ce(IV)–As(III) reaction, Os catalyst	0.6	58
In	Dimethyl-[6-(dimethylamino)xanthen-3-xylidene]ammonium chloride	5.0	88
	8-Hydroxyquinoline	0.04	29
	8-Hydroxyquinaldine	0.2	29
	2,2′-Pyridylbensimidazole	0.1	79
	Rhodamine S	0.5	89
Ir	2,2′,2″-Terpyridine	2.0	90
Li	Dibenzothiazolylmethane	1.0	91
	8-Hydroxyquinoline	0.2	42
Lu	Thenolytfluoroacetone	~100.0	60

Element	Reagent	Limit-of-Detection (μg cm^{-1})	Reference
Mg	N,N'-Bissalicylidene ethylenediamine	0.00001	92
	N,N'-Bissalicylidene-2,3-diaminebenzofuran	0.002	93
	2-Hydroxy-3-sulfo-5-chlorophenylazo barbituric acid	0.004	80
Mn	8-Hydroxyquinoline	0.002	94
Mo	Carminic acid	0.1	95
N	Resorcinol plus H_2SO_4	0.3	32
Nb	Lumogallion	0.1	56
Nd	Calcium tungstate	5.0	32
Ni	Al-1-(2-Pyridylazo)-2-naphthol	0.00006	51
O(O_2)	Trypaflavine	0.0001	35
Os	4,6-Bis(methylthio-3-amino-pyrimidine)	0.05	57
P	Al–morin complex	0.05	58
	Glycogen, phosphatase, TPN$^+$	0.0000006	59
Pb	Morin	5.0	32
Pr	Calcium tungstate	0.5	32
	Thenolyltrifluoroacetone	100.0	60
Ru	5-Methyl-1,10-phenanthroline	1.0	42
S(H_2S)	Fluorescein	0.0002	61
S(S^{2-})	Fluorescein	0.001	62
S(SO_3^{2-})	Quinine plus H_2O_2 plus acid	10.0	32
Sb	Luminol	0.05	56
	Rhodamine 6 Zh	0.1	63
Sc	Morin plus antipyrine	0.01	64
	Salicyladehyde semicarbazone	0.002	65

Elements	Reagent	Limit-of-Detection ($\mu g\ ml^{-1}$)	Reference
Se	2,3-Diaminobenzidine	0.01	66
	2,3-Diaminonaphthalene	0.005	67
Si	Benzoin	0.08	68
Sm	Calcium tungstate	0.1	32
	Sodium tungstate	0.5	69
	Thenoyltrifluoroacetone	2.0	60
Sn	Flavanol	0.1	35
Tb	Bis(1,2-pyridyl-3-methyl-5-pyrazalonyl)-4,4′-methane	0.1	70
	Calcium tungstate	5.0	32
	EDTA plus sulfosalicylic acid	0.006	71
	HCl	3.0	42
	Sodium tungstate	0.1	69
	Thenoyltrifluoroacetone	100.0	60
Te	Butyl-rhodamine B	0.2	35
Th	1-Amino-4-hydroxyanthraquinone	0.04	35
	Morin	0.02	42
	Quercetin	0.5	72
Tl	HBr (77 K)	0.02	73
	HCl (77 K)	0.02	74
	Rhodamine B	0.1	29
Tm	Calcium tungstate	10.0	32
U	Morin	0.5	42
	Rhodamine B	0.01	35
V	Resorcinol	2.0	42
W	Carminic acid	0.04	29
Y	8-Hydroxyquinoline	0.02	42
	5,7-Dibromohydroxyquinoline	0.1	75
Zn	Benzothiazoylmethane	0.002	76
	2,2′-Methylenebibenzothiazol	0.002	77
	Picolinealdehyde-2-quinoylhydrazone	0.03	78

TABLE 7B.3. (*Continued*)

Element	Reagent	Limit-of-Detection (μg ml^{-1})	Reference
	2, 2'-Pyridylbensimidazole	0.01	79
	p-Tosylaminoquinoline	0.02	80
	1, 1, 3-Tricyano-2-amino-1-propene	0.02	55
Zr	Morin	0.02	42

Table 7B.4 is a cross-reference of reagents and the elements for which they have been used. This table includes not only elements that have been determined at analytically useful levels but also elements that have been reported to react, and hence may be potential sources of interference.

These lists should be considered an introduction to the detailed literature of luminescence analysis. The original papers should be consulted for procedural information, applicability, and potential interferences. It is strongly recommended that several alternative approaches to the analysis of any given element be considered before a specific procedure is adopted.

TABLE 7B.4. **Comparison of Reagents and Elements Determinable by Molecular Luminescence Spectrometry (Fluorimetry)**

Reagent	Element	Reference
Acetylsalicylic acid	Be, B	38, 34
Alizarin garnet R	Al, F	35
Acridine	Au, Cr, Se, Te	97
1-Amino-4-hyrdoxyanthraquinone	B, Be, Li, Th	35
7-Amino-3-nitronaphthalenesulfonic acid	Sn	98
Benzamido (p-Dimethyibenzylidene) acetic acid	Cu	52
8-(Benzene sulfoamino)-quinoline	Cd, Zn	80, 47
Benzoin	B, Be, Cu, Ge, Sb, Zn, Si	35, 68
Benzothiazoylmethane	Li, Zn	91, 76
p-Benzoquinone	CN$^-$	40
2, 2'-Bipyridine	Ir, rare earths	99
4, 6'bis(methylthio-3-amino-pyridine)	Os	57
N, N'-Bis(salicylidene)-2, 3-diamino benzofuran	Mg	35
N, N'-Bissalicylidene ethylenediamine	Mg	35
Butyl rhodamine B	B, Tn, Te	35
Calcein	Be, Ca, Mg, Sn	100
Carminic acid	B, Co, Cu, Fe, Ga, Mn, Mo, Ni Tl, W, Zr	29
Chloramine T plus nicotinamide	CN$^-$	101

Reagent	Element	Reference
4'-Chloro-2 hydroxy-4-methoxybensophenone	B	102
Curcumin	Ba, Mg	103
Datiscetin (3,5,7,2'-tetrahydroxyflavone)	Al, Ga, In, Zr	35
Diaminobenzidine	Se, Sc	66
2,3-Diaminonaphthalene (DAN)	Sc	35
Dibenzoylmethane	B, Eu, Sm, Tb	36
Dibensothiazolylmethane	Li, Zn	77
5,7-Dibromo-8-hydroxyquinoline	Ga, La, Lu, Sc, Y	75
Dicarboxymethylaminomethyl-2,6-dihydroxynaphthalene	Al, Ba, Be, La, Mg, Sr	45
1,4-Dihydroxyanthraquinone (quinizarin)	Be	35
o,o'-Dihydroxyazobenzene	Al, Ga, Mg	104
Dihydroxy-2,4-benzophenone	B	105
2-2'-Dihydroxy-(1-azo-1')-4-naphthalenesulfonic acid (superchrome blue black Z, pontachrom blue black R, acid chrome blue black)	Al, Ga	35
Dimethyl-[6-(dimethylamino)xanthen-3-xylidene]ammonium chloride	In	88
2,9-Dimethyl-4,7-diphenyl-1,10-phenanthroline	Cu	53
4-[(2,4-Dihydroxyphenyl)azo]-3-hydroxy-1-naphthalenesulfonic acid	Ce	48
2-Ethyl-3-methyl-5-hydroxychromon	Be	106
Eosin	Ag, Hg, Pb	107
Eosin plus 1,10-phenanthroline	Ag	24, 108
Eriochrome red B	F	29
Flavanol	Al, Hf, Sn, W, Zr	35
Fluorescein (disodium salt)	Ag, Cr	32
Fluorexone[bis-di(carboxymethyl)amino-methylfluorescein]; see calcein		
Glycogen, phosphatase TPN$^-$	P	59
HBr(77-298 K) (also HCl)	Bi, Ce, Cu, Pb, Sb, Sn, Te, Tl	73
2-Hydroxy-4-methoxy-4'-chlorobenzophenone	B	109
2-Hydroxy-3-naphthoic acid	Al, Be, Hf, In, Mg, Se, Th, Y, Zr$^-$	38
2-Hydroxy-1-naphthaldehyde benzoyl hydrazone	Al	110
8-Hydroxyquinaldine	Ba, Ga, In	29
8-Hydroxyquinoline	Al, Be, Ca, Cd, Cs, Ga, In, K, Li, Mg, Mn, Na, Rb, Sn, Zn, Zr	35

Reagent	Element	Reference
8-Hydroxyquinoline-5-sulfonic acid	Ag, Cd, Mg, Zn	25
2-Hydroxy-3-sulfo-5-chlorophenyl- azobarbituric acid	Mg	80
Kojic acid	Au	111
Luminol	Sb	56
Lumogallion [2,2′,4′-tribydroxy-5-chloro-1, 1′-azobenzene-3-sulfonic acid]	Al, Ga, Nb	56, 84
Lumonomagneson	Mg	112
2,2′-Methylenebibenzothiazol	Zn	77
Mordant blue 31	Al, Ga, Sc	113
Mordant blue 9	Al	28
Morin	Al, Be, Ca, Cd, Ga, Ge, In, Mo, Pb, Sb, Sc, Sn, Sr, Th, W, Zn, Zr	35
2,3-Naphthotriazole	Ag	26
o-(*p*-Nitrobenzenesulfonyl)quinone monoxime	CN^-	41
6-Nitro-2-naphthylasmine-8-sulfonic acid	Sn	98
1,10-Phenanthroline	Dy, Eu, Ir, Sm, Tb	99
Picolinealdehyde-2-quinolylhydrazone	Cd, Zn	78
Pontachrome BBR, CI 202	Al	35
Pontachrome VSW, CI 169	Al	35
1-(2-Pyridylazo)-2-naphthol (PAN)	Al, Co, Ni	31
2,2′-Pyridylbensimidazole	Ga, In, Zn	79
Quercetin	Li, Al, B, Ce, Eu, Ge, Hg, Hf, Sc, Th, Zn	72, 86
Quinizarin	B, Be, P	35
8-Quinolyhydrazone of 8-hydroxyquinaldehyde	Ca	46
Resacetophenone	B, Ge	114
Resorcinol	N, V	32, 42
Rezarson	Ge	115
Rhodamine B	Au, Co, Fe, Ga, Hg, In, Mn, Sb, Sn, Te, Tl, U, W	35
Rhodamine 6G	Ag, Ga, In, Re, Sb, Ta	63
Rhodamine S	Au, Ga, Hg, In, Sb, Te, Tl	89
Rose bengal plus 1,10-phenanthroline	Cu	116

TABLE 7B.4. (*Continued*)

Reagent	Element	Reference
Salicylaldehyde	Al	117
Salicylalazine	Cu	32
Salicyclic acid plus 1,10-phenanthroline	Eu, Tb	99
Salicylidene-*o*-aminophenol	Al, Ga	30
N-(Salicylidene)-*o*-hydroxybenzylamine	Al, Be, Zn	118
N-Salicylidene-2-amino-3 hydroxyfluorene	Al	31
Sulfonaphtholazoresorcinol	Co, Ga,	119
5-Sulfo-8-hydroxyquinoline	S, CN$^-$	42
Sulfosalicyclic acid	Ni, Tb	71
Superchrome blue black extra	Co	120
Superchrome garnet Y	Al, Ga	121
Salicylidene-2-amino-3,5-dimethyl phenylarsonic acid	Be	122
2,2′,2″-Terpyridine	Ir	90
2-Thenoyltrifluoroacetone	Eu, Sm, Tb, other rare earths	60, 81
Thiamine	Cu	54
3,4′,7-Trihydroxyflavone	Sb, Sn	123, 124
1,1,3-Tricyano-2-amino-1-propene	Cu, Zn	55
2,2′,4′-Trihydroxy-3-arseno-5-chlorobenzene	Ge	85
2,2′,4′-Trihydroxy-5-chloro-1,1′-azobenzene-3-sulfoninic acid	Ga	80
8-(*p*-Tosylsulfonamido)quinoline	Cd, Zn	80
Trypaflavine	O$_2$	35
Sodium tungstate	Eu, Dy	69, 70
Calcium tungstate	Dy, Gd	32
Calcium tungstate	Eu, En, Nd, Pr	32
Uranyl nitrate	Ag, Tl	32

The tables are compiled from the literature and must not be construed as specific recommendations of any specific procedure.

The lack of a generally agreed upon system for the estimation of limits of detection results in a wide disparity of estimates. The values given should only be considered general indicators of the potential usefulness of each reagent.

BIBLIOGRAPHY

I. B. Berlman, *Handbook of Fluorescence Spectra of Aromatic Molecules*, Academic Press, New York, 1965.

J. G. Calvert and J. N. Pitts, Jr., *Photochemistry*, John Wiley, New York, 1966.

F. Daniels, Ed., *Photochemistry in the Liquid and Solid States*, John Wiley, New York, 1960.

G. F. J. Garlick, *Luminescence, Handbuch der Physik*, Vol. 26, Springer Verlag, Berlin, 1958.

G. G. Guilbault, Ed., *Fluorescence*, Marcel Dekker, New York, 1967.

D. M. Hercules, Ed., *Fluorescence and Phosphorescence Analysis*, Wiley-Interscience, New York, 1966.

H. Kallman and G. M. Spruch, Eds., *Luminescence of Organic and Inorganic Materials*, John Wiley, New York, 1962.

M. A. Konstantinova-Schlezinger, *Fluorometric Analysis*, N. Kaner, Transl., Davey, New York, 1966.

J. N. Murrell, *The Theory of the Electronic Spectra of Organic Molecules*, Methuen, London, 1963.

W. A. Noyes, Jr., G. S. Hammond, and J. N. Pitts, Jr., *Advances in Photochemistry*, Interscience, New York, Vol. 1 (1963), Vol. 2 (1964), Vol. 3 (1964), Vol. 4 (1966).

C. A. Parker, *Photoluminescence of Solutions*, Elsevier, New York, 1968.

R. A. Passwater, *Guide to Fluorescence Literature*, Plenum Press, New York, 1967.

C. N. R. Rao, *Ultra-violet and Visible Spectroscopy*, Butterworths, London, 1961.

C. Reid, *Excited States in Chemistry and Biology*, Butterworths, London, 1957.

S. Udenfriend, *Fluorescence Assay in Biology and Medicine*, Academic Press, New York, 1962.

S. Udenfriend, *Fluorescence Assay in Biology and Medicine*, Vol. II, Academic Press, New York, 1969.

T. S. West, *Trace Characterization; Chemical and Physical*, NBS Monograph 100, U. S. Govt. Printing Office, Washington, D. C., 1967.

C. E. White and A. Weissler, in *Handbook of Analytical Chemistry*, L. Meites, Ed., McGraw-Hill, New York, 1964.

C. E. White and R. J. Argauer, *Fluorescence Analysis*, Marcel Dekker, New York, 1970.

J. D. Winefordner, Ed., *Spectrochemical Methods or Analysis, Advances in Analytical Chemistry and Instrumentation*, C. N. Reilley and F. W. McLafferty, Eds., Vol. 9, Wiley-Interscience, New York, 1971.

J. D. Winefordner, W. J. McCarthy, and P. A. St. John, in *Methods of Biochemical Analysis*, David Glick, Ed., Vol. XV, Wiley-Interscience, New York, 1967.

J. D. Winefordner, S. G. Shulman, and T. C. O'Haver, in *Chemical Analysis*, P. J. Elving and I. M. Kolthoff, Eds., Vol. 38, Wiley-Interscience, New York, 1972.

A. B. Zahlan, Ed., *The Triplet State*, Beirut Symposium, Cambridge University Press, Cambridge, 1967.

M. Zander, *Phosphorimetry*, T. H. Goodwin, Transl., Academic Press, New York, 1968.

REFERENCES

1. R. A. Passwater, *Guide to Fluorescence Literature*, Plenum Press, New York, 1967.
2. A. Weissler, *Anal. Chem.*, **46**, 500R (1974).
3. D. M. Hercules, Ed., *Fluorescence and Phosphorescence Analysis*, Wiley-Interscience, New York, 1966.
4. C. A. Parker, *Photoluminscence of Solutions*, Elsevier, Amsterdam, 1968.
5. J. D. Winefordner, S. G. Shulman, and T. C. O'Haver, in *Chemical Analysis*, P. J. Elving and I. M. Klothoff, Eds., Vol. 38, Wiley-Interscience, New York, 1972.
6. M. Zander, *Phosphorimetry*, T. H. Goodwin, Transl., Academic Press, New York, 1968.
7. H. M. Stevens, *Anal. Chim. Acta*, **20**, 389 (1959).
8. M. Katyal, *Talanta*, **15**, 95 (1968).
9. G. A. Crosby, R. E. Whan, and R. M. Alire, *J. Chem. Phys.*, **34**, 743 (1961).
10. K. De Armand and L. S. Forster, *Spectrochim. Acta*, **19**, 1393, 1403, 1687 (1963).
11. H. Diehl and J. L. Ellingbo, *Anal. Chem.*, **28**, 882 (1956).
12. D. F. H. Wallach, D. M. Surgenor, J. Soderberg, and E. Delano, *Anal. Chem.*, **31**, 456 (1959).
13. J. B. Hill, *Clin. Chem.*, **11**, 122 (1965).
14. I. Landa and J. C. Kremen, *Anal. Chem.*, **46**, 1694 (1974).
15. C. W. Sill, *Anal. Chem.*, **33**, 1584 (1961).
16. J. D. Winefordner and P. A. St. John, *Anal. Chem.*, **35**, 2211 (1963).
17. R. J. Lukasiewicz, P. A. Rozynes, L. B. Sanders, and J. D. Winefordner, *Anal. Chem.*, **44**, 237 (1972).
18. C. A. Parker and W. T. Rees, *Analyst*, **85**, 587 (1960); **87**, 83 (1967).
19. J. B. Birks and I. H. Munro, *Progress in Reaction Kinetics*, Vol. 4, Pergamon Press, Oxford, 1967, p. 239.
20. T. Förster and K. Kasper, *Z. Electrochem.*, **59**, 976 (1955).
21. B. Stevens and E. Hutton, *Nature*, **186**, 1045 (1960).
22. B. L. Funt and E. Neparko, *J. Phys. Chem.*, **60**, 257 (1956).
23. W. R. Ware, *J. Phys. Chem.*, **66**, 445 (1962).
24. M. T. El-Ghamry, W. Frei, and G. W. Higgs, *Anal Chim. Acta*, **47**, 41 (1969).
25. D. E. Ryan and B. K. Pal, *Anal Chim. Acta*, **44**, 385 (1969).
26. M. P. Grigor'eva, E. N. Stepanova, and G. A. Sapozhnikova, *Vop. Pitan.* **28** (3), 65 (1969).
27. S. S. Brown and D. C. Williams, *Anal. Brochem.*, **11**, 199 (1965).
28. J. P. F. DeAlbinati, *An Asoc. Quim. Argent*, **53**, 61 (1965).

29. T. S. West, in *Trace Characterization—Chemical and Physical*, W. W. Meinke and B. F. Scribner, Eds., NBS Monograph 100, U. S. Govt. Printing Office, Washington, D. C., 1967.

30. R. M. Dagnell, R. Smith, and T. S. West, *Talanta*, **13**, 609 (1966).

31. C. E. White, H. C. E. McFarlane, J. Fugt, and R. Fuchs, *Anal. Chem.*, **39**, 367 (1967).

32. M. A. Konstantinova-Shlezinger, *Fluorimetric Analysis*, N. Kaner, Transl., Israel Program for Scientific Translations, Jerusalem, 1965.

33. J. Marienko and I. May, *Anal. Chem.*, **40**, 1137 (1968).

34. V. N. Podchainova, L. V. Skornyakova, and B. L. Dvinyaninoy, *Izv. Vyssh. Ucheb. Zaved. Khim. Khim. Tekhnol.*, **11**, 241 (1968).

35. C. E. White and R. J. Argauer, *Fluorescence Analysis—A Practical Approach*, Marcel Dekker, New York, 1970.

36. M. Marcantonatos, G. Gamba, and D. Monnier, *Helv. Chim. Acta*, **52**, 538 (1969).

37. A. Holme, *Acta Chem. Scand.*, **21**, 1679 (1967).

38. G. F. Kirkbright and W. I. Stephen, *Anal. Chim. Acta*, **32**, 544 (1965).

39. R. Sivori and A. H. Guerro, *An. Asoc. Quim. Argent.*, **55**, (3-4), 157 (1967).

40. G. G. Guilbault and D. N. Kramer, *Anal. Chem.*, **37**, 918 (1965).

41. G. G. Guilbaut and D. N. Kramer, *Anal. Chem.*, **37**, 1395 (1965).

42. A. Weissler and C. E. White, in *Handbook of Analytical Chemistry*, L. Meites, Ed., McGraw-Hill, New York, 1963.

43. G. F. Kirkbright, T. S. West, and C. Woodward, *Anal. Chim. Acta*, **36**, 298 (1966).

44. H. Diehl, *Calcein Calmagite, and 0,0-Dihydroxyazobenzene: Titrametric, Colorimetric, and Fluorometric Reagents for Calcium and Magnesium*, G. F. Smith Chemical, Columbus, Ohio, 1964.

45. B. Budesinsky and T. S. West, *Talanta*, **16**, 399 (1969).

46. E. A. Bozhevol'nev, L. F. Fedorova, I. A. Krasavin, and V. M. Dziomko, *J. Anal. Chem. USSR*, **24**, 399 (1969).

47. D. E. Ryan, A. E. Pitts, and R. M. Cassidy, *Anal. Chim. Acta*, **34**, 491 (1966).
48. C. Huu, A. I. Volkova, and T. E. Get'man, *Zh. Anal. Chim.*, **24**, 688 (1969).

49. A. K. Babko, A. V. Terletskaya, and. L. I. Dubovenko, *Ukr. Khim. Zh.*, **32**, 728 (1966).

50. E. A. Bozhevol'nov and S. U. Kreingold, *Tr. Vses. Nauchn. Issled. Inst. Khim. Reakt. Osab. Chist. Khim. Veshchestv.*, **26**, 204 (1966).

51. G. H. Schenk, K. P. Dilloway, and J. S. Coulter, *Anal. Chem.*, **41**, 510 (1969).

52. J. B. Allred and D. G. Guy, *Anal. Biochem.*, **29**, 293 (1969).

53. D. A. Britton and J. C. Guyon, *Anal. Chim. Acta*, **44**, 397 (1967).

54. Y. Yamane, Y. Yamada, and S. Kunihiro, *Bunseki Kagaku*, **17**, 973 (1968).

55. K. Ritchie and J. Harris, *Anal. Chem.*, **41**, 163 (1969).

56. O. I. Komley and V. K. Zinchuk, *Visn. l'viv. Univ. Ser. Khim.*, **1967** (9), 50.

57. A. S. Burchett, *Dissertation Abstr.*, **B27**, 1384 (1966).

58. D. B. Land and S. M. Edmonds, *Mikrochim. Acta*, **1966**, 1013.

59. D. W. Schultz, J. V. Passoneau, and O. H. Lowry, *Anal. Biochem.*, **19**, 300 (1967).

60. E. C. Stanley, B. I. Kinneberg, and L. P. Varga, *Anal. Chem.*, **38**, 1362 (1966).

61. H. D. Axelrod, J. H. Cary, J. E. Bonelli, and J. P. Lodge, Jr., *Anal. Chem.*, **41**, 1856 (1969).

62. A. Gruenert and G. Toelg, *Talanta*, **18**, 881 (1971).

63. A. I. Ivankova and D. P. Shcherbov, *Issled. Razrab. Fotometrich. Metod Opred. Mikrokolichestv, Elem. Miner. Syr'e* **1967**, 138.

64. V. A. Nazarenko and V. P. Antonvich, *J. Anal, Chem. USSR*, **24**, 254 (1969).

65. G. F. Kirkbright, T. S. West, and C. Woodward, *Analyst*, **91**, 23 (1966).

66. M. Costa, *Rev. Port. Quim.*, **8** (3), 136 (1966).

67. R. C. Ewan, C. A. Baumann, and A. L. Pope, *J. Agr. Food. Chem.*, **16** (2), 212 (1968).

68. G. Elliot and J. A. Radley, *Analyst*, **33**, 1623 (1961).

69. G. Alberti and M. A. Massucci, *Anal. Chem.*, **38**, 214 (1966).

70. E. Butter, U. Kolowos, and H. Holzapfel, *Talanta*, **15**, 901 (1968).

71. R. M. Dagnall, R. Smith, and T. S. West, *Analyst*, **92**, 358 (1967).

72. A. K. Babko, T. H. Chan, A. I. Volkova, and T. E. Gef'man, *Ukr. Khim. Zh.*, **35**, 642 (1969).

73. E. A. Solov'ev, A. P. Golovina, E. A. Bozhevlo'nov, and I. M. Plotnikova, *Vestn. Mosk. Gos. Univ., Ser. Khim.* **1966** (5), 89.

74. M. U. Belyi and I. Ya. Kushnirenko, *Zh. Prikl. Spektrosk.*, **9**, 272 (1968).

75. A. I. Krillov, R. S. Lauer, and N. J. Poluektov, *J. Anal. Chem. USSR*, **22**, 1123 (1967).

76. E. Sawicki and R. A. Carnes, *Anal. Chim. Acta*, **41**, 178 (1968).

77. R. R. Trenholm and D. E. Ryan, *Anal. Chim. Acta*, **32**, 317 (1965).

78. E. R. Jensen and R. T. Pflaum, *Anal. Chem.*, **38**, 1268 (1966).

79. L. S. Bark and A. Rixon, *Anal. Chim. Acta*, **45**, 425 (1969).

80. E. A. Bozhevol'nov, *Oesterr. Chemik.-Z.* **66**, 74 (1965).

81. V. Skramovsky and P. Heaberle, *Clin. Chim. Acta*, **22**, 161 (1968).

82. M. Laamaa, M. L. Allaalu, and H. Kokk, *Tartu Riikliku Ulti Kaali Toim.*, **219**, 199 (1968).

83. A. K. Babko and I. E. Kalinichenko, *Ukr. Khim. Zh.*, **31**, 1316 (1965).

84. M. A. Matveets and D. P. Shcherbov, *Issled. Razrab. Fotometrich. Metod*

Opred. Mikrokolichestv. Elem. Miner. Syr'e, **1967**, 122.

85. A. M. Lukin, O. A. Efremenko, and G. S Petrova, *J. Anal. Chem. USSR*, **22**, 1040 (1967).

86. A. Brookes and A. Townsend, *Chem. Commun.*, **24**, 1660 (1968).

87. H. Imai, *Nippon Kagaka Zasshi*, **90**, 275 (1969).

88. A. Bordea, *Bul. Inst. Politch. Iasi.*, **13**, 209 (1967).

89. E. P. Mulikovskaya, *Nov. Metody Anal. Khim. Sustava Podzemn. Vod.* **1967**, 78.

90. D. W. Fink and W. E. Ohnesorge, *Anal. Chem.*, **41**, 39 (1969).

91. A. E. Pitts and D. E. Ryan, *Anal. Chim. Acta*, **37**, 460 (1967).

92. S. J. Weisman, *J. Chem. Phys.*, **10**, 214 (1942).

93. R. M. Dagnall, R. Smith, and T. S. West, *Analyst*, **92**, 20 (1967).

94. B. K. Pal and D. E. Ryan, *Anal. Chim. Acta*, **47**, 35 (1969).

95. G. F. Kirkbright, T. S. West, and C. Woodward, *Talanta*, **13**, 1637 (1966).

96. F. Will, *Anal. Chem.*, **33**, 1360 (1961).

97. K. P. Stolyarov, N. N. Grigor'ev. and G. A. Khomenok, *Vestn. Leningrad. Univ., Fiz., Khim.* **1972** (4), 120; *Chem. Abstr.* **1973**, 78, 79233v.

98. J. R. A. Anderson and S. L. Lowy, *Anal. Chim. Acta.*, **15**, 246 (1956).

99. A. A. Schilt, *Analytical Application of* 1, 10-*phenanthroline and Related Compounds*, Pergamon Press, Oxford, 1969.

100. S. Udenfriend, Fluorescence Assay in Biology and Medicine, Vol. II, Academic Press, New York, 1969.

101. J. S. Hawker, R. M. Gamson, and H. Klapper, *Anal. Chem.*, **29**, 879 (1957).

102. D. Monnier, C. A. Menzinger, and M. Marcantonatos, *Anal. Chim. Acta*, **60**, 233 (1972).

103. R. A. Moore, *Proc. Soc. Anal. Chem.*, **9**, 35 (1972).

104. H. Diehl, R. Olsen, G. I. Spielholtz, and R. Jensen, *Anal. Chem.*, **35**, 1238 (1963).

105. P. V. Kristalev and Ya. F. Shevchenko, *Sb. Nauch. T. Perm. Politekh. Inst.*, **71**, 38 (1970).

106. T. Ito and A. Murata, *Jap. Anal.*, **20**, 335 (1971); *Z. Anal. Chem.*, **256**, 303 (1971).

107. Sh. T. Talipov, L. E. Zel'tsen, and A. T. Tashkhodzhaev, *Izv. Vyssh. Ucheb. Zaved. Khim. Tekhnol.*, **16** (2), 299 (1973); *Chem. Abstr.*, **78**, 168179z (1973).

108. D. P. Shcherbov and D. N. Lisitsyna, *Zavod. Lab.*, **39** (6), 656 (1973); *Chem. Abstr.*, **79**, 111397k (1973).

109. B. Liebich, D. Monnier, and M. Marcantonatos, *Anal. Chim. Acta*, **52**, 305 (1970).

110. T. Uno and H. Taniguchi, *Bunseki Kagaku*, **20**, 1123 (1971); *Chem. Abstr.* **75**, 147470k (1971).

111. N. K. Podberezskaya, D. P. Shcherbov, E. A. Shilenko, and V. A. Sushkova, *Issled. Tsvel. Fluoreslsent. Reakts., Opred. Blagorod. Metal.*, **1969**, 108.

112. L. Coutelier, *Experientia*, **29**, 192 (1973).

113. K. Hiraki, *Bull. Chem. Soc. Jap.*, **45**, 789 (1972).

114. N. A. Raju and G. C. Rao, *Nature*, **174**, 400 (1954).

115. D. P. Shcherbov, R. N. Plotnikova, and I. N. Astaf'eva, *Zavod. Lab.*, **36**, 528 (1970); *Chem Abstr.*, **73**, 83538n (1970).

116. B. W. Bailey, R. M. Dagnall, and T. S. West, *Talanta*, **13**, 1661 (1966).

117. L. Ben-Dor, and E. Jungreis, *Isr. J. Chem.*, **8**, 951 (1970).

118. Z. Holzbecher and P. Holler *Collect. Czech. Chem. Commun.*, **37**, 2557 (1972).

119. C. T. Hjou, A. I. Volkova, and T. E. Get'man, *J. Anal. Chem. USSR*, **24**, 534 (1969).

120. S. B. Zamochnick and G. A. Rechnitz, *Z. Anal. Chem.*, **199**, 424 (1964).

121. K. Hiraki, *Bull. Chem. Soc. Jap.*, **46**, 2438 (1973).

122. Z. Holzbecher and K. Volka, *Collect Czech. Chem. Commun.*, **35**, 2925 (1970).

123. T. D. Filer, *Anal. Chem.*, **43**, 1753 (1971).

124. T. D. Filer, *Anal. Chem.*, **43**, 725 (1971).

NUCLEAR METHODS

M. L. PARSONS

Department of Chemistry
Arizona State University
Tempe, Arizona

A. ACTIVATION ANALYSIS (1–6)

Nuclear methods can be broadly divided into two types: (a) *activation analysis* in which a sample is irradiated with suitable particles to produce radioactive or excited states of elements in the sample; and (b) *radioactive tracer methods* in which radioactive tracers for an element in a compound are utilized.

1. NUCLEAR INTERACTIONS (1–6)

Simple nuclear interactions may be represented:

$$A + a \rightarrow A + a \tag{1}$$

$$A + a \rightarrow A^* + a \tag{2}$$

$$A + a \rightarrow B + b \tag{3}$$

where A and B are the target or residual nuclei, A^* is an excited species, and a and b are the bombarding or emitted particles or photons. Processes 1 and 2 are elastic and inelastic scattering types, respectively. In process 1, the kinetic energy (KE) of the reactants and products is the same, whereas in process 2 the KE of the products is less than that of the reactants; the difference in energy is used to raise the nucleus A to an excited state. Process 3 is the transmutation of element A to another element B with one or more particles or γ-photons being emitted; the nucleus B can be either in the ground state or in an excited state, the latter decaying rapidly to the ground state with emission of γ-photons; also, B may be unstable and decay to the ground or excited states of another nucleus, for example, C, by emission of α- or β-radiation. The lifetime of the excited species, if very short, results in prompt emission and, if long, results in delayed emission.

By use of energy balance relations as calculated from the masses of reactants and products, it is possible to determine whether the reaction is exoergic or endoergic. Exoergic reactions can in principle, occur even if the particle has zero energy, while endoergic reactions require an amount of KE from the incoming particle to balance the loss of energy in the reaction. Of course, the energy of the incoming particles must actually be greater because of the bombarding particle imparting momentum to the compound (reaction) nucleus.

The general principles (also see Chapter 4) of activation analysis have not changed since they were formulated in a few key papers of the late 1940s and early 1950s. The basic equation of activation analysis is:

$$R_0 = \epsilon \sigma_Q \rho w_Q \phi \frac{N_A}{M_Q} (1 - e^{-\lambda t_i}) \tag{4}$$

where σ_Q = activation cross section for nuclear reaction of concern (in m^2)

ϵ = fraction of the total counting rate in the measured photopeak (dimensionless)

ρ = fractional abundance of particular isotope of element concerned (dimensionless)

w_Q = weight of element Q irradiated (in kg)

ϕ = flux of particles used in irradiation (in $m^{-2} s^{-1}$)

M_Q = atomic weight of element Q concerned (in $kg\ mol^{-1}$)

N_A = Avogadro's number (in mol^{-1})

λ = decay constant of induced radionuclide (in s^{-1})

t_i = irradiation time (in s)

and R_0 is the induced activity* in the product R (counting rate) due to the bombarding particles (in disintegrations s^{-1}), evaluated at the termination of irradiation. Following the termination of irradiation, the induced activity decays with a characteristic half-life decreasing by a factor of 2 for each half-life period, $t_{1/2,R}(\lambda = 1/t_{1/2,R})$, that is,

$$R_t = R_0 e^{-\lambda t_a} \tag{5}$$

where t_a is the time after termination of irradiation.

There are several limiting cases of equation 4. For example, the term $1 - e^{-\lambda t_i}$ reaches a value of unity for all practical purposes after an irradiation time of $10/t_{1/2R}$, where $t_{1/2R}$ is the half-life of the radioactive species of concern R, and so R_0 becomes:

$$R_{0(t_i \, \widetilde{>} \, 10t_{1/2,R})} = \epsilon \sigma_Q \rho w_Q \phi \frac{N_A}{M_Q} \tag{6}$$

*The term R is used for activity (count rate) rather than A.

If the irradiative time t_i is much shorter than $t_{1/2,R}$ $(t_i \stackrel{\sim}{<} \frac{1}{10} t_{1/2,R})$,

$$R_0\left(t_i \stackrel{\sim}{<} \frac{1}{10} t_{1/2}\right) = \epsilon \sigma_Q \rho w_Q \phi \frac{N_A}{M_Q} \lambda t_i \tag{7}$$

Generally, in applying activation analysis to quantitative analysis, the above equations are not utilized directly, but rather the sample and a known mass of standard are irradiated under identical experimental conditions, so that a simple comparison of the induced activity gives the mass of the element to be determined, that is,

$$\frac{R_s}{R_r} = \frac{w_s}{w_r} \tag{8}$$

where the subscripts s and r refer to the sample and the reference (standard).

Most commonly, irradiated samples are measured shortly after the irradiation (by removal of the irradiated sample and measurement of its activity with a γ-ray spectrometer), and as long as $t_a < t_{1/2,R}$, $R_t = R$. However, if the half-life is very short $(t_{1/2,R} < 1$ ms and even as small as 1 ns), it is necessary to measure the prompt emission (γ-photons) during the irradiation. For this latter case, where "prompt counting" is carried out, the number of measured counts C in a counting time of t_c is

$$C = K\epsilon \sigma_Q \rho w_Q t_c \frac{N_A}{M_Q} \tag{9}$$

where K is a constant accounting for the efficiency and geometry of the detector and the proportion of the decaying species measured. Even though, it seems that C could be made finite even if w_Q is extremely small by simply increasing t_c; unfortunately, a lower limit of C is imposed by background and matrix interferences. Finally, by pulsed irradiation methods, it is possible to improve considerably the ratio of analyte counts to fluctuation background counts (the signal-to-noise ratio).

2. EXPERIMENTAL METHOD

The experimental method in a typical activation analysis (1) can be summarized:

1. Irradiate weighed sample and standard in suitable containers for sufficient time t_i to give adequate R_0.
2. After irradiation, dissolve the samples and standards and add a

known weight of the element being determined as a carrier for the small amount of analyte irradiated.

3. Treat the sample in such a way that the carrier and analyte element are in the same chemical form.

4. Perform chemical separations to isolate the element for a suitable compound free from all other radionuclides.

5. Determine the chemical yield by any conventional method (this step makes unnecessary 100% separation in step (4)), because the final measured activity is corrected for losses via the chemical yield measure.

6. Compare the activities of the sample and standard under identical counting conditions, making corrections where needed for self-absorption dead-time losses, decay, and so on.

7. Determine the radiochemical purity of the isolated active compound by measuring the (a) rate of decay from which $t_{1/2}$ can be determined, (b) activity through various thicknesses of aluminum for which the energy of the rays or particles can be determined, and (c) γ-ray energy, using either a scintillation counter or a solid-state γ-ray spectrometer.

Under favorable conditions, steps 2 to 5 can be omitted. The mass of the unknown can then be determined from equation 8, that is,

$$w_s = \frac{R_s}{R_r} \times w_r \tag{10}$$

If the activation fluxes for the standard and reference differ, equation 10 must be changed by multiplying $(R_s/R_r) \times C_R$ by ϕ_R/ϕ_s and, if charged particle activation is used, it is also necessary to multiply by r_r/r_s, where the ϕ's are the fluxes, the r's are the particle ranges, and the subscripts r and s refer to the standard (reference) and sample, respectively.

3. ESTIMATION OF SENSITIVITY AND LIMIT OF DETECTION

The sensitivity of measurement m (the slope of the analytical curve of C versus w) is given by:

$$m = \frac{C}{w} = \frac{KR_t t_c}{w} = \frac{K\epsilon\sigma\rho\phi N_A t_c (1 - e^{-\lambda t_i}) e^{-\lambda t_a}}{M} \tag{11}$$

and if $t_i \gtrsim 10 t_{1/2}$ and $t_a \lesssim \frac{1}{10} t_{1/2}$,

$$m = \frac{C}{w} = \frac{K\epsilon\rho\phi N_A t_c}{M} \tag{12}$$

For example, for ^{56}Mn [produced via ^{55}Mn(n, γ) ^{56}Mn] for 1 μg of ^{55}Mn and for a 1 hr irradiation with a thermal neutron flux of $\phi = 10^{17}$ neutrons m^{-2} s^{-1} and a cross section of $\sigma = 13.3 \times 10^{-28}$ m^2, a half-life $t_{1/2}$ of 2.56 hr, an immediate counting of the irradiation, that is, $t_a = 0$ s, a counting time of 100 s, and a 10% efficiency of counting ($K \approx 0.10$) result in an activity R_0 of 3.43×10^5 s^{-1} and a slope of 3.43×10^{15} s^{-1} kg^{-1}. If we assume a detectable number of counts to be 100 s^{-1}, the limiting detectable amount w_L is found to be 3×10^{-14} kg or 3×10^{-11} g or 30 pg. If the sample weight is assumed to be 3 g, the limiting detectable concentration c_L is 10^{-5} ppm (under the most favorable conditions). For long-lived isotopes, such as ^{59}Fe and ^{60}Co, an irradiation time of a few hours is too short to obtain optimal sensitivity and detection limits. For some elements, such as sodium, only one isotope (^{23}Na) is important, but for other elements, such as iron, the most abundant isotope (^{56}Fe) which has an abundance of 92% cannot be activated by thermal neutrons, whereas, Fe which neutron activation is based on is only present to ~0.3% in natural iron, and so the detection limit for ^{58}Fe is quite poor, for example, ~10 μg.

In Table 8.1, limits of detection estimated from equation 11 (with the assumptions listed at the end of the table) are given for elements activated by thermal neutrons; also, the atomic number, stable isotopes, abundance of stable isotopes, type of activation reaction, half-lives of the various reaction products, and potential primary and secondary interferences are given. In Table 8.2 similar information and detection limits are given for elements activated by fast neutrons.

4. MEANS OF IRRADIATION

a. Slow (Thermal) Neutrons

The most generally useful (1, 6) and readily available source for high sensitivity in the determination of trace elements is certainly the *nuclear reactor* with its abundant flux of *slow (thermal) neutrons* (10^{15} to 10^{18} neutrons m^{-2} s^{-1}). The major advantage of slow neutrons (reactors contain thermal neutrons with energies from 0.02 eV to fast neutrons with energies exceeding 10 MeV) is that usually only one nuclear reaction, namely, R(n, γ)P, occurs, and the cross section for the reaction is generally quite high. Although no special effort is made to select the energy range of the neutrons, it is essential to have a knowledge of the energy spectrum of the neutrons for irradiation of the sample and to introduce and remove samples from the reactor reproducibly. In Table 8.1, calculated detection limits for elements irradiated by thermal neutrons are given; the most

TABLE 8.1A. Estimated Limits of Detection for Thermal Neutron Activation [(n,γ) Reactions][a,b]

Element	Atomic number	Isotopes	Abundance (%)	Reaction (1, 2)	Half-Life of Reaction Product (2)	Potential Interfering Elements Primary	Secondary	Estimated Detection Limits $\mu g(2, 3)$
Ac	89	No stable isotopes	—	—	—	—	—	—
Al	13	^{27}Al	100	^{27}Al(n,γ)^{28}Al	2.31 min	^{28}Si, ^{31}P	^{26}Mg	0.02
Sb	51	^{121}Sb	57.25	^{121}Sb(n,γ)^{122}Sb	3.3 min, 2.75 days	^{122}Te	—	0.005
		^{123}Sb	42.75	^{123}Sb(n,γ)^{124}Sb	21 min, 1. 3 min, 60.1 days	^{124}Te, ^{127}I	^{122}Sn	0.1
Ar	18	^{36}Ar	0.337	^{36}Ar(n,γ)^{37}Ar	34.3 days	^{40}Ca	—	0.2
		^{38}Ar	0.063	—	—	—	—	—
		^{40}Ar	99.600	^{40}Ar(n,γ)^{41}Ar	1.827 hr	^{41}K, ^{44}Ca	—	0.01
				^{41}Ar(n,γ)^{42}Ar	32.9 yr	—	—	—
As	33	^{75}As	100	^{75}As(n,γ)^{76}As	26.8 hr	^{76}Se, ^{79}Br	^{74}Ge	0.002
At	85	No stable isotopes	—	—	—	—	—	—
Ba	56	^{130}Ba	0.101	^{130}Ba(n,γ)^{131}Ba	11.52 days	—	—	2.0
		^{132}Ba	0.097	—	—	—	—	—
		^{134}Ba	2.42	—	—	—	—	—
		^{135}Ba	6.59	—	—	—	—	—
		^{136}Ba	7.81	—	—	—	—	—
		^{137}Ba	11.32	—	—	—	—	—
		^{138}Ba	71.66	^{138}Ba(n,γ)^{139}Ba	82.9 min	^{139}La, ^{142}Ce	—	0.04

Element	Z	Isotope	% Abundance	Reaction	Half-life			Sensitivity
				$^{139}Ba(n,\gamma)^{140}Ba$	12.8 days	—	—	—
				$^{140}Ba(n,\gamma)^{141}Ba$	18.3 min	—	—	—
Be	4	^{9}Be	100	^{10}Be Too stable for analyses	2.7×10^{6} yr	None apparent	—	—
Bi	83	^{209}Bi	100	$^{209}Bi(n,\gamma)^{210}Bi$	5 days	—	—	1
B	5	^{10}B	19.61	$^{10}B(n,\alpha)^{7}Li$	Stable (α-counting used)	^{6}Li	—	—
		^{11}B	80.39	—		—	—	—
Br	35	^{79}Br	50.537	$^{79}Br(n,\gamma)^{80}Br$	4.37 hr, 17.55 min	^{80}Kr	^{80}Se	0.002
		^{81}Br	49.463	$^{81}Br(n,\gamma)^{82}Br$	35.344 hr	$^{82}Kr, ^{85}Rb$	—	0.007
Cd	48	^{106}Cd	1.24	—	—	—	—	—
		^{108}Cd	0.87	—	—	—	—	—
		^{110}Cd	12.32	—	—	—	—	—
		^{111}Cd	12.67	—	—	—	—	—
		^{112}Cd	24.15	—	—	—	—	—
		^{113}Cd	12.21	—	—	—	—	—
		^{114}Cd	28.93	$^{114}Cd(n,\gamma)^{115}Cd$	43 days, 2.2 days	$^{115}In, ^{118}Sn$	—	0.05
		^{116}Cd	7.61	$^{116}Cd(n,\gamma)^{117}Cd$	3.0 hr	^{120}Sn	—	0.1
Ca	20	^{40}Ca	96.97	—	—	—	—	—
		^{42}Ca	0.64	—	—	—	—	—
		^{43}Ca	0.145	—	—	—	—	—
		^{44}Ca	2.06	$^{44}Ca(n,\gamma)^{45}Ca$	161.4 days	$^{45}Sc, ^{48}Ti$	—	7
		^{46}Ca	0.0033	—	—	—	—	—
		^{48}Ca	0.185	$^{48}Ca(n,\gamma)^{49}Ca$	8.75 min	—	—	3

TABLE 8.1A. (*Continued*)

Element	Atomic number	Isotopes	Abundance (%)	Reaction (1, 2)	Half-Life of Reaction Product (2)	Potential Interfering Elements Primary	Secondary	Estimated Detection Limits μg(2, 3)
C	6	^{12}C	98.893	—	—	—	—	—
			1.107}	^{13}C(n,γ)^{14}C	5685 yr	^{14}N, ^{17}O	—	1700
				^{14}C(n,γ)^{15}C	2.48 s			
Ce	58	^{136}Ce	0.193	—	—	—	—	—
		^{138}Ce	0.250					—
		^{140}Ce	88.48	^{140}Ce(n,γ)^{141}Ce	32.5 days	^{141}Pr, ^{144}Nd	^{139}La	0.007
		^{142}Ce	11.07	^{142}Ce(n,γ)^{143}Ce	33.4 hr	^{146}Nd	—	0.2
				^{143}Ce(n,γ)^{144}Ce	284.3 days	—	—	—
Cs	55	^{133}Cs	100	^{133}Cs(n,γ)^{134}Cs	3.15 hr, 2.05 yr	^{134}Ba	—	0.01
Cl	17	^{35}Cl	75.53	—	—	—	—	—
		^{37}Cl	24.47	^{37}Cl(n,γ)^{38}Cl	37.12 min	^{38}Ar, ^{41}K	^{36}S	0.04
Cr	24	^{50}Cr	4.31	^{50}Cr(n,γ)^{51}Cr	27.8 days	^{54}Fe	—	0.005
		^{52}Cr	83.76					—
		^{53}Cr	9.55					
		^{54}Cr	2.33	^{54}Cr(n,γ)^{55}Cr	3.52 min	^{55}Mn, ^{58}Fe	—	0.8
Co	27	^{59}Co	100	^{59}Co(n,γ)^{60}Co	10.47 min, 5.26 yr	^{60}Ni, ^{63}Cu	^{60}Co	0.0005
Cu	29	^{63}Cu	69.09	^{63}Cu(n,γ)^{64}Cu	12.88 hr	^{64}Zn	^{62}Ni	0.003
		^{65}Cu	30.91					—

Element	Z	Isotope	Abundance	Reaction	Half-life			
Dy	66	^{156}Dy	0.0524					
		^{158}Dy	0.0902					
		^{160}Dy	2.294					
		^{161}Dy	18.9					
		^{162}Dy	25.53					
		^{163}Dy	24.97					
		^{164}Dy	28.18	^{164}Dy$(n, \gamma)^{165}$Dy	1.25 min, 139 min	^{165}Ho, ^{168}Er		0.002
Er	68	^{162}Er	0.136					
		^{164}Er	1.56					
		^{166}Er	33.41					
		^{167}Er	22.94					
		^{168}Er	27.07	^{168}Er$(n, \gamma)^{169}$Er	9.6 days	^{169}Tm, ^{172}Yb		0.03
		^{170}Er	14.88	^{170}Er$(n, \gamma)^{171}$Er	7.52 hr	^{174}Yb		0.02
Eu	63	^{151}Eu	47.82	^{151}Eu$(n, \gamma)^{152}$Eu	9.2 hr	^{152}Gd	^{152}Eu	0.00003
		^{153}Eu	52.18	^{153}Eu$(n, \gamma)^{154}$Eu	16 yr			
				^{154}Eu$(n, \gamma)^{155}$Eu	1.811 yr			
				^{155}Eu$(n, \gamma)^{156}$Eu	14 days			0.04
F	9	^{19}F	100	^{19}F$(n, \gamma)^{20}$F	11.36 s	^{23}Na, ^{20}Ne		0.3
Fr	87	No stable isotopes	—					
Gd	64	^{152}Gd	0.205					
		^{154}Gd	2.23					
		^{155}Gd	15.10					
		^{156}Gd	20.60					
		^{157}Gd	15.70					

TABLE 8.1A. (Continued)

Element	Atomic number	Isotopes	Abundance (%)	Reaction (1, 2)	Half-Life of Reaction Product (2)	Potential Interfering Elements Primary	Potential Interfering Elements Secondary	Estimated Detection Limits $\mu g(2, 3)$
		^{158}Gd	24.50	^{158}Gd$(n, \gamma)^{159}$Gd	18.0 hr	^{159}Tb, ^{162}Dy	—	0.02
		^{160}Gd	21.60	^{160}Gd$(n, \gamma)^{161}$Gd	3.73 min	^{164}Dy	—	0.1
Ga	31	^{69}Ga	60.4	^{69}Ga$(n, \gamma)^{70}$Ga	21.37 min	^{70}Ge	^{68}Zn	0.009
		^{71}Ga	39.6	^{71}Ga$(n, \gamma)^{72}$Ga	14.12 hr	^{72}Ge, ^{75}As	^{70}Zn, ^{70}Ge	0.005
Ge	32	^{70}Ge	20.52	—	—	—	—	—
		^{72}Ge	27.43	—	—	—	—	—
		^{73}Ge	7.77					
		^{74}Ge	36.51	^{74}Ge$(n, \gamma)^{75}$Ge	49 s, 79 min	^{75}As, ^{78}Se	—	0.05
		^{76}Ge	7.76	^{76}Ge$(n, \gamma)^{77}$Ge	53.6 s, 11.3 hr	^{80}Se	—	0.4
Au	79	^{197}Au	100	^{197}Au$(n, \gamma)^{198}$Au	2.7 days	^{198}Hg	^{196}Pt, ^{198}Au	0.0003
				^{199}Au$(n, \gamma)^{200}$Au	48.4 min			—
Hf	72	^{174}Hf	0.18	^{174}Hf$(n, \gamma)^{175}$Hf	70 days	—	—	0.06
		^{176}Hf	5.20					
		^{177}Hf	18.50	—	—	—	—	—
		^{178}Hf	27.14	^{178}Hf$(n, \gamma)^{179}$Hf	19 s	—	—	—
		^{179}Hf	13.75	^{179}Hf$(n, \gamma)^{180}$Hf	5.5 hr	^{182}W	—	—
		^{180}Hf	35.24	^{180}Hf$(n, \gamma)^{181}$Hf	42.4 days	^{181}Ta, ^{184}W	—	0.03
He	2	^{3}He	0.000137	Product stable	—	—	—	—

Element	Z	Isotope	Abundance (%)	Reaction	Half-life	Product	Product	Cross section
		^{4}He	99.999863	No reaction	—	—	—	—
Ho	67	^{165}Ho	100	^{165}Ho$(n,\gamma)^{166}$Ho	27.3 hr	^{166}Er, ^{169}Tm	^{164}Dy	0.0003
H	1	^{1}H	99.985	—	—	—	—	—
		^{2}H	0.015	^{2}H$(n,\gamma)^{3}$H	12.262 yr	^{6}Li, ^{3}He	—	3×10^{3}
In	49	^{113}In	4.28	^{113}In$(n,\gamma)^{114}$In	1.2 min, 50 days	^{114}Sn	—	0.1
		^{115}In	95.72	^{115}In$(n,\gamma)^{116}$In	13.4 s, 54 min	^{116}Sn	—	0.03
I	53	^{127}I	100	^{127}I$(n,\gamma)^{128}$I	25 min	^{128}Xe	—	0.003
Ir	77	^{191}Ir	37.3	^{191}Ir$(n,\gamma)^{192}$Ir	1.45 min, 74.4 days	^{192}Pt	^{190}Os	0.002
		^{193}Ir	62.7	^{193}Ir$(n,\gamma)^{194}$Ir	19.7 hr	—	—	0.0001
Fe	26	^{54}Fe	5.84	—	—	—	—	—
		^{56}Fe	91.68	—	—	—	—	—
		^{57}Fe	2.17	—	—	—	—	—
		^{58}Fe	0.31	^{58}Fe$(n,\gamma)^{59}$Fe	46.5 days	^{59}Co, ^{62}Ni	—	10
Kr	36	^{78}Kr	0.35	^{78}Kr$(n,\gamma)^{79}$Kr	34.92 hr	—	—	1
		^{80}Kr	2.27	—	—	—	—	—
		^{82}Kr	11.56	—	—	—	—	—
		^{83}Kr	11.55	—	—	—	—	—
		^{84}Kr	56.90	^{84}Kr$(n,\gamma)^{85}$Kr	4.5 hr, 10.76 yr	^{85}Rb, ^{88}Sr	—	0.1
		^{86}Kr	17.37	^{86}Kr$(n,\gamma)^{87}$Kr	1.3 hr	^{87}Sr	—	1
				^{87}Kr$(n,\gamma)^{88}$Kr	2.805 hr	—	—	—
La	57	^{138}La	0.089	—	—	—	—	—

TABLE 8.1A. (*Continued*)

Element	Atomic number	Isotopes	Abundance (%)	Reaction (1, 2)	Half-Life of Reaction Product (2)	Potential Interfering Elements Primary	Secondary	Estimated Detection Limits µg(2,3)
		^{139}La	99.911	^{139}La(n,γ)^{140}La	40.27 hr	^{140}Ce	^{138}Ba	0.002
				^{140}La(n,γ)^{141}La	3.85 hr	—	—	—
Pb	82	^{204}Pb	1.37	^{204}Pb(n,2n)^{203}Pb	52.1 hr	—	—	—
		^{206}Pb	25.0	—		—	—	—
		^{207}Pb	21.2	—	—	—	—	—
		^{208}Pb	52.4	^{208}Pb(n,γ)^{209}Pb	3.31 hr	^{209}Bi	—	90
Li	3	^{6}Li	7.42	^{6}Li(n,α)^{3}H	12.262 yr	^{2}H, ^{3}He	—	—
		^{7}Li	92.58	^{7}Li(n,γ)^{8}Li	0.847 s	—	—	0.02
Lu	71	^{175}Lu	97.40	^{175}Lu(n,γ)^{176}Lu	3.71 hr	^{176}Hf	—	—
		^{176}Lu	2.60	^{176}Lu(n,γ)^{177}Lu	6.8 days	^{176}Yb, ^{177}Hf ^{180}Ta, ^{180}Hf	—	0.002
Mg	12	^{24}Mg	78.70			—	—	—
		^{25}Mg	10.13	—	—	—	—	—
		^{26}Mg	11.17	^{26}Mg(n,γ)^{27}Mg	10.0 min	^{27}Al, ^{30}Si	—	1
				^{27}Mg(n,γ)^{28}Mg	20.88 hr			
Mn	25	^{55}Mn	100	^{55}Mn(n,γ)^{56}Mn	2.586 hr	^{56}Fe, ^{59}Co	^{54}Cr	0.0006

290

Element	Z	Isotope	Abundance (%)	Reaction	Half-life	Product		Value
Hg	80	^{196}Hg	0.146	^{196}Hg$(n,\gamma)^{197}$Hg	24 hr, 2.7 days	—	—	0.006
		^{198}Hg	10.02	—	—	—	—	—
		^{199}Hg	16.84	—	—	—	—	—
		^{200}Hg	23.13	—	—	—	—	—
		^{201}Hg	13.22	—	—	—	—	—
		^{202}Hg	29.80	^{202}Hg$(n,\gamma)^{203}$Hg	46.9 days	^{203}Tl, ^{206}Pb	—	0.02
		^{204}Hg	6.85	^{204}Hg$(n,\gamma)^{205}$Hg	5.1 min	^{205}Tl, ^{208}Pb	—	0.9
Mo	42	^{92}Mo	15.84	—	—	—	—	—
		^{94}Mo	9.04	—	—	—	—	—
		^{95}Mo	15.53	—	—	—	—	—
		^{96}Mo	16.53	—	—	—	—	—
		^{97}Mo	9.46	—	—	—	—	—
		^{98}Mo	23.78	^{98}Mo$(n,\gamma)^{99}$Mo	67 hr	^{102}Ru	—	0.1
		^{100}Mo	9.63	—	—	—	—	—
Nd	60	^{142}Nd	27.11	—	—	—	—	—
		^{143}Nd	12.17	—	—	—	—	—
		^{144}Nd	23.85	—	—	—	—	—
		^{145}Nd	8.30	—	—	—	—	—
		^{146}Nd	17.20	^{146}Nd$(n,\gamma)^{147}$Nd	11.06 days	^{150}Sn	—	0.01
		^{148}Nd	5.73	^{148}Nd$(n,\gamma)^{149}$Nd	1.8 hr	^{152}Sn	—	0.1
		^{150}Nd	5.62	—	—	—	—	—
Ne	10	^{20}Ne	90.92	—	—	—	—	—
		^{21}Ne	0.26	—	—	—	—	—
		^{22}Ne	8.82	^{22}Ne$(n,\gamma)^{23}$Ne	37.5 s	^{23}Na, ^{26}Mg	—	1
Np	93	No stable isotopes	—	—	—	—	—	—

TABLE 8.1A. (*Continued*)

Element	Atomic number	Isotopes	Abundance (%)	Reaction (1, 2)	Half-Life of Reaction Product (2)	Potential Interfering Elements Primary	Secondary	Estimated Detection Limits μg(2,3)
Ni	28	58Ni	67.88	58Ni(n,p)58Co	9.2 hr, 71.3 days	—	—	—
		60Ni	26.23	60Ni(n,p)60Co	5.26 yr	—	—	—
		61Ni	1.19	—	—	—	—	—
		62Ni	3.66	—	—	—	—	—
		64Ni	1.08	64Ni(n,γ)65Ni	2.553 hr	65Cu, 68Zn	63Cu	0.4
				65Ni(n,γ)66Ni	54.8 hr	—	—	—
Nb	41	93Nb	100	93Nb(n,γ)94Nb	6.29 min	94Mo	—	0.01
				93Nb(n,α)90Y	64.3 hr	89Y, 90Zr	—	—
N	7	14N	99.6337	—	—	—	—	—
		15N	0.3663	15N(n,γ)16N	7.352 s	—	—	2×10⁴
Os	76	184Os	0.018	—	—	—	—	—
		186Os	1.59	—	—	—	—	—
		187Os	1.64	—	—	—	—	—
		188Os	13.3	—	—	—	—	—
		189Os	16.1	—	—	—	—	—
		190Os	26.4	190Os(n,γ)191Os	14.6 days	191Ir, 194Pt	—	0.003
		192Os	41.0	192Os(n,γ)193Os	31.0 hr	193Ir, 196Pt	193Os	—
				193Os(n,γ)194Os	1.9 yr	—	—	—
O	8	16O	99.59	—	—	—	—	—

		^{17}O	0.037	—	—	—	—	—
		^{18}O	0.204	$^{18}O(n,\gamma)^{19}O$	29.1 s	—	—	6×10^3
Pd	46	^{102}Pd	0.69	—	—	—	—	—
		^{104}Pd	10.97	—	—	—	—	—
		^{105}Pd	22.23	—	—	—	—	—
		^{106}Pd	27.33	—	—	—	—	—
		^{108}Pd	26.71	$^{108}Pd(n,\gamma)^{109}Pd$	4.69 min, 13.45 hr	^{109}Ag, ^{112}Cd	—	0.005
		^{110}Pd	11.81	—	—	—	—	—
P	15	^{31}P	100	$^{31}P(n,\gamma)^{32}P$	14.5 days	^{32}S, ^{35}Cl	^{30}Si	0.04
Pt	78	^{190}Pt	0.0127	—	—	—	—	—
		^{192}Pt	0.78	$^{192}Pt(n,\gamma)^{193}Pt$	4.4 days	^{196}Hg	—	0.04
		^{194}Pt	32.9	$^{194}Pt(n,\gamma)^{195}Pt$	3.5 days	^{198}Hg	^{193}Ir	0.4
		^{195}Pt	33.8					
		^{196}Pt	25.3	$^{196}Pt(n,\gamma)^{197}Pt$	18.0 hr	^{197}Au, ^{200}Hg	—	0.1
		^{198}Pt	7.21	$^{198}Pt(n,\gamma)^{199}Pt$	30 min	—	—	0.1
				$^{199}Pt(n,\gamma)^{200}Pt$	11.5 hr	—	—	—
Pu	94	No stable isotopes	—		—	—	—	—
Po	84	No stable isotopes	—		—	—	—	—
K	19	^{39}K	93.10					
		^{40}K	0.012					
		^{41}K	6.88	$^{41}K(n,\gamma)^{42}K$	12.258 hr	^{42}Ca, ^{45}Sc	^{40}Ar	0.07

TABLE 8.1A. (*Continued*)

Element	Atomic number	Isotopes	Abundance (%)	Reaction (1, 2)	Half-Life of Reaction Product (2)	Potential Interfering Elements Primary	Secondary	Estimated Detection Limits μg(2, 3)
Pr	59	^{141}Pr	100	^{141}Pr(n, γ)^{142}Pr	19.0 hr	^{142}Nd	—	0.002
				^{142}Pr(n, γ)^{143}Pr	13.659 days	—	—	—
				^{143}Pr(n, γ)^{144}Pr	17.27 min	—	—	—
Pm	61	No stable isotopes	—	—	—	—	—	—
Pa	91	No stable isotopes	—	—	—	—	—	—
Ra	88	No stable isotopes	—	—	—	—	—	—
Rn	86	No stable isotopes	—	—	—	—	—	—
Re	75	^{185}Re	37.07	^{185}Re(n, γ)^{186}Re	3.7 days	^{186}Os	—	0.0007
		^{187}Re	62.93	^{187}Re(n, γ)^{188}Re	16.74 hr	^{188}Os, ^{191}Ir	^{186}W	0.0006
				^{188}Re(n, γ)^{189}Re	150 days	—	—	—
Rh	45	^{103}Rh	100	^{103}Rh(n, γ)^{104}Rh	4.41 min, 44 s	^{104}Pd, ^{107}Ag	—	0.001
				^{104}Rh(n, γ)^{105}Rh	45 s, 36.5 hr	—	—	—
Rb	37	^{85}Rb	72.15	^{85}Rb(n, γ)^{86}Rb	18.66 days	^{86}Sr, ^{89}Y	^{89}Kr	0.03

294

Element	Z	Isotope	Abundance (%)	Reaction	Half-life	^{88}Sr	^{86}Kr	
		^{87}Rb	27.85	^{87}Rb(n,γ)^{88}Rb	17.7 hr	—	—	0.3
				^{88}Rb(n,γ)^{89}Rb	14.9 min	—	—	—
Ru	44	^{96}Ru	5.51	—	—	—	—	—
		^{98}Ru	1.87	—	—	—	—	—
		^{99}Ru	12.72	—	—	—	—	—
		^{100}Ru	12.62	—	—	—	—	—
		^{101}Ru	17.07	—	—	—	—	—
		^{102}Ru	31.61	^{102}Ru(n,γ)^{103}Ru	39.4 days	^{103}Rh, ^{106}Pd	—	0.1
		^{104}Ru	18.58	^{104}Ru(n,γ)^{105}Ru	4.44 hr	^{108}Pd	—	0.1
				^{105}Ru(n,γ)^{106}Ru	1 yr	—	—	—
Sm	62	^{144}Sm	3.09	—	—	—	—	—
		^{147}Sm	14.97	—	—	—	—	—
		^{148}Sm	11.24	—	—	—	—	—
		^{149}Sm	13.83	—	—	—	—	—
		^{150}Sm	7.44	—	—	—	—	—
		^{152}Sm	26.72	^{152}Sm(n,γ)^{153}Sm	47.0 hr	^{153}Eu, ^{156}Gd	—	0.0005
		^{154}Sm	22.71	^{154}Sm(n,γ)^{155}Sm	21.9 min	^{158}Gd	—	0.02
Sc	21	^{45}Sc	100	^{45}Sc(n,γ)^{46}Sc	19.5 s, 83.9 days	^{46}Ti	—	0.0006
				^{45}Sc(n,α)^{42}K	12.358 hr	—	—	—
Se	34	^{74}Se	0.87	^{74}Se(n,γ)^{75}Se	120.4 days	^{78}Kr	—	0.5
		^{76}Se	9.02	—	—	—	—	—
		^{77}Se	7.58	—	—	—	—	—
		^{78}Se	23.52	—	—	—	—	—
		^{80}Se	49.82	^{80}Se(n,γ)^{81}Se	62.0 min, 18.6 min	^{81}Br, ^{84}Kr	—	0.04
		^{82}Se	9.19	^{82}Se(n,γ)^{83}Se	69 s, 25 min	^{86}Kr	—	2

TABLE 8.1A. (*Continued*)

Element	Atomic number	Isotopes	Abundance (%)	Reaction (1, 2)	Half-Life of Reaction Product (2)	Potential Interfering Elements Primary	Potential Interfering Elements Secondary	Estimated Detection Limits μg(2,3)
Si	14	^{28}Si	92.21	—	—	—	—	—
		^{29}Si	4.70		—	—	—	
		^{30}Si	3.09	^{30}Si(n,γ)^{31}Si	2.64 hr	^{31}P, ^{34}S	—	0.1
Ag	47	^{107}Ag	51.35	^{107}Ag(n,γ)^{108}Ag	2.42 min	^{108}Cd	—	0.0001
		^{109}Ag	48.65	^{109}Ag(n,γ)^{110}Ag	253 days, 24.5 s	^{110}Cd, ^{113}In	^{108}Pd	0.3
Na	11	^{23}Na	100	^{23}Na(n,γ)^{24}Na	15.05 hr	^{24}Mg, ^{27}Al	—	0.006
Sr	38	^{84}Sr	0.56	^{84}Sr(n,γ)^{85}Sr	1.17 hr, 63.9 days	—	—	10
		^{86}Sr	9.86	^{86}Sr(n,γ)^{87}Sr	2.80 hr	^{90}Zr	—	0.1
		^{87}Sr	7.02	^{88}Sr(n,γ)^{89}Sr	53.6 days	^{89}Y, ^{92}Zr	^{87}Rb	3
		^{88}Sr	82.56	^{89}Sr(n,γ)^{90}Sr	28 yr	—	—	—
				^{90}Sr(n,γ)^{91}Sr	9.67 hr	—	—	
S	16	^{32}S	95.0	—	—	—	—	—
		^{33}S	0.76			—	—	—
		^{34}S	4.22	^{34}S(n,γ)^{35}S	86.73 days	^{35}Cl, ^{38}Ar	—	50
		^{36}S	0.0136	^{36}S(n,γ)^{37}S	5.07 min	^{37}Cl, ^{40}Ar	—	1
Ta	73	^{180}Ta	0.0122	—	—	^{182}W, ^{185}Re	^{180}Hf	—
		^{181}Ta	99.9878	^{181}Ta(n,γ)^{182}Ta	115.1 days			0.1

		No stable isotopes	^{182}Ta(n,γ)^{183}Ta	5 days			
Tc	43	—	—	—	—	—	—
Te	52	^{120}Te 0.089	—	—	—	—	—
		^{122}Te 2.46	—	—	—	—	—
		^{123}Te 0.89	—	—	—	—	—
		^{124}Te 4.74	—	—	—	—	—
		^{125}Te 7.03	—	—	—	—	—
		^{126}Te 18.72	^{126}Te(n,γ)^{127}Te	105 days, 9.35 hr	^{127}I, ^{130}Xe	—	0.1
		^{128}Te 31.75					
		^{130}Te 34.72	^{130}Te(n,γ)^{131}Te	1.20 days, 24.8 hr	^{134}Xe	—	0.2
Tb	65	^{159}Tb 100	^{159}Tb(n,γ)^{160}Tb	72.3 days	^{160}Dy	—	0.003
			^{160}Tb(n,γ)^{161}Tb	7.20 days	—	—	—
Tl	81	^{203}Tl 29.50	^{203}Tl(n,γ)^{204}Tl	2.78 yr	^{204}Pb	^{202}Hg	0.01
		^{205}Tl 70.50	^{205}Tl(n,γ)^{206}Tl	4.26 min	^{206}Pb, ^{209}Bi		0.3
Th	90	^{232}Th 100	^{232}Th(n,γ)^{233}Th	22.12 min	^{236}U	—	0.004
			^{233}Th(n,γ)^{234}Th	24.1 days	—	—	—
Tm	69	^{169}Tm 100	^{169}Tm(n,γ)^{170}Tm	129 days	^{170}Yb	^{168}Er	0.002
			^{170}Tm(n,γ)^{171}Tm	1.9 yr	—	—	—
Sn	50	^{112}Sn 0.96	—	—	—	—	—
		^{114}Sn 0.66					
		^{115}Sn 0.35					

TABLE 8.1A. (*Continued*)

Element	Atomic number	Isotopes	Abundance (%)	Reaction (1, 2)	Half-Life of Reaction Product (2)	Potential Interfering Elements Primary	Secondary	Estimated Detection Limits µg(2, 3)
		^{116}Sn	14.30	—	—	—	—	—
		^{117}Sn	7.61	—	—	—	—	—
		^{118}Sn	24.03	—	—	—	—	—
		^{119}Sn	8.58	—	—	—	—	—
		^{120}Sn	32.65	^{120}Sn$(n,\gamma)^{121}$Sn	28.2 hr, > 5 yr	^{121}Sb, ^{124}Te	—	10
		^{122}Sn	4.72	—	—	—	—	—
		^{124}Sn	5.94	—	—	—	—	—
Ti	22	^{46}Ti	7.93	^{46}Ti$(n,p)^{46}$Sc	83.9 days	^{45}Sc	^{46}Ca	—
		^{47}Ti	7.28	—	—	—	—	—
		^{48}Ti	73.94	^{48}Ti$(n,p)^{48}$Sc	1.833 days	^{57}V	^{48}Ca	—
		^{49}Ti	5.51	—	—	—	—	—
		^{50}Ti	5.34	^{50}Ti$(n,\gamma)^{51}$Ti	5.80 min	^{51}V, ^{54}Cr	—	0.1
W	74	^{180}W	0.14	—	—	—	—	—
		^{182}W	26.41	—	—	—	—	—
		^{183}W	14.40	—	—	—	—	—
		^{184}W	30.64	—	—	—	—	—
		^{186}W	28.41	^{186}W$(n,\gamma)^{187}$W ^{187}W$(n,\gamma)^{188}$W	24.0 hr 69.5 days	^{187}Re, ^{190}Os	—	0.002
U	92	^{234}U	0.0056	—	—	—	—	—
		^{235}U	0.7205	—	—	—	—	—
		^{238}U	99.2739	^{238}U$(n,\gamma)^{239}$U	6.75 days	—	—	0.01

		Isotope	Abundance	Reaction (n,γ)	Half-life			
V	23	50V	0.24	—	—	—	50Ti	—
		51V	99.76	51V(n,γ)52V	3.77 min	52Cr, 55Mn	—	0.002
Xe	54	124Xe	0.096	—	—	—	—	—
		126Xe	0.090					
		128Xe	1.919					
		129Xe	26.44					
		130Xe	4.08					
		131Xe	21.18					
		132Xe	26.89	132Xe(n,γ)133Xe	5.65 days	133Ce, 136Ba	—	0.003
		134Xe	10.44	134Xe(n,γ)135Xe	9.13 hr	138Ba		0.9
				135Xe(n,γ)136Xe				
		136Xe	8.87	136Xe(n,γ)137Xe	3.8 min		—	0.01
Yb	70	168Yb	0.135	—	—	—	—	—
		170Yb	3.191					
		171Yb	14.40					
		172Yb	21.90					
		173Yb	16.2					
		174Yb	31.6	174Yb(n,γ)175Yb	4.2 days	175La, 178Hf	—	0.001
		176Yb	12.60					
Y	39	89Y	100	89Y(n,γ)90Y	64.3 hr	90Zr, 93Nb	—	1
				90Y(n,γ)91Y	58.8 days			
Zn	30	64Zn	48.89	64Zn(n,γ)65Zn	245 days	—	63Cu	1
		66Zn	27.81			—	—	
		67Zn	4.11		—			

TABLE 8.1A. (*Continued*)

Element	Atomic number	Isotopes	Abundance (%)	Reaction (1, 2)	Half-Life of Reaction Product (2)	Potential Interfering Elements Primary	Secondary	Estimated Detection Limits μg(2, 3)
Zn		^{68}Zn	18.57	^{68}Zn(n,γ)^{69}Zn	13.9 hr, 58.5 min	^{69}Ga, ^{72}Ge	—	—
		^{70}Zn	0.62	—	—	—	—	—
Zr	40	^{90}Zr	51.46	—	—	—	—	—
		^{91}Zr	11.23	—	—	—	—	—
		^{92}Zr	17.11	—	—	—	—	—
		^{94}Zr	17.40	^{94}Zr(n,γ)^{95}Zr	65.2 days	^{98}Mo	—	5
		^{96}Zr	2.80	^{96}Zr(n,γ)^{97}Zr	17.0 hr	^{100}Mo	—	5

References:

1. R. C. Koch, *Activation Analysis Handbook*, Academic Press, New York, 1960.
2. A. I. Aliev, V. I. Drynkin, D. I. Leipunskaya, and V. A. Kasatkin, *Handbook of Nuclear Data for Neutron Activation Analysis*, B. Bruch, Transl., Israel Program for Scientific Translations, Jerusalem, 1970.
3. W. S. Lyon, Jr., Ed., *Guide to Activation Analysis*, D. Van Nostrand, New York, 1964.

[b]Table 8.1A lists the elements in alphabetical order, their atomic numbers, their isotopes with relative abundances, nuclear reactions possible with thermal neutrons which are suitable for analysis, half-lives of the reaction products which are suitable for counting, any potential interferences, and estimated sensitivities (in μg) for the reaction.

Potential interferences are divided into two classes. Primary interferences are isotopes capable of nuclear reaction with thermal neutrons to produce the same daughter product as the element of interest, and secondary interferences are those capable of nuclear reaction with thermal neutrons to produce a product that is in turn capable of reaction to produce a product identical to the element of interest.

The estimated sensitivities are based on the following assumptions: (1) A thermal neutron flux of 10^{16} neutrons m^{-2} s^{-2}; (2) a detectable counting rate of 40 disintegrations s^{-1}; (3) a saturation factor of 0.5, that is, $1 - e^t = 0.5$; (4) σ values from reference 2 when no information was available in reference 3; (5) K is 1.0, that is, 100% efficiency in measurement of the disintegration; (6) $t_a = 0$, that is, the irradiated sample is measured immediately on irradiation.

The expression used to calculate the detection limits is equation 12 in the text.

TABLE 8.1B. Estimated Limits of Detection for Thermal Neutron Activation [(n, particle) reactions][a]

Element	Reaction (1,2)	Half-Life of Reaction Product (2)	Energy of γ-Rays (MeV)(2)	Potential Interferences (1,2)[b]	Limit of Detection (mg) (1)
Ac	No stable isotopes	—	—	—	—
Al	^{27}Al(n,p)^{27}Mg	10.0 min	0.834	Ir, Rb, Pb, As, Ge	0.055[c]
			1.013	Cr, Ti, Ga, Rb	0.15[c]
	^{27}Al(n,α)^{24}Na	15.05 hr	1.37	Mg, Ni, F, Eu	0.85[c]
			2.75	Mg, Ar, Fe	1.0[c]
Sb	^{121}Sb(n,2n)^{120}Sb	16.2 min	0.511	[d]	0.4[e]
	^{123}Sb(n,2n)^{122}Sb	3.3 min	0.061	Lu, Sm, Ho, U, Ni	0.2[e]
			0.075	Lu, Sm, Ho, U, Ni	—
Ar	^{40}Ar(n,p)^{40}Cl	1.42 min	1.46	Eu, Cr, Mn, Rb, Zn	[f]
			2.75	Fe, Al, Mg	—
			6.0	B, F, O	—
	^{40}Ar(n,α)^{37}S	5.07 min	3.09	Cl	—
	^{40}Ar(n,d)^{39}Cl	55.5 min	0.264	As, Ge, Se	—
			1.266	Lu, Rh, Si	—
			1.52	Rb, Zn, Sc, Zr	—
As	^{75}As(n,p)^{75m}Ge	49 s	0.138	Ge, Pd	0.75[e]
	^{75}As(n,p)^{75}Ge	79 min	0.265	Ar, Ge, Se	4.2[c]
	^{75}As(n,2n)^{74}As	17.74 days	0.511	[d]	8.5[c]
			0.596	Sm, Mg, Zr, Ir, Rh, Ge	13[c]
	^{75}As(n,α)^{72}Ga	14.12 hr	0.84	Ir, Rb, Pb, Ge, Al Co, Fe, Eu	21[c]

TABLE 8.1B. (*Continued*)

Element	Reaction (1,2)	Half-Life of Reaction Product (2)	Energy of γ-Rays (MeV)(2)	Potential Interferences (1,2)[b]	Limit of Detection (mg) (1)
At	No stable isotopes	—	—	—	—
Ba	^{138}Ba(n, 2n)^{137}Ba	2.5 min	0.622	Ag, Br, Se, Mo, I	—
	^{138}Ba(n, n'*)^{137m}Ba	2.57 min	0.622	Ag, Br, Se, Mo, I	0.11
Be	No γ-rays from products	—	—	—	—
Bi	Products too stable	—	—	—	—
B	^{11}B(n, p)^{11}Be	13.57 s	2.12	Ca, Cl, S, Co, Fe	1.6[e]
			5.86	O, F, Ar	1.1[e]
			6.76	O, F, Ar	
Br	^{79}Br(n, 2n)^{78}Br	6.4 min	0.511	[d]	0.009[e]
			0.612	Ag, Rb	—
	^{81}Br(n, 2n)^{80}Br	17.55	0.62	Ag, Rb	0.15[c]
	^{79}Br(n, n'*)^{79m}Br	4.8 s (1)	0.21 (1)	U, Ag, Pd, Nb, Hf	0.18[e]
Cd	^{110}Cd(n, 2n*)^{111m}Cd	48.7 min	0.15	Au, Sr, Sn	0.18[e]
	^{111}Cd(n, n'*)^{111m}Cd	48.7 min	0.15	Au, Sr, Sn	0.18[e]
			0.25 (1)	None apparent	0.07[e]
	^{106}Cd(n, p)^{106}Ag	24.0 min	0.512	[d]	1.2[e]
	^{106}Cd(n, 2n)^{105}Cd	54.7 min	0.511	[d]	—
Ca	^{44}Ca(n, α)^{41}Ar	1.827 hr	1.283	Ti, Rb, Nd, Rh, Ar, Lu, K	23[e]
	^{44}Ca(n, p)^{44}K	22.3 min	1.13 (1)	Ge, Ru, Sc, Cl	11[c]

			2.07 (1)	Cs, Cd, B, S, Co, Fe	22[e]
C	$^{12}C(n,2n)^{11}C$	20.74 min	0.511	[d]	[f]
Ce	$^{140}Ce(n,2n)^{139m}Ce$	60 s	0.74	Nd, Re	[f]
	$^{142}Ce(n,2n)^{141}Ce$	32.5 days	0.142	Ti, Cl, As, Ge	—
Cs	$^{133}Cs(n,2n)^{132}Cs$	6.48 days	—	[d]	[f]
			0.67	Nd, I, Se, Mo, Ba	—
			1.98	Ca, Y, Sr	—
Cl	$^{35}Cl(n,2n)^{34}Cl$	32.4 min	0.146	Ti, Ce, Au, Cd, Sr	0.75[c]
			0.51	[d]	0.32[c]
			1.17	Ca, Sc	8[c]
			2.14	B, S, Co, Fe	2[c]
	$^{37}Cl(n,p)^{34}S$	5.07 min	3.09	Ar	0.6[c]
	$^{37}Cl(n,\alpha)^{34}P$	12.40 s	2.13	B, S, Co, Fe	3[e]
Cr	$^{50}Cr(n,2n)^{49}Cr$	41.7 min	0.511	[d]	7[e]
	$^{53}Cr(n,p)^{53}V$	2.0 min	1.00	Eu, Mg, Ti, Lu, Al	4[e]
	$^{52}Cr(n,p)^{52}V$	3.77 min	1.434	Eu, Mn, Ar, Rb	0.65[e]
Co	$^{59}Co(n,\alpha)^{56}Mn$	2.586 hr	0.846	Fe, Eu, Ir, Rb, Pb, As, Ge, Al	1.3[c]
			1.811	P, Si, Fe, Sr, Y	4.6[c]
			2.111	Ca, B,Cl, S, Fe	9[c]
Cu	$^{63}Cu(n,2n)^{62}Cu$	9.76 min	0.511	[d]	0.007[c]
	$^{65}Cu(n,2n)^{64}Cu$	12.88 hr	0.511	[d]	—
	$^{65}Cu(n,\alpha)^{62}Co$	13.91 min	1.17	Sc, Ca, Cl, Ni	1.8[c]

TABLE 8.1B. (Continued)

Element	Reaction (1, 2)	Half-Life of Reaction Product (2)	Energy of γ-Rays (MeV)(2)	Potential Interferences (1, 2)[b]	Limit of Detection (mg) (1)
Dy	Poor sensitivity	—	—	—	—
Er	^{168}Er (n, 2n) ^{168m}Er	2.5 s	0.208	Hf, U, Ag, Pd, Br, Nb	f
Eu	^{151}Eu (n, 2n) ^{150}Eu	14.0 hr	0.33	Hf, V, Au, Pt, Ru	f
			1.66	Na, Mg	—
	^{153}Eu (n, 2n) ^{152}Eu	9.2 hr	0.122	Re	—
			0.854	Al, Co, Fe	—
			0.976	Zn, Tl, Mg, Ti, Lu	—
			1.39	Ni, F, Al, Mg	—
F	^{19}F (n, p) ^{19}O	29.1 s	0.199	Er, Ir	0.07[e]
			1.36	Ti, Ni, Al, Mg, Eu	0.48[e]
	^{19}F (n, 2n) ^{18}F	1.58 hr	0.511	[d]	1.7[e]
	^{19}F (n, α) ^{16}N	7.352 s	6.134	O, Ar, B	0.15[e]
			7.121	O	3[e]
Fr	No stable isotopes	—	—	—	—
Gd	^{160}Gd (n, 2n) ^{159}Gd	18 hr	0.057	Ni, Lu	f
			0.364	Ir, Hg, Pb, U	—
Ga	^{69}Ga (n, 2n) ^{68}Ga	67.7 min	0.511	[d]	0.16[c]
	^{69}Ga (n, α) ^{66}Cu	5.10 min	1.038	Al, Ti, Rb, Pb	0.8[c]

Element	Reaction	Half-life	Interferences		
Ge	^{76}Ge(n, 2n)^{75m}Ge	49 s	As, Re, Ti, Ce, Pd	0.138	0.15^e
	^{76}Ge(n, 2n)^{75}Ge	79 min	As, Se, Ar	0.265	0.95^e
	^{70}Ge(n, 2n)^{69}Ge	1.65 days	^{a}	0.511	1.8^c
	^{74}Ge(n, p*)^{74}Ga	8.0 min	Zr, Ir, Rh, As, Sm	0.60	1.1^c
			Rb, Ru, Sc, Ca	1.10 (1)	7^c
			K, Sc	2.35 (1)	10^c
	^{72}Ge(n, p)^{72}Ga	14.12 hr	Ir, Rb, Pb, As, Al	0.84	7^c
Au	^{197}Au(n, n'*)^{197m}Au	7.4 s (1)	Pb, Lu, Yb	0.28 (1)	0.13^e
	^{197}Au(n, 2n)^{196m}Au	9.7 hr	Cl, Cd, Sr	0.149	4^c
			Pd, Ir	0.188	5^c
	^{197}Au(n, p)^{197}Pt	80 min	Eu, Pt, Ru	0.34	1.8^c
Hf	^{179}Hf(n, 2n)^{178m}Hf	4.8 s	Ag, Ho, Nd	0.089	f
			Ag, Ta, Ho, Nd	0.093	—
			U, Er, Ag, Pd, Br, Nb	0.214	—
			Th, V, Eu, Au, Pt, Ru	0.326	—
			Nd, Mg, Na, Tl	0.427	—
	^{180}Hf(n, 2n)^{179m}Hf	19.0 s	Hg, Se, W, Ir	0.16	—
			U, Ag, Pd, Br, Nb	0.217	—
He	No useful reactions	—	—	—	—
Ho	^{165}Ho(n, 2n)^{164m}Ho	36.5 min	None apparent	0.037	f
			Sm, Sb	0.073	—
			Hf, Nd, Ag, Ta	0.091	—
H	Stable products	—	—	—	—
In	^{113}In(n, 2n)^{112}In	11 min	^{a}	0.511	0.46^c

TABLE 8.1B. (*Continued*)

Element	Reaction (1,2)	Half-Life of Reaction Product (2)	Energy of γ-Rays (MeV)(2)	Potential Interferences (1,2)[b]	Limit of Detection (mg) (1)
I	^{127}I(n, 2n)^{126}I	13.1 days	0.65	Te, Se, Mo, Ba, Cs	15[c]
Ir	^{191}Ir(n, 2n)^{190m}Ir	3.2 hr	0.36	Gd, Hg, Pb, U	[f]
			0.52	Tl[d]	—
			0.6	Rh, Ge, As, Sm, Zr	—
	^{193}Ir(n, α)^{190}Re	2.8 min	0.19	Au, Pd	—
			0.39	Sr, Lu, Yb, Mg	—
			0.57	Pb, Rh, Mg	—
			0.83	Rb, Pb, As, Ge, Al	—
Fe	^{57}Fe(n, p)^{57}Mn	1.75 min	0.117	Eu, Re, W, Pd, Mn	18[e]
			0.134	Eu, Re, W, Pd, Mn	
	^{54}Fe(n, 2n)^{53}Fe	8.50 min	0.511	[d]	1.5[c]
	^{56}Fe(n, p)^{56}Mn	2.586 hr	0.846	Al, Co, Eu	3.6[e]
			1.811	P, Si, Co, Sr, Y	24[c]
			2.111	Ca, B, Cl, S, Co	21[c]
			2.65 (1)	Sc, Al, Mg, Ar	20[c]
Kr	No information	—	—	—	—
La	^{139}La(n, p)^{139}Ba	82.9 min	0.167	W	—
Pb	^{208}Pb(n, 2n)^{207m}Pb	0.797 s	0.57	Rh, Ir, Mg	3.2[e]
	^{207}Pb(n, n'*)^{207m}Pb	0.797 s	0.57	Rh, Ir, Mg	3.2[e]
			1.06	Ti, Ga, Rb	5.5[e]

306

Element	Reaction	Half-life		Interferences	
	$^{204}Pb(n,2n)^{203m}Pb$	6.7 s (1)	0.83 (1)	Ir, Rb, As, Ge, Al	3.6[e]
	$^{204}Pb(n,2n)^{203}Pb$	52.1 hr	0.28	Au, Lu, Yb	24[c]
		68 min	0.37 (1)	Ir, Gd, Hg, U	21[c]
	$^{204}Pb(n,n'*)^{204m}Pb$		0.91 (1)	Rb, Y, Sr, Zr	20[c]
Li	Stable products	—	—		—
Lu	$^{175}Lu(n,p)^{175}Yb$	4.2 days	0.114	Yb, Tb, Fe	f
			0.282	Au, Pb, Yb	—
			0.396	Sr, Ir, Yb, Mg	—
	$^{175}Lu(n,2n)^{174}Lu$	165 days	0.045	None apparent	—
			0.059	Gd, Ni, U, Sb	—
			0.067	Sm	—
			0.077	Sb	—
			0.99	Ti, Mg, Eu, Cr, Al	—
			1.23	Ar, Rh, Cu, Ni	—
Mg	$^{25}Mg(n,p)^{25}Na$	1.0 min	0.40	Lu, Yb, Sr, Ir, Nd	0.75[e]
	$^{25}Mg(n,\alpha)^{23}Ne$	37.5 s	0.45	Hf, Na, Tl, Te	—
	$^{25}Mg(n,p)^{25}Na$	1.0 min	0.58	Pb, Ir, Zr	3.6[e]
			0.98	Eu, Ti, Lu	3[e]
			1.61	K, Na, Eu	7.5[e]
	$^{25}Mg(n,\alpha)^{23}Ne$	37.5 s	1.65	K, Na, Eu	—
	$^{24}Mg(n,p)^{24}Na$	15.05 hr	1.37	F, Ni, Al	0.6[c]
			2.75	Fe, Ar, Al	0.6[c]
Mn	$^{55}Mn(n,\alpha)^{52}V$	3.77 min	1.434	Cr, Ar	1.9[e]
Hg	$^{200}Hg(n,2n*)^{199m}Hg$	42 min	0.37	Ir, Gd, Pb, U, Cd, Cl, N	0.6[c]

TABLE 8.1B. (*Continued*)

Element	Reaction (1, 2)	Half-Life of Reaction Product (2)	Energy of γ-Rays (MeV)(2)	Potential Interferences (1,2)[b]	Limit of Detection (mg) (1)
	^{199}Hg$(n, n'*)$ ^{199m}Hg	42 min	0.158 (1)	Sm, Se, Hf	0.11[c]
Mo	^{92}Mo$(n, 2n)$ ^{91m}Mo	64 s	0.511	[d]	—
	^{92}Mo$(n, 2n)$ ^{91}Mo	15.7 min	0.511	[d]	1.6[e]
	^{92}Mo$(n, 2n)$ ^{91m}Mo	64 s	0.658	Se, Ba, I, Cs	16[e]
Nd	^{142}Nd$(n, 2n)$ ^{141}Nd	2.42 hr	0.42	Mg, Hf	[f]
			1.3	Ar, Ca, K, Rb, Ti	—
	^{142}Nd$(n, 2n)$ ^{141m}Nd	63.9 s	0.76	Re	—
	^{148}Nd$(n, 2n)$ ^{147}Nd	11.06 days	0.091	Ho, Hf, Ag, Ta	—
			0.533	Tl, Rh	—
			0.69	Cs, Rb	—
Ne	No information	—	—	—	—
Np	No stable isotopes	—	—	—	—
Ni	^{60}Ni(n, p) ^{60m}Co	10.47 min	0.059	Lu, Gd, U, Sb	2.4[c]
	^{62}Ni(n, p) ^{62}Co	13.91 min	1.17	Sc, Cu	9.5[c]
	^{58}Ni$(n, 2n)$ ^{57}Ni	36.5 hr	1.367	Ti, F, Al, Mg	20[c]
Nb	^{93}Nb$(n, 2p*)$ ^{92}Y	3.53 hr	0.21 (1)	Er, U, Ag, Pd, Br, Hf	4.6[c]
			0.94	Zr, Zn	4.6[c]
N	^{14}N$(n, 2n)$ ^{13}N	9.956 min	0.511	[d]	0.075[c]

308

Element	Reaction	Half-life	Energy (MeV)	Interference	Sensitivity
Os	Poor sensitivity	—	—	—	—
O	^{16}O(n,p)^{16}N	7.352 s	6.134 / 7.121	Ar, F, B / F, B	0.075[e] / 1.5[e]
Pd	^{105}Pd(n,p*)^{105m}Rh	45 s	0.130	W, Mn, Fe	1[e]
	^{108}Pd(n,2n)^{107m}Pd	21.3 s	0.22	U, Ag, Br, Nd, Hf	0.08[e]
	^{110}Pd(n,2n*)^{109m}Pd	4.69 min	0.188 (1)	Au, Ir[d]	0.17[c]
P	^{31}P(n,2n)^{30}P	2.497 min	0.511	[d]	0.8[e]
	^{31}P(n,α)^{28}Al	2.31 min	1.78	Eu, Si, Co, Fe[d]	0.19[e]
Pt	^{198}Pt(n,2n)^{197m}Pt	80 min	0.34	Eu, Ru, Au, Hf	4.2[c]
Pu	No stable isotopes	—	—	—	—
Po	No stable isotopes	—	—	—	—
K	^{39}K(n,2n)^{38}K	7.65 min	0.511	[d]	0.36[c]
	^{41}K(n,p)^{41}Ar	1.827 hr	1.283	Si, Ar, Rh, Ca, Nd	4.6[c]
	^{41}K(n,α)^{38}Cl	37.12 min	1.60	Mg	4.4[c]
	^{39}K(n,2n)^{38}K	7.65 min	2.16	Cl, S, Co, Fe, Zr	—
	^{34}K(n,α)^{38}Cl	37.12 min	2.15	Cl, S, Co, Fe, Zr[d]	—
Pr	^{141}Pr(n,2n)^{140}Pr	3.4 min	0.511	[d]	[f]
Pm	No stable isotopes	—	—	—	—
Pa	No stable isotopes	—	—	—	—

TABLE 8.1B. (*Continued*)

Element	Reaction (1, 2)	Half-Life of Reaction Product (2)	Energy of γ-Rays (MeV)(2)	Potential Interferences (1, 2)[b]	Limit of Detection (mg) (1)
Ra	No stable isotopes	—	—	—	—
Rn	No stable isotopes	—	—	—	—
Re	$^{187}Re(n,2n)^{186}Re$	3.7 days	0.123	Ru, Fe	f
			0.137	Mn, Fe, As, Ge	—
			0.768	Nd, Rb	—
Rh	$^{103}Rh(n,p)^{103}Ru$	39.4 days	0.053	None apparent	—
	$^{103}Rh(n,\alpha)^{100}Tc$	16 s	0.56	Rb, Pb, Ir	—
	$^{103}Rh(n,\gamma)^{104}Rh$	44 s	0.54	Rb, Pb, Ir	—
			0.56	Rb, Pb, Ir	—
			1.24	Ar, Lu, Cu, Ni, Ca, K	—
	$^{103}Rh(n,\alpha)^{100}Tc$	16 s	0.6	Ir, Ge, As, Sm, Rb	—
Rb	$^{85}Rb(n,\alpha)^{82}Br$	35.349 hr	0.554	Rh, Nd, Pb, Ir	f
			0.617	Ag, Rh, Ir, Ge, As, Sm, Br	—
			0.7	Nd	—
			0.78	Nd, Re	—
			0.83	Ir, Pb, As, Ge, Al	—
			1.32	Ca, K, Ti	—
			1.48	Ar, Zn	—
			1.04	Al, Ti, Ga, Pb	—
	$^{87}Rb(n,2n)^{86}Rb$	18.66 days	1.079	Al, Ti, Ga, Pb[d]	—
	$^{85}Rb(n,2n)^{84}Rb$	33 days	0.511	d	—
			0.89	Sc, Y, Sr, Pb	—

310

Element	Reaction	Half-life	Energy	Interferences	
Ru	^{96}Ru(n, 2n)^{95}Ru	1.65 hr	0.34	Au, Pt, Eu, Hf[d]	f
			0.511	d	—
			1.1	Pb, Rb, Ge, Sc, Ca, Cl	—
Sm	^{144}Sm(n, 2n)^{143}Sm	8.6 min	0.511	d	f
	^{154}Sm(n, 2n)^{153}Sm	47.0 hr	0.07	Lu, Ho	—
			0.607	Zr, Ir, Rh, Ge, As, Rb, Br	—
			0.103	Se, W, Ta, Tb	—
Sc	^{45}Sc(n, α)^{42}K	12.358 hr	1.524	Zn, Ar, Zr[d]	f
	^{45}Sc(n, 2n)^{44}Sc	3.92 hr	0.511	d	—
			1.16	Cl, Ca, Cu, Ni	—
			2.51	Ge, Fe	—
	^{44}Sc(n, 2n)^{44m}Sc	2.44 days	0.888	Rb, Y	—
			1.119	Ge, Ru, Ca	—
Se	^{78}Se(n, 2n*)^{77m}Se	17.5 s	0.162	Hg, Hf, W	0.05^e
	^{77}Se(n, n'*)^{77m}Se	17.5 s	0.162	Hg, Hf, W	0.05^e
	^{80}Se(n, p*)^{80}As	15.3 (1)	0.66 (1)	Mo, Ba, I, Cs	3.6^e
	^{82}Se(n, 2n)^{81}Se	62.0 min	0.103	Ta, W, Sm	0.19^e
	^{78}Se(n, α)^{75}Ge	79 min	0.265	As, Ge, Ar	4.4^c
	^{74}Se(n, 2n)^{73m}Se	44 min	0.511	d	—
	^{74}Se(n, 2n)^{73}Se	7.1 hr	0.511	d	—
Si	^{29}Si(n, p)^{29}Al	6.56 min	1.28	—	1.4^e
	^{28}Si(n, p)^{28}Al	2.31 min	1.78	—	1.65^e
Ag	^{109}Ag(n, n'*)^{109m}Ag	39.2 s (1)	0.088	Hf, Nd, Ta	—
	^{107}Ag(n, n'*)^{107m}Ag	44.3 s (1)	0.093	Hf, Nd, Ta	0.44^e
	^{107}Ag(n, p)^{107m}Pd	21.3 s	0.22	Er, U, Pd, Br, Nb, Hf	2.4^e
	^{107}Ag(n, 2n)^{106}Ag	24.0 min	0.511	d	0.55^e
	^{109}Ag(n, 2n)^{108}Ag	2.42 min	0.632	Rb, Br	0.9^e

TABLE 8.1B. (*Continued*)

Element	Reaction (1,2)	Half-Life of Reaction Product (2)	Energy of γ-Rays (MeV)(2)	Potential Interferences (1,2)[b]	Limit of Detection (mg) (1)
Na	^{23}Na(n,p)^{23}Ne	37.5 s	0.44	Mg, Tl, Hf	0.42[e]
	^{23}Na(n,α)^{20}F	11.36 s	1.1629	K, Mg, Eu	—
	^{23}Na(n,p)^{23}Ne	37.5 s	1.65 (1)	K, Mg, Eu	0.1[e]
Sr	^{86}Sr(n,2n)^{85m}Sr	1.17 hr	0.150	Au, Cd	1.7[c]
			0.225	None apparent	0.55[c]
	^{88}Sr(n,2n)^{87m}Sr	2.80 hr	0.388	Ir, Lu, Yb, Hg, Pd, U, Hf	0.15[c]
	^{87}Sr(n,n'*)^{87m}Sr	2.80 hr	0.388	Ir, Lu, Yb, Hg, Pd, U, Hf	0.15[c]
	^{88}Sr(n,p)^{88}Rb	17.7 min	0.91 (1)	Rb, Sc, Y, Pb, Zr	4.2[c]
			1.835	Co, Fe, Y	2.8[c]
S	^{34}S(n,p)^{34}P	12.40 s	2.13	B, Cl, Co, Fe	7[e]
Ta	^{181}Ta(n,2n)^{180m}Ta	8.15 hr	0.093	Ag, Hf, W, Se, Sm	—
			0.102	Ag, Hf, W, Se, Sm	2.2[c]
Tc	No stable isotopes	—	—	—	—
Te	^{130}Te(n,2n)^{129}Te	1.2 hr	0.46	Mg, Na, Tl	0.9[c]
Tb	^{159}Tb(n,2n)^{158m}Tb	11 s	0.111	W, Lu, Yb	[f]
Tl	^{203}Tl(n,2n)^{202}Tl	12 days	0.44	Mg, Na, Hf, Te	[f]
			0.523	Ir, Nd	—
			0.965	Zn, Eu, Mg	—

Element	Reaction	Half-life	Energy (MeV)	Interferences	Value
Th	^{232}Th(n,2n)^{231}Th	25.64 hr	0.017 / 0.084 / 0.31	U / Lu, Ag, Hf / V, Hf	f / — / —
Tm	No information	—	—	—	—
Sn	^{124}Sn(n,2n)^{123}Sn	40.0 min	0.153	Cd, Sr, Hg, Cl^d	0.24^c
	^{112}Sn(n,2n)^{111}Sn	35.0 min	0.511	d	1.4^c
Ti	^{46}Ti(n,p*)^{46m}Sc	19.5 s	0.142 (1)	Ce, Cl, As, Ge, Se	0.7^e
	^{46}Ti(n,2n)^{45}Ti	3.06 hr	0.511	d	5^e
	^{48}Ti(n,p)^{48}Sc	1.833 days	0.986 / 1.040 / 1.314	Eu, Mg, Lu, Cr, Al Ga, Rb, Pb	5.5^c / 10^c
W	^{184}W(n,2n*)^{183m}W	5.1 s	0.102 / 0.108	Ta, Se, Sm, Tb, Lu Yb, Fe, Eu, Re	0.34^e
	^{183}W(n,n'*)^{183m}W	5.1 s	0.125	Ta, Se, Sm, Tb, Lu Yb, Fe, Eu, Ra	—
	^{186}W(n,2n*)^{185m}W	1.6 min	0.175	Au, La	1.3^e
U	^{238}U(n,2n)^{237}U	6.75 days	0.026 / 0.06 / 0.21 / 0.37	None apparent / Ni, Lu, Sb / Ag, Pd, Er, Br, Nb, Hf / Ir, Gd, Hg, Pb	f / — / — / —
V	^{51}V(n,p)^{51}Ti	5.79 min (1)	0.319	Th, Hf, Eu	0.8^e
Xe	No information	—	—	—	—

TABLE 8.1B. (*Continued*)

Element	Reaction (1,2)	Half-Life of Reaction Product (2)	Energy of γ-Rays (MeV)(2)	Potential Interferences (1,2)[b]	Limit of Detection (mg) (1)
Yb	^{176}Yb (n, 2n) ^{175}Yb	42 days	0.114	Tb, Lu, Fe	f
			0.282	Au, Pb, Lu	—
			0.396	Sr, Ir, Lu, Mg	—
Y	^{89}Y (n, 2n) ^{88}Y	105 days	0.395	Sr, Ir, Lu, Mg[d]	f
			0.511	[d]	—
			1.853	Co, Fe, Sr	—
	^{89}Y (n, n'γ) ^{89m}Y	16.5 s	0.91	Sr, Pb, Zr, Rb	—
			0.908	Sr, Pb, Zr, Rb[d]	—
Zn	^{64}Zn (n, 2n) ^{63}Zn	38.4 min	0.511	[d]	0.06[c]
	^{64}Zn (n, p) ^{64}Cu	12.88 hr	0.511	[d]	—
	^{64}Zn (n, 2n) ^{63}Zn	38.4 min	0.966	Nb, Tl, Eu, Mg	4[c]
	^{68}Zn (n, α) ^{65}Ni	2.56 hr (1)	1.48 (1)	Ar, Rb	17[c]

Zr	$^{90}Zr(n,2n)^{89m}Zr$	4.18 min	0.588	Mg, Ir, Rh, Ge, As, Sm	0.85[e]
			1.53	Zn, Ar, Sc	25[e]
	$^{94}Zr(n,p)^{94}Y$	20.35 min	0.92	Sr, Pb, Y, Nb	0.55[e]
	$^{91}Zr(n,2n*)^{90m}Zr$	0.83 s (1)	2.32 (1)	Ge	7.5[e]
	$^{90}Zr(n,n'*)^{90m}Zr$	0.83 s (1)	2.32 (1)	Ge	7.5[e]

[a]References:

1. J. Perdijon, *Anal. Chem.*, **39**, 448 (1967).
2. A. I. Aliev, V. I. Drynkin, D. I. Leipunskaya, and V. A. Kasatkin, *Handbook of Nuclear Data for Neutron Activation Analysis*, B. Benny, Transl. Israel Program for Scientific Translations, Jerusalem, 1970.

[b]The interferences in this column are all determined by the ability of the instrumentation to resolve γ-ray emissions that are close. The assumption was made that for emissions above 0.1 MeV the resolution limitation was ±0.1 MeV. This of course varies from facility to facility and also depends on the wavelength range of interest. Interferences of isotopes that produce the same daughter products should be obtained from Table 8.1A.

[c]Limits of detection were measured using the following conditions: irradiation time, 5 min; delay time, 10 min; counting time, 5 min; limits normalized to 10^{13} neutrons m^{-2} s^{-1}; the minimum detectable activity varied between 45 and 200.

[d]The following elements produce γ-radiation in the region between 0.51 and 0.52 MeV: Sb, As, Br, Cd, C, Cs, Cl, Cu, F, Ga, Ge, In, Ir, Fe, Mo, N, K, Pr, Rb, Ru, Sm, Sc, Se, Ag, Sn, Ti, Y, and Zn.

[e]Limits of detection were measured using the following conditions: irradiation time, 30 s; delay time, 3 s; counting time, 30 s; limits were normalized to 10^{13} neutrons m^{-2} s^{-1}; the minimum detectable activity varied between 30 and 85.

[f]Not determined in reference 1. However, the limits of detection should be similar to the corresponding isotopes in Table 8.1A and vary with the limiting detectable activity estimated from reference 1.

315

TABLE 8.2. Estimated Limits of Detection for Fast-Neutron Activation[a]

Element	Atomic Number	Reaction (1, 2, 3)	Half-Life (1)	Interfering Elements (2, 3)		Estimated Detection Limits (mg)
				Primary	Secondary	
Ac	89	No stable isotopes	—	—	—	—
Al	13	^{27}Al(n,p)^{27}Mg	10.0 min	^{26}Mg, ^{30}Si	—	0.06
		^{27}Al(n,α)^{24}Na	15.05 hr	^{24}Mg	^{23}Na	0.03
Sb	51	^{121}Sb(n,2n)^{120}Sb	16.2 min	^{120}Te	^{119}Sn, ^{120}Sn, ^{123}Te	0.03
		^{123}Sb(n,2n)^{122}Sb	2.75 days, 3.3 min	^{122}Te	^{122}Sn, ^{125}Te	0.03
Ar	18	^{40}Ar(n,p)^{39}Ar	1.42 min	—	—	0.3
		^{40}Ar(n,α)^{37}S	5.07 min	—	—	0.3
		^{40}Ar(n,d)^{39}Cl	55.5 min	—	—	1.8
As	33	^{75}As(n,p)^{75}Ge	79 min, 49 s	^{78}Se, ^{74}Ge, ^{76}Ge	—	0.6
		^{75}As(n,α)^{72}Ga	14.12 hr	^{72}Ge, ^{71}Ga	—	1.0
		^{75}As(n,2n)^{74}As	17.74 days	^{74}Se	^{73}Ge, ^{74}Ge	0.009
At	85	No stable isotopes	—	—	—	—
Ba	56	^{136}Ba(n,2n)^{135}Ba	28.7 hr	—	—	0.3
		^{138}Ba(n,2n)^{137}Ba	2.57 min	—	—	0.02
Be	4	^{9}Be(n,p)^{9}Li	0.169 s	—	—	2.0
		^{9}Be(n,α)^{6}He	0.80 s	—	—	0.1

316

Element	Z	Reaction	Half-life			
Bi	83	$^{209}\text{Bi}(n,p)^{209}\text{Pb}$	3.31 hr	—	—	29.0
		$^{209}\text{Bi}(n,\alpha)^{206}\text{Tl}$	4.26 min	$^{205}\text{Tl}, ^{206}\text{Pb}$	—	32.0
B	5	$^{11}\text{B}(n,p)^{11}\text{Be}$	13.57 s	—	—	0.5
		$^{11}\text{B}(n,\alpha)^{8}\text{Li}$	0.847 s	—	—	0.05
Br	35	$^{79}\text{Br}(n,2n)^{78}\text{Br}$	6.5 min	^{78}Kr	$^{77}\text{Se}, ^{78}\text{Se}$	0.02
		$^{81}\text{Br}(n,p)^{81}\text{Se}$	18.6 min, 62.0 min	—	—	0.8
		$^{81}\text{Br}(n,2n)^{80}\text{Br}$	17.5 min, 4.37 hr	^{80}Kr	$^{80}\text{Se}, ^{83}\text{Se}$	0.04
Cd	48	$^{111}\text{Cd}(n,p)^{111}\text{Ag}$	7.6 days	—	—	5
		$^{112}\text{Cd}(n,p)^{112}\text{Ag}$	3.2 hr	—	—	6
		$^{116}\text{Cd}(n,2n)^{115}\text{Cd}$	2.2 days, 43 days	—	—	0.2
Ca	20	$^{44}\text{Ca}(n,p)^{44}\text{K}$	22.3 min	—	—	8
		$^{44}\text{Ca}(n,\alpha)^{41}\text{Ar}$	1.827 hr	—	—	8
C	6	$^{12}\text{C}(n,2n)^{11}\text{C}$	20.74 min	—	^{11}B	0.3
Ce	58	$^{140}\text{Ce}(n,2n)^{139}\text{Ce}$	60 s	—	—	0.02
		$^{142}\text{Ce}(n,2n)^{141}\text{Ce}$	32.5 days	—	—	0.1
Cs	55	$^{133}\text{Cs}(n,p)^{133}\text{Xe}$	5.65 days	—	—	1
		$^{133}\text{Cs}(n,2n)^{132}\text{Cs}$	6.48 days	—	—	0.01
Cl	17	$^{35}\text{Cl}(n,p)^{35}\text{S}$	86.73 days	$^{34}\text{S}, ^{36}\text{S}, ^{38}\text{Ar}$	—	0.05
		$^{35}\text{Cl}(n,\alpha)^{32}\text{P}$	14.5 days	$^{31}\text{P}, ^{32}\text{S}$	—	0.06
		$^{35}\text{Cl}(n,2n)^{34}\text{Cl}$	32.4 min, 1.588 s	—	—	0.5

TABLE 8.2. (*Continued*)

Element	Atomic Number	Reaction (1, 2, 3)	Half-Life (1)	Interfering Elements (2, 3) Primary	Interfering Elements (2, 3) Secondary	Estimated Detection Limits (mg)
		$^{37}Cl(n,p)^{37}S$	5.07 min	$^{36}S, ^{40}Ar$	—	0.6
		$^{37}Cl(n,\alpha)^{34}P$	12.40 s	—	—	0.4
Cr	24	$^{52}Cr(n,p)^{52}V$	3.77 min	$^{51}V, ^{55}Mn$	—	0.08
		$^{52}Cr(n,2n)^{51}Cr$	27.8 days	—	—	0.03
Co	27	$^{59}Co(n,p)^{59}Fe$	46.5 days	$^{58}Fe, ^{62}Ni$	—	0.1
		$^{59}Co(n,\alpha)^{56}Mn$	2.586 hr	$^{55}Mn, ^{56}Fe$	—	0.3
		$^{59}Co(n,2n)^{58}Co$	9.2 hr, 71.3 days	—	—	0.02
Cu	29	$^{63}Cu(n,\alpha)^{60}Co$	10.47 min	—	—	0.4
		$^{63}Cu(n,2n)^{62}Cu$	9.76 min	—	^{61}Ni	0.03
Dy	66	$^{162}Dy(n,\alpha)^{159}Gd$	18.0 hr	—	—	24
		$^{163}Dy(n,p)^{163}Tb$	7 min	—	—	29
		$^{164}Dy(n,\alpha)^{161}Gd$	3.73 min	—	—	17
		$^{164}Dy(n,\gamma)^{165}Dy$	2.36 hr, 1.25 min	$^{165}Ho, ^{165}Er$	—	10
Er	68	$^{166}Er(n,2n)^{165}Er$	9.5 hr	—	—	0.07
		$^{168}Er(n,2n)^{167}Er$	2.5 s	—	—	0.4
		$^{170}Er(n,2n)^{169}Er$	9.6 days	—	—	0.1
Eu	63	$^{151}Eu(n,2n)^{150}Eu$	14.0 hr	—	—	0.08
		$^{153}Eu(n,p)^{153}Sm$	47 hr	—	—	5

Element	Z	Reaction	Half-life			
		$^{153}Eu(n,\alpha)^{150}Pm$	2.7 hr	—	—	4
		$^{153}Eu(n,2n)^{152}Eu$	9.2 hr	—	—	0.05
F	9	$^{19}F(n,p)^{19}O$	29.1 s	$^{18}O, ^{22}Na$	—	0.1
		$^{19}F(n,\alpha)^{16}N$	7.325 s	—	—	0.08
		$^{19}F(n,2n)^{18}F$	1.87 hr	—	$^{17}O, ^{18}O, ^{21}Ne$	0.06
Fr	87	No stable isotopes	—	—	—	—
Gd	64	$^{160}Gd(n,2n)^{159}Gd$	18 hr	$^{159}Tb, ^{162}Dy$	—	0.06
		$^{160}Gd(n,\gamma)^{161}Gd$	3.73 min	^{164}Dy	—	5
Ga	31	$^{69}Ga(n,p)^{69}Zn$	13.9 hr, 58.5 min	$^{72}Ge, ^{68}Zn, ^{70}Zn$	—	0.6
		$^{69}Ga(n,2n)^{68}Ga$	67.7 min	—	$^{67}Zn, ^{68}Zn$	0.02
		$^{71}Ga(n,2n)^{70}Ga$	21.37 min	^{70}Ge	$^{70}Zn, ^{73}Ge$	0.03
		$^{71}Ga(n,\alpha)^{68}Cu$	30 s	—	—	0.4
Ge	32	$^{70}Ge(n,p)^{70}Ga$	21.37 min	$^{69}Ga, ^{71}Ga$	^{70}Zn	0.8
		$^{70}Ge(n,2n)^{69}Ge$	1.65 days	—	^{69}Ga	0.09
		$^{72}Ge(n,p)^{72}Ga$	14.12 hr	$^{71}Ga, ^{75}As$	—	0.8
		$^{76}Ge(n,2n)^{75}Ge$	49 s, 79 min	$^{78}Se, ^{75}As$	—	0.1
Au	79	$^{197}Au(n,2n)^{196}Au$	6.2 days, 9.7 hr	^{196}Hg	$^{196}Pt, ^{199}Hg$	0.02
Hf	72	$^{179}Hf(n,2n)^{178}Hf$	4.8 s	—	—	0.2
		$^{180}Hf(n,2n)^{179}Hf$	19.0	—	—	0.1
He	2	No fast-neutron reactions	—	—	—	—
Ho	67	$^{165}Ho(n,p)^{165}Dy$	2.36 hr	—	—	0.5

TABLE 8.2. (*Continued*)

Element	Atomic Number	Reaction (1, 2, 3)	Half-Life (1)	Interfering Elements (2, 3) Primary	Secondary	Estimated Detection Limits (mg)
		$^{165}\text{Ho}(n,2n)^{164}\text{Ho}$	36.5 min	—	—	0.008
H	1	Forms stable products	—	—	—	—
In	49	$^{115}\text{In}(n,2n)^{114}\text{In}$	50 days, 1.2 min	—	—	0.01
		$^{115}\text{In}(n,n')^{115}\text{In}$	4.4 hr	—	—	0.04
I	53	$^{127}\text{I}(n,2n)^{126}\text{I}$	13.1 days	^{126}Xe	$^{125}\text{Te},\ ^{126}\text{Te},\ ^{129}\text{Xe}$	0.01
Ir	77	$^{191}\text{Ir}(n,2n)^{190}\text{Ir}$	3.2 hr	—	—	0.2
		$^{193}\text{Ir}(n,\alpha)^{190}\text{Re}$	2.8 min	—	—	4
Fe	26	$^{54}\text{Fe}(n,p)^{54}\text{Mn}$	313.5 days	^{55}Mn	$^{53}\text{Cr},\ ^{54}\text{Cr}$	0.4
		$^{56}\text{Fe}(n,p)^{56}\text{Mn}$	2.586 hr	$^{59}\text{Co},\ ^{55}\text{Mn}$	—	0.07
		$^{56}\text{Fe}(n,2n)^{55}\text{Fe}$	2.60 yr	—	—	0.02
Kr	36	No fast-neutron reactions	—	—	—	—
La	57	$^{139}\text{La}(n,p)^{139}\text{Ba}$	82.9 min	$^{138}\text{Ba},\ ^{142}\text{Ce}$	—	4
Pb	82	$^{208}\text{Pb}(n,2n)^{207}\text{Pb}$	0.797 s	—	—	0.04
Li	3	$^{6}\text{Li}(n,p)^{6}\text{He}$	0.80 s	—	—	0.2
		$^{6}\text{Li}(n,\alpha)^{3}\text{H}$	12.262 yr	$^{2}\text{H},\ ^{3}\text{He}$	—	0.5

Element		Reaction	Half-life			Abundance
Lu	71	$^{175}Lu(n,2n)^{174}Lu$	165 days	—	—	0.02
		$^{175}Lu(n,p)^{175}Yb$	4.2 days	—	—	7
Mg	12	$^{24}Mg(n,p)^{24}Na$	15.05 hr	^{27}Al, ^{23}Na	—	0.02
		$^{25}Mg(n,p)^{25}Na$	1.0 min	—	—	0.6
		$^{26}Mg(n,p)^{26}Na$	1.03 s	—	—	1
		$^{26}Mg(n,\alpha)^{23}Ne$	37.5 s	—	—	0.3
Mn	25	$^{55}Mn(n,p)^{55}Cr$	3.52 min	—	—	0.1
		$^{55}Mn(n,\alpha)^{52}V$	3.77 min	^{51}V, ^{52}Cr	—	0.2
		$^{55}Mn(n,2n)^{54}Mn$	313.5 days	—	—	0.008
Hg	80	$^{200}Hg(n,\alpha)^{197}Pt$	18.0 hr	—	—	68
		$^{200}Hg(n,p)^{200}Au$	48.4 min	—	—	32
Mo	42	$^{92}Mo(n,2n)^{91}Mo$	15.7 min	—	—	0.8
		$^{100}Mo(n,2n)^{99}Mo$	67 hr	^{102}Ru	—	0.07
Nd	60	$^{142}Nd(n,2n)^{141}Nd$	2.42 hr, 63.9 s	—	—	0.05
		$^{148}Nd(n,2n)^{147}Nd$	11.06 days	—	—	0.2
		$^{150}Nd(n,2n)^{149}Nd$	1.8 hr	—	—	8
Ne	10	No fast-neutron reactions	—	—	—	—
Np	93	No stable isotopes	—	—	—	—
Ni	28	$^{58}Ni(n,p)^{58}Co$	9.2 hr, 71.3 days	—	—	0.04
		$^{58}Ni(n,np) +$ $(n,d)+(n,pn)]^{57}Co$	67 days	—	—	0.02

TABLE 8.2. (*Continued*)

Element	Atomic Number	Reaction (1, 2, 3)	Half-Life (1)	Interfering Elements (2, 3) Primary	Secondary	Estimated Detection Limits (mg)
		^{60}Ni(n,p)^{60}Co	10.47 min, 5.26 yr	—	—	0.3
Nb	41	^{93}Nb(n,2n)^{92}Nb	10.16 days	—	—	0.03
		^{93}Nb(n,nα)^{89}Y	16.5 s	—	—	0.05
N	7	^{14}N(n,2n)^{13}N	9.965 min	—	^{13}C, ^{16}O	0.3
Os	76	^{188}Os(n,p)^{188}Re	16.74 hr	—	—	22
		^{190}Os(n,p)^{190}Re	2.8 min	—	—	48
O	8	^{16}O(n,p)^{16}N	7.352 s	^{15}N, ^{19}F	—	0.06
Pd	46	^{104}Pd(n,p)^{104}Rh	44 s	^{107}Ag, ^{103}Rh	^{104}Rh	1
		^{105}Pd(n,p)^{105}Rh	36.5 hr	^{104}Rh	^{104}Rh	0.09
		^{110}Pd(n,2n)^{109}Pd	13.45 hr, 4.69 min	—	—	0.06
P	15	^{31}P(n,p)^{31}Si	2.64 hr	^{30}Si, ^{34}S	—	0.05
		^{31}P(n,α)^{28}Al	2.31 min	^{27}Al, ^{28}Si	—	0.03
		^{31}P(n,2n)^{30}P	2.497 min	—	^{30}Si; ^{29}Si, ^{33}S	0.5
Pt	78	^{198}Pt(n,2n)^{197}Pt	18.0 hr, 80 min	^{197}Au, ^{200}Hg	—	0.2
Pu	94	No stable isotopes	—	—	—	—
Po	84	No stable isotopes	—	—	—	—

Element	Z	Reaction	Half-life			
K	19	^{39}K$(n,2n)^{38}$K	7.66 min, 0.946 s	—	^{38}Ar	2
		^{41}K$(n,p)^{41}$Ar	1.827 hr	^{40}Ar, ^{44}Ca	—	1
		^{41}K$(n,\alpha)^{38}$Cl	37.12 min	^{37}Cl, ^{38}Ar	—	2
Pr	59	^{141}Pr$(n,2n)^{140}$Pr	3.4 min	—	^{140}Ce, ^{143}Nd	0.01
Pm	61	No stable isotopes				—
Pa	91	No stable isotopes				—
Ra	88	No stable isotopes				—
Rn	86	No stable isotopes				—
Re	75	^{187}Re$(n,2n)^{186}$Re	3.7 days	—	—	0.02
Rh	45	^{103}Rh$(n,p)^{103}$Ru	39.4 days	^{100}Ru	—	0.9
		^{103}Rh$(n,\alpha)^{100}$Tc	16 s	^{104}Pd, ^{107}Ag	^{100}Mo	1
		^{103}Rh$(n,\gamma)^{104}$Rh	44 s	—	—	1
Rb	37	^{85}Rb$(n,\alpha)^{82}$Br	35.344 hr	—	—	0.1
		^{85}Rb$(n,2n)^{84}$Rb	33 days	—	—	0.01
		^{87}Rb$(n,2n)^{86}$Rb	18.66 days	^{86}Sr, ^{89}Y	^{86}Kr	0.03
Ru	44	^{96}Ru$(n,2n)^{95}$Ru	1.65 hr	—	—	0.3
		^{102}Ru$(n,p)^{102}$Tc	4.5 min, 5 s	—	—	6
Sm	62	^{144}Sm$(n,2n)^{143}$Sm	8.6 min, 2.3 min	—	—	0.4
		^{154}Sm$(n,2n)^{153}$Sm	47.0 hr	^{153}Eu, ^{156}Gd	—	0.1

TABLE 8.2. (Continued)

Element	Atomic Number	Reaction (1, 2, 3)	Half-Life (1)	Interfering Elements (2, 3)		Estimated Detection Limits (mg)
				Primary	Secondary	
Sc	21	^{45}Sc(n,p)^{45}Ca	161.5 days	—	—	0.1
		^{45}Sc(n,α)^{42}K	12.358 hr	—	—	0.1
		^{45}Sc(n,2n)^{44}Sc	3.92 hr, 2.44 days	—	—	0.04
Se	34	^{76}Se(n,2n)^{75}Se	120.4 days	—	—	0.1
		^{80}Se(n,α)^{77}Ge	11.3 hr	^{76}Ge	—	0.6
		^{80}Se(n,2n)^{79}Se	3.91 min	—	—	0.2
		^{82}Se(n,2n)^{81}Se	18.6 min, 62 min	^{81}Br, ^{84}Kr	—	0.08
Si	14	^{28}Si(n,p)^{28}Al	2.31 min	^{27}Al, ^{31}P	—	0.02
		^{29}Si(n,p)^{29}Al	6.56 min	—	—	0.8
Ag	47	^{107}Ag(n,2n)^{106}Ag	24 min, 8.3 days	^{106}Cd	^{105}Pd, ^{106}Pd	0.05
		^{109}Ag(n,2n)^{108}Ag	2.42 min	^{108}Cd	^{108}Pd, ^{111}Cd	0.04
Na	11	^{23}Na(n,p)^{23}Ne	37.5 s	^{22}Ne, ^{26}Mg	—	0.09
		^{23}Na(n,α)^{20}F	11.36 s	—	—	0.02
		^{23}Na(n,2n)^{22}Na	2.58 yr	—	^{22}Ne, ^{25}Mg	0.1
Sr	38	^{86}Sr(n,2n)^{85}Sr	63.9 days, 1.17 hr	—	—	0.4
		^{88}Sr(n,p)^{88}Rb	17.7 min	^{87}Rb	—	0.8
		^{88}Sr(n,α)^{85}Kr	4.50 hr	^{85}Rb, ^{84}Kr	—	0.2
		^{88}Sr(n,2n)^{87}Sr	2.80 hr	—	—	0.07

Element	Z	Reaction	Half-life			
S	16	$^{32}S(n,p)^{32}P$	14.5 days	$^{31}P, ^{35}Cl$	—	0.02
		$^{32}S(n,t)^{30}P$	2.497 min	—	—	0.4
		$^{34}S(n,p)^{34}P$	12.40 s	^{37}Cl	—	1
		$^{34}S(n,\alpha)^{31}Si$	2.64 hr	$^{30}Si, ^{31}P$	—	0.8
Ta	73	$^{181}Ta(n,2n)^{180}Ta$	8.15 hr	^{180}W	^{180}Hf	0.02
Tc	43	No stable isotopes				—
Te	52	$^{128}Te(n,p)^{128}Sb$	10 hr, 10.3 min	$^{127}I, ^{130}Xe$	—	0.08
		$^{128}Te(n,2n)^{127}Te$	9.35 hr	—	—	0.08
		$^{130}Te(n,2n)^{129}Te$	1.2 hr, 41 days	^{132}Xe	—	0.2
Tb	65	$^{159}Tb(n,2n)^{158}Tb$	11 s	—	—	0.1
Tl	81	$^{203}Tl(n,2n)^{202}Tl$	12 days	^{204}Pb	$^{201}Hg, ^{202}Hg$	0.08
		$^{205}Tl(n,2n)^{204}Tl$	3.78 yr	—	—	0.02
Th	90	$^{232}Th(n,2n)^{231}Th$	25.64 hr	—	—	0.02
Tm	69	No fast neutron reactions	—	—	—	—
Sn	50	$^{118}Sn(n,2n)^{117}Sn$	14 days	—	—	0.05
Ti	22	$^{46}Ti(n,p)^{46}Sc$	83.9 days	^{45}Sc	^{46}Ca	0.3
		$^{48}Ti(n,pn)^{47}Sc$	3.45 days	—	—	0.5
		$^{48}Ti(n,p)^{48}Sc$	1.833 days	^{51}V	^{48}Ca	0.1

TABLE 8.2. (*Continued*)

Element	Atomic Number	Reaction (1, 2, 3)	Half-Life (1)	Interfering Elements (2, 3)		Estimated Detection Limits (mg)
				Primary	Secondary	
W	74	$^{182}W(n,2n)^{185}W$	120 days	—	—	0.04
		$^{184}W(n,2n)^{183}W$	5.1 s	—	—	0.1
		$^{186}W(n,2n)^{185}W$	1.6 min, 75.8 days	—	—	0.04
U	92	$^{238}U(n,2n)^{237}U$	6.75 days	—	—	0.04
V	23	$^{51}V(n,p)^{51}Ti$	5.80 min	^{50}Ti, ^{54}Cr	—	0.3
		$^{51}V(n,\alpha)^{48}Sc$	1.833 days	^{48}Ti	^{48}Cr	0.2
		$^{51}V[(n,n\alpha)+(n,\alpha n)]^{47}Sc$	3.45 days	^{48}Ti	^{47}Ti, ^{46}Ca	0.2
Xe	54	No fast-neutron reactions	—	—	—	—
Yb	70	$^{176}Yb(n,2n)^{175}Yb$	42 days	—	—	0.4
Y	39	$^{89}Y(n,p)^{89}Sr$	53.6 days	—	—	0.5
		$^{89}Y(n,2n)^{88}Y$	105 days	—	—	0.01
		$^{89}Y(n,n',\gamma)^{89}Y$	16.5 s	—	—	0.03

Zn	30	^{64}Zn(n,p)^{64}Cu	^{63}Cu, ^{65}Cu	^{64}Ni	12.88 hr	0.08
		^{64}Zn(n,2n)^{63}Zn	—	^{63}Cu	38.4 min	0.2
		^{66}Zn(n,p)^{66}Cu	^{65}Cu, ^{69}Ga	—	5.10 min	0.4
Zr	40	^{92}Zr(n,p)^{92}Y	—	—	3.53 hr	3
		^{92}Zr(n,α)^{89}Sr	—	—	53.6 days	7
		^{94}Zr(n,p)^{94}Y	—	—	20.35 min	7

[a]References:
1. A. I. Aliev, V. I. Drynkin, D. I. Leipunskaya, and V. A. Kasatkin, *Handbook of Nuclear Data for Neutron Activation Analysis*, B. Benny, Transl., Israel Program for Scientific Translations, Jerusalem, 1970.
2. J. Perdijon, *Anal. Chem.*, **39**, 448 (1967).
3. R. C. Koch, *Activation Analysis Handbook*, Academic Press, New York, 1960.
4. W. S. Lyon, Jr., Ed., *Guide to Activation Analysis*, D. Van Nostrand, New York, 1964.

[b]Table 8.2 lists the elements in alphabetical order, their atomic numbers, possible reactions with fast neutrons, half-lives of the daughter products, possible interferences, and estimated limits of detection (in mg).

Potential interferences are divided into two types. Primary interferences are those arising from isotopes that undergo nuclear reactions with fast neutrons to produce the same daughter product as the element of interest, and secondary interferences are those that undergo nuclear reaction with fast neutrons to produce a product which in turn can undergo nuclear reaction to produce a product that is the same as the element of interest.

The estimated sensitivities are based on the following assumptions: (1) A fast neutron flux of 10^{13} neutrons m^{-2} s^{-1}; (2) a detectable counting rate of 40 disintegrations s^{-1} and a counting time of 1 s; (3) a saturation factor of 0.5, that is, $1 - e^{\lambda t} = 0.5$; (4) σ values from reference 1; (5) the K value is 1.0, that is 100% efficiency in measuring disintegration; (6) $t_a = 0$, that is, the irradiated sample is measured immediately on irradiation.

serious limitations of the estimated detection limits are for species having short-lived nuclides, resulting in radioactivity losses prior to measurement, and for detection systems in which significant losses of counts occur, that is, $K < 1$. Of course, if the neutron flux is less than 10^{16} m^{-2} s^{-1}, a further increase in the detection limits results. Fluxes higher than 10^{16} m^{-2} s^{-1} cause severe problems in handling the increased total activity of the irradiated sample, and so the preliminary sampling steps following irradiation necessitate *special laboratory techniques* and *sample handling equipment*. Sample loading can be done in two major ways: (1) a pneumatic tube to transfer samples from the laboratory to the reactor core for periods of seconds to about an hour; and (2) a facility for long irradiations up to 1 week—which operates more slowly and must be composed of radiation-resistant materials. If it is assumed that the above limitations can be minimized, there are still two other inherent limitations. *Self-shielding* (interaction of the nucleus with bombarding particles is independent of the matrix but, practically, the matrix influences the neutron flux via attenuation of the neutron flux charged particles and α-rays also attenuate) according to Beer's law, that is, $e^{-N\sigma}$, where N is the number of atoms of a given element (in cm^{-3}) and other terms are as defined before. Hogdahl (7) studied the effect of self-shielding on the accuracy of measurement. Kamenoto's (8) simple equation $(\sigma w / m)$ 3×10^{-33} m^2 kg amu^{-1} can be used to determine quickly the need to test for self-shielding effects in more detail. Also (primary) *nuclear reactions* can interfere. Interferences due to nuclear reactions (both thermal and fast neutrons) can often be avoided by proper placement of the sample within the reactor. Fast neutrons can cause interferences of a more subtle nature, namely, interfering reactions producing nuclides with atomic numbers of 1 more or 2 less via (n, p) and (n, α) reactions occur. By surrounding the samples with certain materials, for example, boron and cadmium can be used in the case of manganese determination these interferences can sometimes be avoided (7). In the case of *secondary reaction* leading to an interference of the analyte from the matrix material, there is little that can be done, except simply to accept poorer sensitivity. *Preconcentration* of the impurity prior to irradiation can often be performed with a concomitant increase in time and the possibility of additional interferences, but such a step involving separation of the interferent prior to irradiation removes the one great advantage of activation analysis over other analytical methods, namely, the avoidance of the reagent blank. In Table 8.1, potential *primary and secondary* interferences are listed.

With TRIGA reactors, it is possible to produce high peak fluxes (5×10^{16} neutrons) in a short time (30 ms), which results in increased sensitivity of analysis. Lukens, Yule, and Gunn (10) showed an improvement in detection limits for elements with half-lives less than 50 s by pulsing the reaction. However, small laboratory neutron generators (8, 9) are now available with 10^{13} neutrons m^{-2} s^{-1}. Although the flux of thermal neutrons is too low for most trace analytical studies, it is possible that the combination of fast-neutron and thermal neutron activation may have some possible use in the trace analysis area.

Because neutron activation is a comparative (rather than an absolute method), only one standard generally needs to be irradiated each time to monitor the neutron flux and to correct for geometric problems, whereas several samples can be simultaneously irradiated. A study of errors resulting in irradiation of samples and standard has been made by Giradi, Guzzi, and Pauls (12).

b. Fast Neutrons

Fast neutrons have also been utilized for trace analysis via activation analysis (9, 10, 13, 14). The sensitivity of analysis of at least 6 elements (oxygen, iron, silicon, phophorus, lead, and yttrium) is greater with fast neutrons than with thermal neutrons, and for 15 others it is only slightly less and may actually be better for certain matrices. The three reactions that provide useful neutron flux with reasonably priced neutron generation (acceleration) are:

$$^{3}H + ^{2}H \rightarrow ^{4}He + n \quad (17.6 \text{ meV}) \tag{13}$$

$$^{2}H + ^{2}H \rightarrow ^{3}He + n \quad (3.3 \text{ MeV}) \tag{14}$$

$$^{9}Be + ^{2}H \rightarrow ^{10}B + n \quad (4.4 \text{ Mev}) \tag{15}$$

Reaction 13 occurs in generators capable of 500 kV voltages and results in 14 MeV neutrons which produce $(n, 2n)$, (n, p), and (n, α) reaction products with almost all elements. Reaction 14 is only of limited use in activation analysis. Reaction 15 produces a broad range of neutron energies, up to 5 MeV, and is the most intense source for neutrons of 1 MeV.

Commercial 14 MeV neutron generators have been primarily used for fast-neutron activation studies. One survey of neutron accelerators was reported in 1965 (15); most laboratories utilize 150 to 200 keV generators based on reaction 13; the cost is $15,000 to $25,000; the shielding of neutron generators is often quite expensive, but can sometimes be done simply and inexpensively if the neutron tube can be installed in a small hole in the ground and then covered with plastic chips. Also, tritium targets have a finite lifetime, especially if the target heats up, resulting in loss of tritium; some tritium targets can be regenerated. Most generators have their own transfer systems which accurately locate the samples and standards each time; because the flux varies so much with position, only one sample or one standard is irradiated at a time.

Generally, the detection limits are from a few to several hundred micrograms (see Table 8.2) for calculated limits of detection—calculated from equation 11 assuming $\phi = 10^{13}$ m^{-2} s^{-1}. However, because samples as large as a few grams can be measured, the concentrational detection limits are of the order of parts per million. In Table 8.A.2 not only are detection limits given, but also nuclear reactions, half-lives of the radioactive isotopes, and primary and secondary interferences. Because fast-neutron generators are small and portable and of constant flux for many hours, and because of the capacity for selecting the energy of the neutrons, the use of fast neutrons is rapidly gaining favor among workers in the area of neutron activation analysis.

To improve the precision and accuracy of measurement, the sample can be rotated (16) and γ-ray absorption errors can be corrected (17). Fast neutrons have particularly been found useful for light elements, for example, nitrogen, oxygen, fluorine, silicon, and copper.

c. Isotopic Neutron Sources

Isotopic neutron sources have *not* been extensively used in neutron activation analysis because of their low fluxes. Many combinations of α- and γ-emitters with light elements produce neutron sources, for example, ^{124}Sb–Be, ^{241}Am–Be, ^{210}Po–Be, and ^{228}Th–Be, but unfortunately these sources have fluxes of 10^7 neutrons s^{-1} or less; ^{252}Cf–Be results in a neutron flux of 10^9 neutrons s^{-1} mg^{-1}. The activity of these sources is highly dependent on their temperature. The cost of these sources for maximum activity is generally less than $1500.

d. Charged Particle Sources

The use of charged particle sources has been limited by the small penetration depths of such particles and the need for dissipation of large amounts of heat. However, charged particles have found some use in the activation of light elements, such as carbon, oxygen, nitrogen, and fluorine, which cannot be conveniently determined after activation by reactor neutrons, and in the study of small surface areas. The Coulomb barrier restricts the reaction of low-energy charged particles to elements of low atomic number. As the energy of the particles increases, many nuclear interferences arise. Charged particles can be obtained from Cockcroft–Walton voltage multipliers, Van de Graff generators, cyclotrons, and linear accelerators. Standardization is more difficult with charged particle activation because of the variation in stopping power of the particles as the composition (matrix) occurs.

In summary, charged particle activation analysis produces high sensitivity (and low detection limits) for light elements, and the method is most useful for surfaces and thin layers. However, selectivity is a problem. In the future, small cyclotrons may become available for ^{3}He particles of $\sim$10 MeV, [e.g., Ricci (17) showed that a 100 μA beam of 10 MeV ^{3}He ions impinging on a thick lithium, beryllium, or boron target produces a neutron yield of 2×10^{11} n s^{-1}, comparable to ordinary neutron generators; also, the neutrons are monoenergetic, which could be used to control selectivity of analysis; thus ^{3}He activation could be useful for concentrations of parts per million and below of certain elements].

e. γ-Ray Photons

γ-Photons provide an alternative means as compared to neutrons to induce nuclear reactions. Most γ-ray activations are based on (γ, n) reactions, but particles other than neutrons can be ejected; photoneutron thresholds usually exceed 5 MeV, but photons of higher energy are normally used to obtain good detection limits (see Table 8.3). Samples are irradiated with γ-rays, bremsstrahlung, from betatrons or linear accelerators. The great penetration of γ-rays allows simple encapulation of targets and pneumatic transfer systems (18).

γ-Rays have been used to activate several low-molecular-weight elements, such as oxygen, nitrogen, and carbon. γ-Ray activation has also been used to analyze trace elements in ores and other materials. However, because large accelerators are needed to give adequate photon activity, γ-photon activation will probably be used only as a reference method to supplement more conventional methods of analysis and will seldom be

TABLE 8.3. Limits of Detection for Photon Activation[a,b]

Element	Reaction	Product Half-Life	Limit of Detection (μg) 10 min Irradiation	Limit of Detection (μg) 4 hr Irradiation
Sb	$^{121}Sb(\gamma, n)^{120m}Sb$	5.8 days	43	1.3
	$^{121}Sb(\gamma, n)^{122}Sb$	2.8 days	13	0.5
Ar	$^{40}Ar(\gamma, p)^{30}Cl$	55.5 min	0.8	0.1
As	$^{75}As(\gamma, n)^{74}As$	17.9 days	100	5
Br	$^{79}Br(\gamma, n)^{78}Br$	6.5 min	0.04	0.02
	$^{81}Br(\gamma, n)^{80m}Br$	4.4 hr	7	0.2

TABLE 8.3. (*Continued*)

Element	Reaction	Product Half-Life	Limit of Detection (μg)	
			10 min Irradiation	4 hr Irradiation
Cd	$^{113}Cd(\gamma,p)^{112}Ag$	3.2 hr	200	10
	$^{114}Cd(\gamma,p)^{113}Ag$	5.3 hr	130	8
C	$^{12}C(\gamma,n)^{11}C$	20.5 min	0.6	0.2
Cs	$^{133}Cs(\gamma,n)^{132}Cs$	6.58 days	13	0.6
	$^{140}Cs(\gamma,n)^{139}Cs$	140 days	400	13
Cl	$^{35}Cl(\gamma,n)^{34}Cl$	32.0 min	0.8	0.2
Cr	$^{50}Cr(\gamma,n)^{49}Cr$	42 min	10	1.3
	$^{53}Cr(\gamma,p)^{52}V$	3.76 min	2	2
Co	$^{59}Co(\gamma,n)^{58}Co$	71.3 days	600	20
Cu	$^{63}Cu(\gamma,n)^{62}Cu$	9.9 min	0.08	0.04
	$^{65}Cu(\gamma,n)^{64}Cu$	12.8 hr	8	0.4
Dy	$^{158}Dy(\gamma,n)^{157}Dy$	8.1 hr	1000	40
Er	$^{166}Er(\gamma,n)^{165}Er$	10.3 hr	4	0.13
F	$^{19}F(\gamma,n)^{18}F$	1.83 hr	5	0.4
Gd	$^{160}Gd(\gamma,n)^{159}Gd$	18 hr	8	0.4
Ge	$^{76}Ge(\gamma,n)^{75}Ge$	82 min	2	0.2
	$^{70}Ge(\gamma,n)^{69}Ge$	38 hr	40	1.3
Au	$^{197}Au(\gamma,n)^{196}Au$	6.2 days	10	0.5
Ho	$^{165}Ho(\gamma,n)^{164}Ho$	37 min	0.06	0.01
In	$^{115}In(\gamma,n)^{114}In$	50 days	100	5
	$^{115}In(\gamma,\gamma')^{115m}In$	4.5 hr	10	0.6
I	$^{127}I(\gamma,n)^{126}I$	13.2 days	20	1
Ir	$^{191}Ir(\gamma,n)^{190}Ir$	11 days	60	2
Fe	$^{54}Fe(\gamma,n)^{53}Fe$	8.5 min	1.3	0.8

332

TABLE 8.3. (*Continued*)

| Element | Reaction | Product Half-Life | Limit of Detection (μg) | |
			10 min Irradiation	4 hr Irradiation
Pb	$^{204}Pb(\gamma,n)^{203}Pb$	52 hr	200	10
Lu	$^{175}Lu(\gamma,n)^{174}Lu$	300 days	1300	70
Mg	$^{25}Mg(\gamma,p)^{24}Na$	15.0 hr	100	5
	$^{26}Mg(\gamma,p)^{25}Na$	60 s	1.3	1.3
Mn	$^{55}Mn(\gamma,n)^{54}Mn$	303 days	2000	70
Hg	$^{198}Hg(\gamma,n)^{197}Hg$	65 hr	50	2
Mo	$^{92}Mo(\gamma,n)^{91}Mo$	15 5 min	0.7	0.2
	$^{100}Mo(\gamma,n)^{99}Mo$	66 hr	80	4
Nd	$^{142}Nd(\gamma,n)^{141}Nd$	2.5 hr	1	0.08
	$^{150}Nd(\gamma,n)^{149}Nd$	1.8 hr	4	0.2
Ni	$^{58}Ni(\gamma,n)^{57}Ni$	36 hr	60	2
Nb	$^{93}Nb(\gamma,n)^{92}Nb$	10.2 days	40	1.3
N	$^{14}N(\gamma,n)^{13}N$	10.0 min	0.2	0.1
Os	$^{192}Os(\gamma,n)^{191}Os$	15 days	70	4
O	$^{16}O(\gamma,n)^{15}O$	21 min	0.13	0.13
Pd	$^{104}Pd(\gamma,n)^{103}Pd$	17 days	1500	20
P	$^{31}P(\gamma,n)^{30}P$	2.50 min	0.13	0.1
K	$^{39}K(\gamma,n)^{38}K$	7.71 min	0.4	0.2
Pr	$^{141}Pr(\gamma,n)^{140}Pr$	3.4 min	0.01	0.01
Re	$^{187}Re(\gamma,n)^{186}Re$	90 hr	10	0.5
Rh	$^{103}Rh(\gamma,n)^{102}Rh$	206 days	800	40
Ru	$^{96}Ru(\gamma,n)^{95}Ru$	99 min	5	0.4
	$^{104}Ru(\gamma,n)^{103}Ru$	40 days	700	20

TABLE 8.3. (*Continued*)

Element	Reaction	Product Half-Life	Limit of Detection (μg)	
			10 min Irradiation	4 hr Irradiation
Rb	^{87}Rb(γ,n)^{86}Rb	18.7 days	200	10
Sm	^{144}Sm(γ,n)^{143}Sm	8.9 min	0.8	0.5
	^{154}Sm(γ,n)^{153}Sm	47 hr	20	0.8
Si	^{29}Si(γ,p)^{28}Al	2.31 min	1.3	1.3
	^{30}Si(γ,p)^{29}Al	6.6 min	5	4
Ag	^{107}Ag(γ,n)^{106}Ag	24 min	0.1	0.04
Na	^{23}Na(γ,n)^{22}Na	2.6 yr	20,000	800
Sr	^{86}Sr(γ,n)^{85}Sr	64 days	4000	130
S	^{32}S(γ,n)^{31}S	2.7 s	0.2	0.2
Ta	^{181}Ta(γ,n)^{180}Ta	8.1 hr	0.6	0.04
Tb	^{159}Tb(γ,n)^{158}Tb	150 y	130,000	5000
Tl	^{203}Tl(γ,n)^{202}Tl	12 days	70	4
Tm	^{169}Tm(γ,n)^{168}Tm	85 days	200	8
Sn	^{112}Sn(γ,n)^{111}Sn	35 min	7	1.3
	^{124}Sn(γ,n)^{123}Sn	125 days	600	200
Ti	^{46}Ti(γ,n)^{45}Ti	3.09 hr	40	2
	^{48}Ti(γ,p)^{47}Sc	3.4 days	200	7
W	^{186}W(γ,n)^{185}W	75 days	500	20
Yb	^{168}Yb(γ,n)^{167}Yb	18 min	20	8
	^{176}Yb(γ,n)^{175}Yb	4.2 days	70	4
Y	^{89}Y(γ,n)^{88}Y	108 days	500	20
Zn	^{64}Zn(γ,n)^{63}Zn	38.4 min	0.4	0.06
Zr	^{90}Zr(γ,n)^{89}Zr	78.4 hr	20	1

TABLE 8.3. (*Continued*)

[a] References:
1. G. J. Lutz, *Anal. Chem.*, **43**, 93 (1974).
2. G. J. Lutz, *Anal. Chem.*, **41**, 424 (1969).
[b] Table 8.4 lists the elements in alphabetical order, possible reactions, half-lives of the daughter products, and estimated detection limits (in μg) for two different times of irradiation and for an electron energy of 30 MeV.

The estimated limits of detection are based on the following assumptions: (1) The excitation source is bremsstrahlung from an electron beam current of 100 mA striking a tungsten target 0.6 cm thick for an electron energy of 30 MeV. Calculations are from Lutz (1,2). (2) A correction was made by Lutz for attenuation of photons in the target. The angular distribution was calculated from the Schiff equation. It was assumed the sample could be positioned to intercept the beam included in the 50° angle from the forward direction. Irradiation was for 10 min and for 4 hr. (3) A detectable emitting rate of 40 disintegrations s^{-1} and a counting time of 1 s.

The expression used to calculate the detection limits was $w_L = 40 \text{ s}^{-1}/m$, where m is the sensitivity (in disintegrations $s^{-1} \mu g^{-1}$) as determined by Lutz (2). p. 320.

used for trace analytical work. In addition, a further disadvantage of γ-ray activation is the production of positron emitting nuclides, so that γ-ray spectrometry cannot be as useful as it is for neutron activation. However, an advantage over neutron activation is the elimination of self-shielding problems, so that it is possible to measure impurities in high neutron absorption cross-sectional matrices, such as boron, tungsten, cadmium, and holmium.

It should also be mentioned that, instead of measuring the induced radioactivity, the neutron product itself may be measured. Finally, Mössbauer spectroscopy, in which γ-ray irradiation and absorption are used to determine structure, could be listed as a type of γ-ray activation.

f.   Survey of Activation Analysis Techniques

In Table 8.4 a comparison of the above activation sources is given.

TABLE 8.4.   Limits of Detection for Photon Activation[a, b]

Type	Bombarding Particle	Main Source	Main Nuclear Reactions
Conventional	Thermal (slow) neutrons	Reactors	(n, γ)
Conventional	Fast neutrons	14 MeV neutron generators, cyclotrons, reactors	$(n, 2n), (n, p), (n, \alpha)$
Conventional	Bremsstrahlung	Betatrons	$(\gamma, n), (\gamma, p)$
Conventional	Charged particles $(p, d, {}^3\text{He})$	Cyclotrons	Several
Prompt	Thermal neutrons	Reactors	(n, γ)
	Charged particles $(p, \ )$	Cyclotrons, linear accelerators	$(p, \alpha), (\alpha, p)$

5. MEASUREMENT OF PROMPT RADIATION

Prompt radiation measurement implies measurement of γ-ray emission while the sample is in its irradiation position. The advantages of this approach are speed of analysis (because counting is performed during the irradiation rather than after it) and the irradiation can last as long as needed to give a statistically useful count. The disadvantages are the impossibility of performing chemical separations, surface etching prior to counting, and the lack of specificity in many cases.

6. ISOTOPIC DILUTION

If it is necessary to deal with a matrix from which it is impossible to isolate the isotope of analytical interest, or if a chemical reaction is a necessary part of the analytical procedure being studied, it is sometimes desirable to utilize the method of isotopic dilution in one of its several variations. This technique is similar to the method of standard[†] addition, except that the standard is a known activity rather than a known amount of the analytical species. The important aspect of this is that the method incorporates the advantages of internal standardization as well as those of standard[†] addition.

The basis of the method can be described as follows. If an unknown number of atoms of the element of analytical interest N_u (which has an unknown number of radioactive isotopes N_u^*) to which is added a known number of atoms of the same element N_a with a known number of the same radioactive isotope N_a^*, the ratio of radioactive isotopes N_t^* to the total number of atoms of that element N_t is given by:

$$\frac{N_t^*}{N_t} = \frac{N_u^* + N_a^*}{N_u + N_a} \tag{16}$$

If this is accomplished prior to some separation procedure or other chemical operation that is less than stoichiometric, the ratio in the separated sample can be given in the same way, that is,

$$\frac{N_s^*}{N_s} = \frac{N_t^*}{N_t} \tag{17}$$

where the subscript s indicates the number of respective atoms (or isotopes) in the separated mixture.

[†]"Standard" and "reference" addition refer to the same process.

If the above situation can be realized when the number of added atoms is very small with respect to the unknown number ($N_u \gg N_a$) and the original amount of radioactive isotopes is negligible ($N_u^* = 0$), equation 16 can be rewritten:

$$N_u = N_s \frac{N_a^*}{N_s} \tag{18}$$

which can also be expressed in terms of activities R:

$$N_u = N_s \frac{R_a}{R_s} \tag{19}$$

If the characteristics of the radioactive isotopes are known, this equation can be related to the weight w and specific activity S (specific activity is activity divided by the weight of the active species) by means of an analogous equation:

$$w_u = w_s \frac{S_a}{S_s} \tag{20}$$

If the added amount of the analytical element is significant when compared to the unknown amount, w_u is given by:

$$w_u = w_a \left(\frac{S_a}{S_s} - 1 \right) \tag{21}$$

This technique is not generally applicable to more than one unknown element at a time. However, if an energy-dispersive readout is available, several unknown elements can be studied simultaneously. There are several other variations of this technique, but they are all quite similar. Generally, these techniques are quite sensitive if the activity of the added isotope is made large. The most important advantage of the method is that the separation techniques do not have to be quantitative for the method to be successful.

7. INTERFERENCES AND SYSTEMATIC ERRORS IN ACTIVATION ANALYSIS

Although specific errors have been discussed in the previous sections with respect to the source of excitation, they are briefly discussed here in a general manner. Interfering reactions and self-shielding constitute the major source of systematic errors.

Interfering nuclear reactions are reactions that modify the linear behavior between the amount of an element to be determined and the amount of the indicator radionuclide produced. *Primary interfering nuclear reactions* are those induced by the primary bombarding radiation on elements other than the element to be determined, but these reactions yield the same indicator radionuclide as do the desired nuclear reactions. Also, reactions producing a radionuclide of a neighboring element which decays into the indicator radionuclide belong to this class. *Secondary interfering nuclear reactions* are those occurring between secondary particles and elements *other than* the ones to be determined, but which produce the indicator radionuclide. Thus, secondary reactions involve the production of particles via the exciting flux and the particles; for example, protons in the case of fast neutrons, neutrons in the case of photons, and neutrons in the case of deuterons result in a nuclear reaction producing the indicator radionuclide. Finally, second-order reactions occurring with the transformation products of the sample constituent can also cause systematic errors; for example, the indicator radionuclide can be greatly decreased if it is further activated by the original bombarding species or radiation, or the amount of nuclide to be activated can be enhanced by the decay of a short-lived radionuclide of a neighboring element.

Self-shielding effects occur when the sample matrix attenuates the exciting radiation. This effect is particularly important in thermal neutron activation and is described by Beer's law:

$$\phi_x = \phi_0 e^{-N\sigma x} \tag{22}$$

where ϕ_0 is the incident flux, ϕ_x is the transmitted flux at a depth x, N is the number of atoms of a given element (in m^{-3}), and σ is the neutron absorption cross section of a given element (in m^2).

Neutron scattering is the major attenuation effect with fast (14 MeV) neutrons. Attenuation of bremstrahlung is usually important. In charged particle activation, self-shielding arises via the rapid decrease in energy of the incident particles as they penetrate into the matrix, and so self-shielding is important in this case.

In Table 8.5, a comparison of errors (systematic and random) and their approximate magnitudes are given. It is apparent from the magnitudes of the systematic errors of the nuclear data why absolute neutron activation analysis is not used for routine trace analysis.

8. RADIOCHEMICAL SEPARATIONS

In most high-sensitivity applications, radiochemical separations are essential. As better resolution detectors become available, radiochemical separations will be needed less but will certainly still be required in the

TABLE 8.5. Comparison of and approximate Magnitudes of Errors in Neutron Activation Analysis (1–4)[a]

Error Type	Magnitude of Error (%)
Chemical manipulations	
Sample weight	±1
Separation efficiency determination	±2
Irradiation	
Self-shielding (correction 50%)	±4
Flux depression	±2
Thermal enhancement	±2
Absolute value of thermal flux	±5
Value of cadmium ratio	±2
Irradiation time (1 min)	±3
Inhomogeneity of neutron flux	±1
Counting	
Detection calibration	±3
Counting rate (10^3 s^{-1})	±4
Geometric factors	±1
Interfering and competing nuclear reactions	variable
Nuclear data	
Half-lives	±2–10
Decay schemes	±2–50
Thermal activation cross sections	±5–30

[a]References:
1. E. Rocci and R. L. Stah, *Anal. Chem.*, **39**, 794 (1967).
2. J. P. Cali, J. R. Weiner, and G. G. Rocco, *Proceedings of the International Conference on Modern Trends in Activation Analysis*, Texas A and M University, College Station, Texas, 1965, p. 253.
3. F. Girardi, G. Gerzzi, and J. Pauly, *Anal. Chem.*, **36**, 1588 (1964).
4. F. Krivan, "Nuclear Data for Activation Analysis," in *Nuclear Data in Science and Technology*, Vol. II, International Atomic Energy Agency, Vienna, 1973, p. 193.

majority of activation analysis experiments. Separation procedures can be as simple as removal of the major activity component, or as complex as removal of groups of elements. A series of monographs on the radio-chemistry of the elements is available from the U.S. National Academy of Sciences Sub-Committee on Radiochemistry (Nuclear Science Series), and these monographs give radiochemical procedures for the separation of a single element from almost any combination of other elements.

Inactive carriers are usually present during solution of the sample to correct for losses during the sample processing steps; when losses are expected to be small, chemical treatment of the sample prior to irradiation is rarely performed.

Precipitation (especially from homogeneous solution) is a widely used separation means for scavenging, for example, separation of one or more nuclides from several others. Extraction is also a widely used means for radiochemical separations (19, 20). Differential migration techniques, inorganic ion exchangers, reversed-phase partition chromatography, paper chromatography, distillation, electrodeposition, amalgam exchange, and the ring oven method are all separation methods that have been applied to radiochemistry.

Separations of short-lived radionuclides must be done rapidly after completion of irradiation, and so special techniques are required (1, 6). Automatic separation systems have also been devised by some workers (21–24).

9.   γ-RAY SPECTROMETERS

γ-Ray spectrometry is the most widely used method of measurement in activation analysis; the technique involves absorption of the γ-rays in a detector, generally a NaI crystal, and then processing the pulses with a multichannel analyzer. γ-Ray spectra of a large number of species have been published by Crothamel (25) and Heath (26). More recently, computer search techniques have been used for identification purposes (27).

Quite recently, higher-resolution semiconductor detectors have become available, such as the lithium-drifted germanium detector. The resolving power (defined as the center peak energy E divided by the peak width ΔE at half-maximum, i.e., $E/\Delta E$) is 10^3 for the 660 keV γ-ray of ^{13}Cs, whereas with a NaI detector the resolving power is $\sim$14. Unfortunately, semiconductive detectors must be operated at liquid nitrogen temperature, and the volume of the detector is small, resulting in lower efficiency than with NaI detectors, for example, $\sim$5 versus $\sim$75 cm^3, resulting in an efficiency of $\frac{1}{100}$ at 200 kV for the semiconductive detector versus the NaI detector; actually the efficiency becomes worse as the energy of the γ-rays increases. Because of the greater resolution of semiconductive detectors, computer methods with their greater number of channels, as well as computational and control features, are gradually replacing multichannel pulse height analyzers.

Energy drift of the γ-ray spectrometer can be controlled by the use of a reference emitter. The use of standards is also necessary (see Sections 2,6,7) to correct for irradiative conditions, counting geometry, and so on.

By means of coincidence counting, several nuclides emitting in cascade can be measured; coincidence methods are plagued with several practical difficulties with semiconductive detectors. The Compton continuum can be compensated for by several approaches (6) involving background subtraction, special instrumentation, and coincidence methods.

The main errors arising in activation analysis are those due to (1) self-shielding, (2) unequal flux at sample and standard position, (3) inaccurate counting procedures, for example, differences in self-scattering, self-absorption, and geometry between sample and standard, and (4) counting statistics. Certainly, if no special problems involving competing nuclear reactions or self-shielding occur, and where the activity and counting time are adequate to give negligible counting errors, the systematic errors and precision can be less than $\pm 10\%$ which is normally obtained (acceptable) in routine activation analysis.

10. APPLICATIONS OF ACTIVATION ANALYSIS

It is not possible to review critically all the fields in which activation analysis has been used for trace analysis. However, activation analysis has made substantive contributions to the following areas: trace constituents in high-purity materials; trace components in biological materials; forensic science; major, minor, and trace elements in geochemical (terrestrial and extraterrestrial) samples; on-line analysis; surface analysis (often via charged particle activation); air and water pollution; and semiconductor analysis.

B. RADIOACTIVE TRACERS (1, 28–30)

The use of radioactive tracers to avoid the necessity for quantitative separations in conventional methods of analysis has been relatively sparse, even though tracers are a superb means of checking the efficiency of separations of small amounts of species, and the equipment and tracer reagents cost little. Tracer methods for determining elements can be divided into three groups: (1) a radioactive tracer of the element to be determined (isotopic labeling); (2) a radioactive tracer of an element different from that being determined (nonisotopic labeling), and (3) radiometric titrations. In *isotopic labeling methods*, the principle is that, if a fraction of the element to be determined can be isolated (that fraction being determined by measuring the radioactivity), and if the mass of the element in that fraction can be measured either via a conventional method or via reaction with an equivalent amount of reagent, the original mass of the element can be estimated. In *nonisotopic labeling methods*, the principle is that, if the element to be determined can be reacted in some quantitative manner with a known excess of a second labeled material, and if the fraction of the labeled material thus released can be measured (via radioactivity), the original mass of the desired element can be estimated. In *radiometric titrations*, the course of the titrations is followed either with a labeled titrand, a labeled titrant, or a labeled indicator element. The measurement of the change in radioactivity within the system often involves a phase change.

However, it is generally recognized that, if irradiation facilities are available, activation analysis is preferred to radiochemical tracer methods, partly because of the lack of a reagent blank in the former method. Nevertheless, where activation analysis is impossible or difficult, because of self-shielding problems, counting problems, too great an overall radioactivity of the sample, or nuclear reactions, and because of the great simplicity and low cost of radioactive tracer methods, radioactive tracer methods have found and will continue to find use in specialized trace analysis problems.

The basic equations for the use of radioactive tracers for analysis were given in Chapter 4.

REFERENCES

1. A. A. Smales, "Radioactivity Techniques in Trace Characterization," *Trace Characterization*, W. W. Meinke and B. F. Scribner, Eds. NBS Monograph 100, U.S. Govt. Printing Office, Washington, D.C.

2a. R. C. Koch, *Activation Analysis Handbook*, Academic Press, New York, 1960.

2b. P. Kruger, *Principles of Activation Analysis*, Wiley-Interscience, New York, 1971.

3a. D. De Soete, R. Gijbels, and J. Hoste, *Neutron Activation Analysis*, Wiley-Interscience, New York, 1972.

3b. W. S. Lyon, *Guide to Activation Analysis*, D. Van Nostrand, New York, 1964.

4. S. S. Nargolwalla and E. P. Przybylewicz, *Activation Analysis with Neutron Generators*, Wiley-Interscience, New York, 1973.

5. H. R. Lukens, *J. Chem. Ed.*, **44**, 668 (1967).

6. R. F. Coleman and T. B. Pierce, *Analyst*, **92**, 1 (1967).

7. T. Hogdahl, *Radiochemical Methods of Analysis*, Vol. 1, International Atomic Energy Agency, Vienna, 1965, p. 23.

8. Y. Kamenoto, *Int. J. Appl. Radiat. Isotopes*, **15**, 447 (1964).

9. D. C. Burg, P. E. Segar, P. Kienle, and L. Campbell, *Int. J. Appl. Radiat. Isotopes*, **11**, 10 (1961).

10. H. R. Lukens, H. P. Yule, and V. P. Guinn, *Nucl. Instrum. Methods*, **33**, 273 (1965).

11. H. P. Yule, H. R. Lukens, and V. P. Guinn, *Nucl. Instrum. Methods*, **33**, 277 (1965).

12. F. Girardi, G. Gazzi, and J. Pauly, *Anal. Chem.*, **36**, 1588 (1964).

13. R. F. Coleman, *Analyst*, **86**, 39 (1961).

14. R. F. Coleman, *Analyst*, **87**, 590 (1962).

15. *Nucleonics*, **23**, 4 (1965).

16. O. U. Anders and D. W. Briden, *Anal. Chem.*, **36**, 287 (1963).

17. E. Ricci, *Proceedings of the International Conference on Modern Trends in Activation Analysis*, Texas A and M University, College Station, Texas, 1965, p. 200.

18. C. Engelmann, *Commissariart a l'Energie Atomique Report* 2559 (1964).

19. G. H. Morrison and H. Freiser, *Solvent Extraction in Analytical Chemistry*, John Wiley, New York, 1962.

20. G. H. Morrison, *Anal. Chem.*, **36**, 93R (1964).

21. F. Girardi, M. Merlini, J. Pauly, and R. Pietra, *Radiochemical Methods of Analysis*, Vol. 2, International Atomic Energy Agency, Vienna, 1965, p. 1.

22. K. Samsahl, *Ab. Atomenergi, Stockholm Rapp.*, AE 159 (1964).

23. K. Samsahl, *Ab. Atomenergi, Stockholm Rapp.*, AE 215 (1966).

24. F. Girardi, *Proceedings of the International Conference on Modern Trends in Activation Analysis*, Texas A and M University, College Station, Texas, 1965, p. 337.

25. C. E. Crouthamel, Ed., *Applied Gamma Ray Spectroscopy*, Pergamon Press, Oxford, 1960.

26. R. L. Heath, U.S. Atomic Energy Commission Report, IDO-16880, 1964.

27. G. D. O'Kelley, Ed., *Applications of Computers to Nuclear and Radiochemistry*, NAP-NS-3107, USAEC, 1962.

28. J. W. McMillan, AERE Report R, 5266 (1966).

29. J. Ruzicka and J. Stary, *At. Energy Rev.*, **2**, 3 (1964).

30. H. A. McKay, *Principles of Radiochemistry*, Butterworths, London, (1971).

X-RAY METHODS

M. L. PARSONS

Department of Chemistry
Arizona State University
Tempe, Arizona

A. X-RAY INSTRUMENTATION

X-ray spectroscopy is more closely associated with nuclear techniques than with UV and visible techniques. The X-ray region of the electromagnetic spectrum covers the wavelength region from about 10^{-5} to about 100 Å; however, the wavelength region of analytical usefulness is from about 0.1 to 20 Å, corresponding to an energy range of approximately 0.6 to 124 keV. This energy range is more suited to ionization-type detectors similar to those used in detecting γ- and β- radiations than to photomultipliers used in optical spectroscopy. Energy-dispersive measuring techniques using multichannel analyzers are becoming more prominent in the field. However, wavelength-dispersive X-ray spectrometers are still used quite extensively, and each type has certain advantages.

1. CHOICE OF X-RAY COMPONENTS IN WAVELENGTH-DISPERSIVE SPECTROMETERS

X-ray spectrometric systems have many features in common: the X-ray tube power supply, the X-ray tube, collimators, sample chambers, the inert gas-flow system or vacuum system, crystals, and detectors with associated electronics. Each is discussed in turn.

a. Power Supplies

The power supply for an X-ray tube consists of a highly regulated (with respect to both current and voltage) high-voltage (50 to 100 kV) DC source with a current flow of at least 0.05%, or 25 V and 25 μA. Further, the power supply must have variable control of both voltage and current. Of course, automatic cutoff features are also valuable and necessary to protect the rather expensive X-ray tubes.

b. X-ray Tubes

The devices used to produce a high-intensity flux of X-rays in analytical work are quite simple; they consist of a tungsten filament which is heated to produce electrons by thermionic emission. These electrons are accelerated to a positive target which is at ground potential with respect to a negative potential of up to 50 kV or more at the filament. X-ray tubes contain associated focusing elements, and the target is water-cooled to prevent the tube from burning up. A low-atomic-number window (usually beryllium) is used to transmit the beam of X-rays to the sample. Of course, the tube must be evacuated to permit electron penetration to the target. Finally, safety precautions must be taken to protect the user from stray X-radiation and/or electrical shock, and the tube itself must have automatic shut-off capabilities in case the cooling water flow stops or the temperature increases past a certain point, because the target will quickly burn away.

The X-rays from this type of source consist of two types—a continuum (bremsstrahlung) and line emission. The continuum is essentially the result of electrons being decelerated in the mass of the target. The continuum has three major characteristics: a low-wavelength (high-energy) cutoff, a maximum which is approximately 1.5 times the low-wavelength cutoff, and a gradual tailing off to the high-wavelength range. The short-wavelength cutoff, λ_0, is given by the Duane–Hunt law:

$$\lambda_0 = \frac{12,400}{V} \tag{1}$$

where λ_0 is the short-wavelength cutoff (in Å), and V is the potential of the power supply (in V). It should also be pointed out that the intensity of the continuum I_c from any particular X-ray tube is proportional to several parameters:

$$I_c \propto iV^2Z \tag{2}$$

where i is the tube current (in A), V is its potential (in V), and Z is the atomic number of the target (no units).

The *line spectra* from a given target of an X-ray tube originate from the relaxation of electrons from a higher quantum level to a lower one in the inner electronic levels for the target atoms. For example, if the impact of the electron beam on the target of an X-ray tube causes an inner electron to be removed from the first major quantum level (K shell), an electron from the second major quantum level (L shell) may fill this vacancy, with the resultant emission of a characteristic X-ray corresponding to the energy difference between the two levels (see Figure 9.4). The major analytically useful emission lines for nearly all elements are tabulated in Tables 9.1 and 9.2. X-ray tubes are available with targets of tungsten, platinum, gold,

TABLE 9.1. Wavelengths (Å) of the Principal X-Ray Spectral Lines of the Elements-*K* Series (1)

Line $\begin{pmatrix} \text{Electron} \\ \text{transition} \\ K \leftarrow \\ \text{Approximate} \\ \text{relative} \\ \text{intensity} \end{pmatrix}$	Wavelength (Å)		
	α	α_1	α_2
		LIII	LII
	150	100	50
3-Li	240	—	—
4-Be	113	—	—
5-B	67	—	—
6-C	44	—	—
7-N	31.603	—	—
8-O	23.707	—	—
9-F	18.307	—	—
10-Ne	14.615	—	—
11-Na	11.909	—	—
12-Mg	9.889	—	—
13-Al	8.339	8.338	8.341
14-Si	7.126	7.125	7.127
15-P	6.155	6.154	6.157
16-S	5.373	5.372	5.375
17-Cl	4.729	4.728	4.731
18-Ar	4.192	4.191	4.194
19-K	3.744	3.742	3.745
20-Ca	3.360	3.359	3.362
21-Sc	3.032	3.031	3.034
22-Ti	2.750	2.749	2.753
23-V	2.505	2.503	2.507
24-Cr	2.291	2.290	2.294
25-Mn	2.103	2.102	2.105
26-Fe	1.937	1.936	1.940
27-Co	1.791	1.789	1.793
28-Ni	1.659	1.658	1.661
29-Cu	1.542	1.540	1.544
30-Zn	1.437	1.435	1.439
31-Ga	1.341	1.340	1.344
32-Ge	1.256	1.255	1.258
33-As	1.177	1.175	1.179
34-Se	1.106	1.105	1.109
35-Br	1.041	1.040	1.044
36-Kr	0.981	0.980	0.984
37-Rb	0.927	0.926	0.930
38-Sr	0.877	0.875	0.880
39-Y	0.831	0.829	0.833
40-Zr	0.788	0.786	0.791

TABLE 9.1. (*Continued*)

Line $\left(\begin{array}{c}\text{Electron} \\ \text{transition} \\ K\leftarrow \\ \text{Approximate} \\ \text{relative} \\ \text{intensity}\end{array}\right)$	Wavelength, Å		
	α	α_1	α_2
		LIII	LII
	150	100	50
41-Nb	0.748	0.747	0.751
42-Mo	0.710	0.709	0.713
43-Tc	0.674	0.673	0.676
44-Ru	0.644	0.643	0.647
45-Rh	0.614	0.613	0.617
46-Pd	0.587	0.585	0.590
47-Ag	0.561	0.559	0.564
48-Cd	0.536	0.535	0.539
49-In	0.514	0.512	0.517
50-Sn	0.492	0.491	0.495
51-Sb	0.472	0.470	0.475
52-Te	0.453	0.451	0.456
53-I	0.435	0.433	0.438
54-Xe	0.418	0.416	0.421
55-Cs	0.402	0.401	0.405
56-Ba	0.387	0.385	0.390
57-La	0.373	0.371	0.376
58-Ce	0.359	0.357	0.362
59-Pr	0.346	0.344	0.349
60-Nd	0.334	0.332	0.337
61-Pm	0.322	0.321	0.325
62-Sm	0.311	0.309	0.314
63-Eu	0.301	0.299	0.304
64-Gd	0.291	0.289	0.294
65-Tb	0.281	0.279	0.284
66-Dy	0.272	0.270	0.275
67-Ho	0.263	0.261	0.266
68-Er	0.255	0.253	0.258
69-Tm	0.246	0.244	0.250
70-Yb	0.238	0.236	0.241
71-Lu	0.231	0.229	0.234
72-Hf	0.224	0.222	0.227
73-Ta	0.217	0.215	0.220
74-W	0.211	0.209	0.213
75-Re	0.204	0.202	0.207
76-Os	0.198	0.196	0.201
77-Ir	0.193	0.191	0.196
78-Pt	0.187	0.185	0.190

TABLE 9.1. (*Continued*)

Line	Wavelength, Å		
	α	α_1	α_2
⎛ Electron			
transition			
$K\leftarrow$		LIII	LII
⎝ Approximate relative intensity	150	100	50
79-Au	0.182	0.180	0.185
80-Hg	0.177	0.175	0.180
81-Tl	0.172	0.170	0.175
82-Pb	0.167	0.165	0.170
83-Bi	0.162	0.161	0.165
84-Po	0.158	0.156	0.161
85-At	0.154	0.152	0.157
86-Rn	0.150	0.148	0.153
87-Fr	0.146	0.144	0.149
88-Ra	0.142	0.140	0.145
89-Ac	0.138	0.136	0.141
90-Th	0.135	0.133	0.138
91-Pa	0.131	0.129	0.134
92-U	0.128	0.126	0.131
93-Np	0.125	0.123	0.128
94-Pu	0.122	0.120	0.125
95-Am	0.119	0.117	0.122
96-Cm	0.116	0.114	0.119
97-Bk	0.113	0.111	0.116
98-Cf	0.110	0.108	0.113

TABLE 9.2. **Wavelengths of the Principal X-Ray Spectral Lines of the Elements L Series (1)**

Line	Wavelength, Å			
	α_1	α_2	β_1	β_2
⎛ Electron				
transition				
LIII$\leftarrow$	MV	MIV		NV
LII$\leftarrow$			MIV	
LI$\leftarrow$				
⎝ Approximate relative intensity	100	10	80	60
16-S	—	—	—	—
17-Cl	—	—	—	—
18-Ar	—	—	—	—

TABLE 9.2. (*Continued*)

Line	α₁	α₂	β₁	β₂
			Wavelength, Å	
$\begin{pmatrix}\text{Electron}\\\text{transition}\\ LIII\leftarrow\\ LII\leftarrow\\ LI\leftarrow\\\text{Approximate}\\\text{relative}\\\text{intensity}\end{pmatrix}$	MV	MIV	MIV	NV
	100	10	80	60
19-K	—		—	—
20-Ca		36.393	36.022	—
21-Sc		31.393	31.072	—
22-Ti		27.445	27.074	—
23-V		24.309	23.898	—
24-Cr		21.713	21.323	—
25-Mn		19.489	19.158	—
26-Fe		17.602	17.290	—
27-Co		16.000	15.698	—
28-Ni		14.595	14.308	—
29-Cu		13.357	13.079	—
30-Zn		12.282	12.009	—
31-Ga		11.313	11.045	—
32-Ge		10.456	10.194	—
33-As		9.671	9.414	—
34-Se		8.990	8.735	—
35-Br		8.375	8.126	—
36-Kr		7.817	7.576	—
37-Rb	7.318	7.325	7.075	—
38-Sr	6.863	6.870	6.623	—
39-Y	6.449	6.456	6.211	—
40-Zr	6.070	6.077	5.836	5.586
41-Nb	5.725	5.732	5.492	5.238
42-Mo	5.406	5.414	5.176	4.923
43-Tc	5.114	5.123	4.887	4.636
44-Ru	4.846	4.854	4.620	4.372
45-Rh	4.597	4.605	4.374	4.130
46-Pd	4.368	4.376	4.146	3.909
47-Ag	4.154	4.162	3.935	3.703
48-Cd	3.956	3.965	3.739	3.514
49-In	3.752	3.781	3.555	3.339
50-Sn	3.600	3.609	3.385	3.175
51-Sb	3.439	3.448	3.226	3.023
52-Te	3.290	3.299	3.077	2.882
53-I	3.148	3.157	2.937	2.751
54-Xe	3.015	3.025	2.803	2.626
55-Cs	2.892	2.902	2.683	2.511
56-Ba	2.776	2.785	2.567	2.404
57-La	2.665	2.674	2.458	2.303
58-Ce	2.561	2.570	2.356	2.208

TABLE 9.2. (*Continued*)

Line	α₁	α₂	β₁	β₂
Electron transition		Wavelength, Å		

Line (Electron transition LIII← LII← LI← Approximate relative intensity)	α_1 MV 100	α_2 MIV 10	β_1 MIV 80	β_2 NV 60
59-Pr	2.463	2.473	2.259	2.119
60-Nd	2.370	2.382	2.166	2.035
61-Pm	2.283	2.292	2.081	1.956
62-Sm	2.199	2.210	1.998	1.882
63-Eu	2.120	2.131	1.920	1.812
64-Gd	2.046	2.057	1.847	1.746
65-Tb	1.976	1.986	1.777	1.682
66-Dy	1.909	1.920	1.710	1.623
67-Ho	1.845	1.856	1.647	1.567
68-Er	1.785	1.796	1.587	1.514
69-Tm	1.726	1.738	1.530	1.463
70-Yb	1.672	1.682	1.476	1.416
71-Lu	1.619	1.630	1.424	1.370
72-Hf	1.569	1.580	1.374	1.327
73-Ta	1.522	1.533	1.327	1.285
74-W	1.476	1.487	1.282	1.245
75-Re	1.433	1.444	1.238	1.206
76-Os	1.391	1.402	1.197	1.169
77-Ir	1.352	1.363	1.158	1.135
78-Pt	1.313	1.325	1.120	1.102
79-Au	1.277	1.288	1.083	1.070
80-Hg	1.242	1.253	1.049	1.040
81-Tl	1.207	1.218	1.015	1.010
82-Pb	1.175	1.186	0.982	0.983
83-Bi	1.144	1.155	0.952	0.955
84-Po	1.114	1.125	0.922	0.929
85-At	1.085	1.097	0.894	0.905
86-Rn	1.057	1.069	0.866	0.881
87-Fr	1.030	1.042	0.840	0.858
88-Ra	1.005	1.017	0.814	0.836
89-Ac	0.980	0.992	0.789	0.814
90-Th	0.956	0.968	0.766	0.794
91-Pa	0.933	0.945	0.742	0.774
92-U	0.911	0.923	0.720	0.755
93-Np	0.889	0.901	0.698	0.736
94-Pu	0.869	0.880	0.678	0.720
95-Am	0.849	0.860	0.658	0.701
96-Cm	0.829	0.841	0.639	0.685
97-Bk	0.810	0.822	0.621	0.669
98-Cf	0.792	0.804	0.603	0.653

molybdenum, chromium, silver, and rhodium. Tungsten is probably used for the most common general-purpose tube, but each has advantages for certain regions of the X-ray spectrum.

c. Collimators

Collimators are the counterpart of lenses in optical spectroscopy; collimators are required to render the X-ray beam as parallel as possible and to focus it in the proper direction. The two most common types of collimators are multiple-tube and multiple-slit arrangements. In all collimator systems, there is a trade-off between intensity of the X-ray beam and the degree of collimation.

d. Sample Chamber

One of the major advantages of X-ray techniques is the variety of sample types that can be directly analyzed. Virtually any type of solid sample can be placed in the sample chamber if the physical dimensions are compatible. Liquids are also analyzed directly if proper precautions are taken to minimize sample evaporation. Fortunately, containers can be prepared from low-atomic-number elements which are "transparent" to X-rays. In fact, X-ray techniques are rivaled only by nuclear techniques with regard to versatility of sample preparation. The sample chamber therefore must be compatible with the overall system and capable of containing a variety of sample holders.

e. System Atmosphere

Air is a suitable atmosphere for an X-ray spectrometer for a limited number of elements. Remember that the wavelengths of X-rays from atomic species vary regularly with atomic number from over 100 Å for beryllium ($Z = 4$) to about 0.1 Å for uranium ($Z = 92$). The high-wavelength (low-energy) cutoff is determined by the atmosphere or lack of it in the spectrometer. Air is satisfactory for elements above vanadium ($Z = 23$). With a helium atmosphere, all elements with an atomic number of 19 (potassium) or greater can be analyzed. An evacuated system permits the measurement of all elements with an atomic number of 9 (fluorine) or greater. Of course, the use of an evacuated system presents certain problems with some sample types, for example, with liquid containment.

f. Crystals

In wavelength-dispersive X-ray spectrometers, the crystal is the dispersive device (based on diffraction) analogous to the diffraction grating is the same device in optical spectrometry. The lattice parameter (*d*-spacing) of the crystal determines the wavelength range of the crystal, and Bragg's law can be directly applied [$\lambda_{upper} = (2d_{upper} \sin\theta)/m$; $\lambda_{lower} = (2d_{lower}\sin\theta)/m$]. Further, the maximum wavelength that can be diffracted is $2d$. In Table 9.3, the $2d$-spacings and useful wavelength ranges for several crystals in wavelength-dispersive X-ray spectrometry are listed.

TABLE 9.3.   Some Properties of Common Diffraction Crystals (1, 2)

Crystal	Value, $2d$(Å)	Wavelength Useful Range (Å) $2\theta = 10°-145°$
α-Quartz	1.624	0.142–1.55
	2.750	0.240–2.62
	6.687	0.583–6.38
	8.510	0.742–8.12
Topaz	2.712	0.236–2.59
Lithium floride	2.848	0.248–2.72
	4.028	0.351–3.84
Fluorite	3.862	0.337–3.68
	6.306	0.550–6.01
Aluminum	4.048	0.353–3.86
	4.676	0.408–4.46
Rock salt	5.641	0.492–5.38
Calcite	6.071	0.529–5.79
Silicon	6.276	0.547–5.98
Germanium	6.532	0.569–6.23
Pentaerythritol (PET)	8.742	0.762–8.34
Ethylenediamine *d*-tartrate (EDDT)	8.808	0.768–8.40
	10.640	0.927–10.15
Ammonium dihydrogen phosphate (ADP)		
Gypsum	15.185	1.32–14.48
Mica (muscorite), cleavage plane	19.84	1.73–18.92
Rock sugar (sucrose)	20.12	1.75–19.19
Potassium Hydrogen Phthalate (KHP)	26.632	2.32–25.40
Octadecyl hydrogen maleate (OHM)	63.5	5.53–60.56
Lead myristate	80.5	7.02–76.77
Lead stearate	100.4	8.75–95.75
Lead lignocerate	130	11.35–124

g. Detectors

In the normal wavelength range used by X-ray spectrometers, proportional, scintillation, and semiconductor detectors are most often used. These have been mentioned in Chapter 8 as they are also the common detectors for nuclear radiations. Many wavelength-dispersive X-ray spectrometers utilize both a proportional detector and a scintillation detector. The useful lower wavelength of the proportional detector is about 0.3 Å, whereas, the scintillation detector allows measurement of X-rays as short as 0.1 Å. The combined signal of both counters offers more sensitivity and is often used.

In energy-dispersive discrimination systems, semiconductor detectors with associated multichannel analyzers are most commonly used; these systems are also discussed in more detail in Chapter 8.

2.   CHOICE OF CONDITIONS

In general, the choice of experimental conditions in X-ray spectrometry involves (a) the form of the sample, (b) the X-ray tube to be used, (c) the voltage and current supplying the tube, (d) the atmosphere in the system, (e) the crystal used in dispersive systems, (f) the detector, (g) the counting and delay times, (h) the pulse height discrimination settings, and (i) the area of the sample to be used.

The amount of flexibility must of necessity depend on the sampling facilities and instrumental system being used. The form of the sample is dictated by its nature and matrix. It may be desirable to measure the sample as it is received, to grind and size it and run it as a solid powder, to make a pellet out of it, or to put it into a solution prior to measurement.

The choice of X-ray tube target depends on the analytical wavelength of the element to be studied. The choice of power (X-ray tube) conditions depends on the sensitivity (as well as detection limit) required.

The crystal type (dispersive spectrometer) is determined by the required wavelength range and resolution needed to isolate the transition of interest. The detector and counting conditions are likewise determined by the wavelength range and sensitivity (as well as detection limit) requirements. Many of these points are elaborated further in the ensuing sections.

3.   ENERGY-DISPERSIVE X-RAY SPECTROMETERS

There are two ways to isolate X-ray atomic line radiation. One is to use a crystal and disperse the wavelengths by crystal diffraction (wavelength-dispersive X-ray spectrometry) as in conventional X-ray spectroscopy; the other is to use a semiconductor detector and pulse height analysis (energy-dispersive X-ray spectrometry). Energy-dispersive X-ray spectrometry offers some advantages over wavelength-dispersive spectrometry. The most important is probably the potential for rapid, simultaneous multielement analysis. The energy spectrum is seen at once if a multichannel analyzer is used, thus eliminating the need for a scanning technique (necessary with wavelength-dispersive systems). The potential for reducing the radiation path is also important in that a greater flux of X-rays can be directed into the sample. This allows one to use radioactive isotopes, as well as X-ray tubes in exciting species. Instruments that incorporate radioisotopes to excite X-ray fluorescence can be made fairly portable. Energy-dispersive X-ray spectrometers have come about because of the development of semiconductor detectors that offer the required increased sensitivity as compared to conventional ionization detectors.

The major disadvantage of the energy-dispersive system compared to the wavelength-dispersive system is the loss in resolution (see Section B.2.c). In energy-dispersive equipment, all the components are identical with those for wavelength-dispersive systems, with the exception that pulse height analysis is required and radioactive isotopes can be used instead of X-ray tubes for excitation.

B. X-RAY FLUORESCENCE SPECTROMETRY

X-ray fluorescence spectrometry or, as some prefer to call it, X-ray secondary emission spectrometry, is now by far the most popular form of X-ray spectrometric analysis.

1. BASIS OF METHOD

Although absolute quantitative analysis of elements by X-ray fluorescence is rarely performed, it is instructive to consider the fundamental fluorescence irradiance expression in order to determine the important parameters and factors influencing the fluorescence signal. From Chapter 4, Section B.3,

$$E_{XF} = E_L G_1 (\csc \phi_1)(\sigma_A \beta n_A l)\left(1 - \frac{1}{J_K}\right) Y_K \theta \alpha_a \left(\frac{1 - e^{-\mu^* \rho l}}{\mu^* \rho l}\right) \tag{3}$$

assuming front surface (90°) excitation-fluorescence measurement and assuming no enhancement effects. All terms have been defined in Chapter 4, Section B.3, and so are not redefined here. It should be stressed that equation 3 for X-ray fluorescence is similar to the analogous equation for atomic fluorescence and molecular luminescence, for example, the linearity between the X-ray fluorescence irradiance Wm^{-2} and the x-ray fluorescence quantum yield Y_K, the photoelectric cross section σ_A, the irradiance of the exciting source E_L, and the concentration n_A of analyte. In addition, it should be stressed that the final term, $[1 - \exp(-\mu^* p l)]/\mu^* p l$ accounts for matrix absorption in the direction of the exciting beam. Thus an analyte species has an analytical calibration curve (the ordinate is a function of E_{XF}, and the abscissa is a function of n_A) that depends on the matrix constituents and their concentrations; in addition, the limiting detectable concentration also depends on the matrix constituents and their concentrations.

Just as in Chapter 4, Section B.3, there are several limiting cases for equation 3; If the sample is very thin,

$$E_{XF} = E_L G_1 (\csc \phi_1)(\sigma_A n_A l)\left(1 - \frac{1}{J_K}\right) Y_K \theta \alpha_a \tag{4}$$

and, if the sample is infinitely thick,

$$E_{XF} = E_L G_1 (\csc \phi_1)\left(\frac{\sigma_A n_A}{\mu^* \rho}\right)\left(1 - \frac{1}{J_K}\right) Y_K \theta \alpha_a \tag{5}$$

The above three expressions apply to X-ray fluorescence spectrometry whatever the dispersive detection device (wavelength dispersive versus energy dispersive).

2.   INSTRUMENTATION

a.   Components and Instrumental Conditions

A typical crystal X-ray fluorescence spectrometer is shown in Figure 9.1. It has the same components as described in Section A. These instruments are designed to produce the largest X-ray flux (primary X-rays) possible, which is focused on the sample; X-ray tubes are generally operated at the maximum voltage and current to ensure maximum flux. Many X-ray fluorescence procedures combine the counts from the proportional and scintillation counters in order to maximize the counting sensitivity. The use of filters or pulse height discriminators depends on the matrix involved and the need to minimize interferences.

Energy-dispersive systems can be incorporated into the same configuration; however, because there is no need for a wavelength-dispersive device in the system, the instrumental system can be greatly simplified (see Figure 9.2); this type of arrangement allows one to increase the geometry to maximize the solid angle of viewing. The use of large semiconductor detectors may improve the sensitivity of the measurements and detection limits of elements.

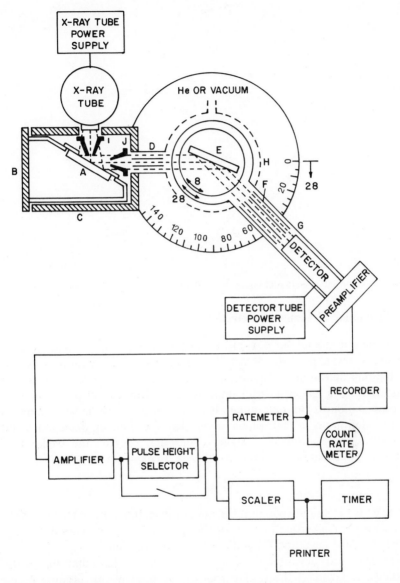

Fig. 9.1 Schematic diagram of typical X-ray fluorescence wavelength-dispersive spectrometer: (*A*) Sample; (*B* and *C*) sample compartment; (*D*) collimator; (*E*) diffraction crystal; (*F*) collimator; (*G*) detector system; (*H*) goniometer; (*I*) collimator; (*J*) collimator.

357

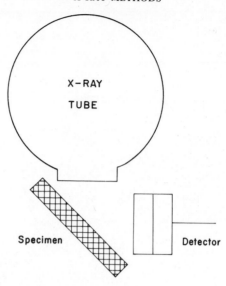

Fig. 9.2 Schematic diagram of typical X-ray fluorescence energy-dispersive spectrometer.

b. Estimation of Limits of Detection

Whereas it should be possible by means of equation 3, and a rather detailed knowledge of the detector's capabilities, to calculate theoretically the limiting detectable concentration of a given element in a given sample measured by a given instrumental system, X-ray spectroscopists have utilized standards to determine detection limits. The most commonly applied definition is that which states that the limit of detection is the concentration resulting in a signal count that exceeds the background count by three standard deviations, that is, 3σ; and so:

$$\sigma_L = C_{BG} + 3\sigma_{BG} \qquad (6)$$

where σ_L is the analyte count at the limit of detection (also refer to Chapter 1), C_{BG} is the background count for the specified counting period, and σ_{BG} is the standard deviation (in counts) of the background.

The standard deviation in X-ray spectrometry, like that in nuclear methods, is taken as the square root of the total count, because of the nature of the random processes, that is, Poisson statistics; and so:

$$\sigma_C = C^{1/2} \qquad (7)$$

where C is the total count during the specified counting period (also see Chapters 1 and 13). It is apparent from the above that the time of counting has a direct bearing on the limit of detection, that is, total counts equals count rate times counting time). This is true for most spectroscopic methods; however, in the case of X-ray spectrometry, there is no sample consumption; therefore measurement time is not sample-limited, as it is, for example, in flame spectrometric methods, and so counting times of several minutes are not uncommon. Consider three counting times—10, 100, and 1000—with a background rate of 5 counts s^{-1}:

$$10 \text{ s} \times 5 \text{ counts s}^{-1} = 50 \text{ counts}; \ \sigma = 7.1; \ RSD = 14.2\%$$

$$100 \text{ s} \times 5 \text{ counts s}^{-1} = 500 \text{ counts}; \ \sigma = 22.4; \ RSD = 4.5\%$$

$$1000 \text{ s} \times 5 \text{ counts s}^{-1} = 5000 \text{ counts}; \ \sigma = 70.7; \ RSD = 1.4\%$$

where RSD is the relative standard deviation of the total background counts. The percentage of the total count (background plus signal) that can be attributed to the analyte signal decreases from 30% for the 10 count to only 4.1% for the 1000 count. This constitutes an important advantage for X-ray fluorescence spectroscopy over several other techniques.

However, the detection limit is constrained not only by the counting statistics of the background in the "real world" situation, but also by other sources of random error. The total standard deviation of an X-ray spectrometer measurement can be expressed:

$$\sigma_T^2 = \sigma_C^2 + \sigma_I^2 + \sigma_O^2 + \sigma_S^2 \tag{8}$$

where σ_T is the total standard deviation, σ_C is the counting standard deviation, σ_I is the standard deviation due to instrumental errors, (e.g., electronic noise, source noise, and drift), σ_O is the standard deviation due to the operator (i.e., reading errors, improper optimization of instrumental conditions, and errors in sample position), and σ_S is the standard deviation due to sample preparation errors. Only in the ideal case does counting statistics dominate the total standard deviation. Instrumental variation should be quite low, unless some instrumental component is not operating correctly (σ_I). This can be checked by utilizing statistical tests (3). Operational variation is dependent on the number and type of operator manipulations of the sample specimen and instrumental settings (σ_O). This too can be checked by performing the complete analytical procedure on the same sample specimen several times and performing a statistical evaluation. Often the dominant source of variation is in the sampling and/or sample (also standards) preparation steps (σ_S). In fact, this source of variation can often completely dominate the overall precision of the technique. This can happen because the value of the analyte signal is determined in part by the matrix elements. Further, the sample form has an effect; therefore, if a solid sample is analyzed, the particle size variation from sample to sample can result in random errors.

c. Resolving Power and Spectral Interferences

In X-ray fluorescence spectrometry, as in optical spectroscopy, one must consider the potential of spectral interferences, that is, the overlap of spectral lines originating from two (or more) different elements. Different crystals have different resolving capabilities in wavelength-dispersive spectrometers, and different detectors have different resolving capabilities in energy-dispersive spectrometers. In wavelength-dispersive spectrometers, the resolution expression and dispersive expressions are identical to those in optical spectrometry:

$$R = \frac{\bar{\lambda}}{\Delta\lambda} = \frac{\bar{\theta}}{\Delta\theta} \tag{9}$$

and

$$\frac{d\theta}{d\lambda} = \frac{m}{2d\cos\theta} \tag{10}$$

where R is the resolving power of the system that can just resolve two line wavelengths (or angles) separated by a wavelength difference of $\Delta\lambda$ with an average wavelength $\bar{\lambda}$ (average angle, $\bar{\theta}$), $d\theta/d\lambda$ is the change in θ, the diffraction angle, with respect to the change in wavelength λ, m is the order of diffraction, and d is the d-spacing of the crystal.

In X-ray spectroscopy, there are several terms that are analogous to several in optical spectrometry but are called by different names, for example:

Divergence is the measure of the intensity distribution of the diffracted peak whether in wavelength or energy units and is identical to the *line profile* in optical spectroscopy.

$W_{1/2}$ is defined as the full width at half-maximum which is the same as the *half-intensity line width* in optical spectroscopy.

The *rocking curve* is the intensity distribution (as a function of wavelength) of the crystal used as the diffraction grating and is similar to the *spectral efficiency curve* in optical spectroscopy.

In X-ray spectroscopy, the rocking curve depends on the nature of the crystal, the nature of the collimator(s), and the nature of the transition.

In energy-dispersive systems, as in wavelength-dispersive systems, the resolving power may be expressed by:

$$R = \bar{E}/\Delta E \tag{11}$$

where ΔE is the energy difference between two transitions that can just be resolved with an average energy $\bar{E}$ between the two transitions. However, most X-ray spectroscopists do not express resolving power in terms of equation 9, but rather:

$$R = \frac{W_{1/2}}{V_{av}} \qquad (12)$$

where $W_{1/2}$ has been defined above (in keV), and V_{av} is the average pulse height (in keV) in the energy distribution being observed. In X-ray spectroscopy, resolution is usually expressed as a percent, that is, $R \times 100$.

For a wavelength-dispersive spectrometer, the resolving power ($R = \bar{\lambda}/ \Delta\lambda = \bar{\theta}/\Delta\theta$) is approximately $\sim$230 for two X-ray lines having an average $\bar{\theta}$ of 70° and half-widths of 0.15° or a separation of 2×0.15° and is approximately 37 for two X-ray lines having an average $\bar{\theta}$ of 11° and a similar half-width of 0.15°. It should be stressed that the useful angular range for crystal spectrometers is about 11 to 70°, which results in the above estimates.

For an energy-dispersive spectrometer, the resolving power ($R = \bar{E}/\Delta E$) is determined by random variation in the number of primary electrons produced in the detector, that is, with a proportional detector each 4 keV photon produces $\sim$160 primary electrons $\pm\sqrt{160}$, whereas with solid-state semiconductive detectors (GeLi or SiLi) each 4 keV photon produces $\sim$1600 primary electrons $\pm\sqrt{1600}$. Thus, with proportional detectors, the energy-dispersive resolving power R is $\sim$12, and with the semiconductive detector R is $\sim$40. Even though the resolving power with the semiconductive detector is greater than with the proportional detector, it is still considerably less than that of the wavelength-dispersive system.

The most common spectral interferences originate from three sources: (a) first-order transitions of the same series from elements adjacent to the analyte species in atomic number; (b) first-order transitions of different series, that is, the L series of element A interfering with the K series of element B; and (c) first-order transitions of the analyte species with higher-order transitions of matrix elements.

3.   QUANTITATIVE ANALYSIS

a.   Preparation of Samples and Use of Standards

X-ray intensities depend greatly on the matrix elements; however, in X-ray spectroscopy, there is one added problem which involves the physical nature of the sample. The fluorescence intensity of a certain amount of an analyte in a graphite briquet differs from the same amount of analyte on a filter paper, or the same amount in a fused-silica matrix, and so on. Further, a problem is apparent if differing sizes of particles of the solid matrices are used. Therefore the preparation of samples and standards must be carefully controlled in both chemical composition and in physical nature.

On the more positive side, in X-ray fluorescence, virtually any type of sample, that is, solids of any nature or liquids, can be simply prepared for analysis. Standards must closely match the samples if a direct comparison of the sample and standard fluorescence intensity is to be made. With respect to analytical techniques, almost every quantitative method has been utilized, from the "absolute" method to that of standard additions. The various techniques that have been used for quantitative analysis of elements in X-ray fluorescence spectroscopy are (a) absolute calculation method, (b) direct comparison with similar standards, (c) reference method (standard additions), (d) standard dilutions, (e) internal standardization, (f) thin-film methods, (g) X-ray scatter standardization, (h) mathematical correction of absorption-enhancement effects; and (i) experimental correction of absorption-enhancement effects. Of these techniques, the first five are common and need no further comment (see Chapter 2), but the last four are more specialized to X-ray methods and are briefly described.

(1) Thin-Film Methods

By means of placing the sample in a thin-film matrix, the absorption-enhancement effects of other elements can be minimized. The fact that the X-ray fluorescence intensity is directly proportional to analyte concentration in a thin film of constant thickness can be used to measure thickness in samples of varying thickness if they have a constant composition.

(2) X-Ray Scatter Standardization

Scattered X-rays can be utilized, like the background emission in the DC arc, as a measure of background intensity. The variation in the intensity of analyte fluorescence compared to variation in the intensity of scattered X-rays should remain nearly constant (at constant analyte concentration) even with a varying sample matrix.

(3) Experimental Correction of Absorption-Enhancement Effects

In this technique, a correction factor is determined experimentally for each interfering matrix element as a function of its concentration. Then the combined correction factors are applied to the measured signal of the analyte according to the approximate concentration of the interferring element, and the "true" concentration is then calculated.

(4)   Mathematical Correction of Absorption-Enhancement Effects

Similarly to the previous technique, correction factors can be calculated from the mass absorption coefficients (values tabulated in Table 9.5). In this technique, a mathematical matrix is developed with correction factors as a function of concentration for each interfering element. This method generally incorporates an iteration technique and results in determination of the concentration of each element in the matrix.

b.   Limits of Detection and Useful Analytical Range (5)

Because of the rather large dependence of fluorescence intensity of the analyte on the sample matrix, the limit of detection of a given element is greatly affected by the form and type of the matrix. However, several generalizations can be made. In favorable cases, it is possible to analyze from about 1 μg of analyte per gram of sample (0.0001% by weight) to essentially 100% of the analyte (a pure analyte sample). Few analytical techniques have such a large analytical working range. The sensitivity falls off rapidly as the atomic number decreases below 25 (manganese) with an air path, 19 (potassium) with a helium path, and 13 (aluminum) with a vacuum path. A limit of detection of about 0.1% by weight is all that can be attained for sodium ($Z = 11$) in a sample placed in a vacuum path instrument. Limits of detection have been reported in the range 1 to 50 μg g^{-1} and from 0.1 to 1 μg (ppm of analyte in the sample) for all elements with atomic numbers greater than 15 (phosphorus). One should be cautioned that rather long counting times are often required to realize these detection limits, that is, 2 to 30 m. Furthermore, most tabulated limits of detection are calculated using the definition given in Section B.2.b, which states that the analyte concentration that produces a signal count of three times the standard deviation count is the limit of detection. These detection limits generally do not include deviations caused by matrix or sampling variations, and so on. Therefore *no tabulation of experimental detection limits* is given here, because such figures of merit for elements measured by X-ray fluorescence spectrometry vary greatly with the matrix.

For thin layers, for example, 0.1 μm thick with $\sim$1 mg of sample, the powdered sample can be homogeneously distributed on Mylar foil. A similar process can be applied to solutions on filter paper or on another substrate. In the case of a thin film, the concentration of analyte is related linearly to the fluorescence irradiance as shown in equation 4.

It should be stressed that in quantitative analytical studies it is important that the following parameters be kept constant during the entire analytical procedure, including the measurement of standards, if systematic and random errors are to be minimized: (1) the source (primary X-ray intensity) irradiance; (2) the geometric factor, including the angles between the exciting X-ray beam and the sample surface, the absorption path length, the atmosphere around the sample, and the angle between the measured fluorescence X-rays and the detector; (3) the constancy of the crystal spectrometer-detector or energy-dispersive detector system. If a reference element is used, it is also important to account for counting efficiency of the detector and the efficiency of the crystal spectrometer (if one is used) at the two wavelengths.

If the analytical range of the standards chosen for a given analysis is small, the systematic errors will generally also be small. However, if a heavy element is to be measured in a matrix consisting of light elements, the analytical calibration curve is concave toward the concentration axis and may actually be parallel to it, so that any analysis is impossible, for example, lead in glass. In such a case, the relevant sample must be ground to a powder, mixed with a heavy element in the compound, and then measured, that is, μ^* then remains constant between standards and samples. However, if a light element is in a matrix of heavy elements, analytical calibration curves act more normally, except that in some instances essentially complete absorption by the heavy element can occur.

c. Applications (1–5)

X-ray fluorescence spectrometry has found widespread application for the determination of elements in samples from the steel and iron industry, the geochemical field, the light-metal industry, the plating industry, the cement industry, the petroleum and coal industries, and the chemical industry. The greatest number of applications has been for elements having concentrations greater than 1%; however, a great deal of work has also been performed at the trace levels, that is, below 0.01% by weight.

C. X-RAY ABSORPTION SPECTROMETRY

Just as in most electromagnetic radiation techniques, it is possible to measure X-ray absorption and to relate the absorption to the concentration of the species being measured. X-ray absorption techniques are straightforward, except that the X-ray instrumentation is not as versatile as that used in the measurement of X-ray fluorescence.

1. BASIS OF METHOD

X-ray absorption is described by an equation similar to the Beer absorption equation in molecular absorption spectrometry. It is generally given in the form:

$$I = I_0 e^{-\mu_m \rho t} \tag{13}$$

where I is the transmitted X-ray intensity, I_0 is the incident X-ray intensity (in intensity units), μ_m is the mass absorption coefficient (in $cm^2 \ g^{-1}$), ρ is the density of the absorbing species (in $g \ cm^{-3}$), and t is the thickness of the absorbing species (in cm).

The more familiar expression is derived from equation 13 and is:

$$\ln\left(\frac{I_0}{I}\right) = \mu_m \rho t \tag{14}$$

where all terms have been defined above. Thus the absorption of X-rays is directly proportional to the concentration of analyte present in the sample, just as in molecular absorption spectrometry. The wavelength of the absorption edges for the K and L transitions of most elements are given in Table 9.4. Further, mass absorption coefficients for some elements and some molecules are listed in Table 9.5 as a function of wavelength. The mass absorption coefficient μ_m varies in a regular way with both atomic number and wavelength:

$$\mu_m = KZ^4\lambda^3 \tag{15}$$

where K is a constant which changes abruptly at each absorption edge, Z is the atomic number of the absorbing species, and λ is the wavelength of observation.

The X-ray absorption edge corresponds to the wavelength beyond which electrons cannot be ejected from the element under X-ray bombardment. The resultant disruption is diagrammed in Figure 9.3. The absorption edge always occurs at a lower wavelength (higher energy) than the characteristic lines of the series.

TABLE 9.4. Wavelengths of the X-Ray Absorption Edges of the Chemical Elements

| Atomic Number | Element | Wavelength (Å) | | | |
		K	LI	LII	LIII
3	Li	226.950	—	—	—
4	Be	107.200	—	—	—
5	B	65.604	—	—	—
6	C	43.887	—	—	—
7	N	31.220	—	—	—
8	O	23.233	—	—	—
9	F	17.897	—	—	—
10	Ne	14.170	—	—	—
11	Na	11.475	—	—	—
12	Mg	9.512	197.300	—	220.534

TABLE 9.4. (*Continued*)

Atomic Number	Element	Wavelength, (Å)			
		K	*L*I	*L*II	*L*III
13	Al	7.951	142.500	163.022	163.942
14	Si	6.745	105.000	122.028	122.761
15	P	5.787	81.000	95.070	95.801
16	S	5.018	64.100	75.314	75.868
17	Cl	4.397	52.100	60.739	61.222
18	Ar	3.871	43.200	50.025	50.446
19	K	3.436	36.400	41.754	42.179
20	Ca	3.070	30.700	35.797	36.173
21	Sc	2.757	26.800	30.943	31.296
22	Ti	2.497	23.400	26.936	27.290
23	V	2.269	19.803	23.877	24.229
24	Cr	2.070	17.840	21.294	21.637
25	Mn	1.896	16.138	19.086	19.417
26	Fe	1.743	14.650	17.188	17.504
27	Co	1.608	13.333	15.545	15.843
28	Ni	1.488	12.201	14.104	14.387
29	Cu	1.380	11.172	12.841	13.113
30	Zn	1.283	10.262	11.725	11.987
31	Ga	1.196	9.416	10.728	10.930
32	Ge	1.116	8.692	9.840	10.089
33	As	1.045	8.067	9.056	9.298
34	Se	0.980	7.456	8.347	8.584
35	Br	0.920	6.920	7.721	7.952
36	Kr	0.866	6.444	7.158	7.377
37	Rb	0.816	5.995	6.642	6.861
38	Sr	0.770	5.591	6.173	6.387
39	Y	0.728	5.226	5.752	5.961
40	Zr	0.689	4.889	5.378	5.565
41	Nb	0.653	4.595	5.031	5.230
42	Mo	0.620	4.323	4.716	4.913
43	Tc	0.589	4.068	4.431	4.623
44	Ru	0.560	3.837	4.169	4.358
45	Rh	0.534	3.623	3.928	4.113
46	Pd	0.509	3.425	3.706	3.889
47	Ag	0.486	3.243	3.501	3.680
48	Cd	0.464	3.073	3.312	3.488
49	In	0.444	2.916	3.137	3.311
50	Sn	0.425	2.648	2.975	3.146
51	Sb	0.407	2.634	2.824	2.994
52	Te	0.390	2.508	2.685	2.853
53	I	0.374	2.390	2.555	2.721
54	Xe	0.358	2.277	2.434	2.597
55	Cs	0.345	2.175	2.321	2.482
56	Ba	0.331	2.078	2.214	2.374

TABLE 9.4.   (*Continued*)

Atomic Number	Element	Wavelength, (Å)			
		K	LI	LII	LIII
57	La	0.318	1.988	2.115	2.273
58	Ce	0.306	1.902	2.022	2.178
59	Pr	0.295	1.822	1.934	2.089
60	Nd	0.284	1.747	1.852	2.007
61	Pm	0.274	1.675	1.775	1.928
62	Sm	0.265	1.608	1.703	1.855
63	Eu	0.256	1.545	1.634	1.785
64	Gd	0.247	1.485	1.569	1.719
65	Tb	0.238	1.428	1.507	1.656
66	Dy	0.230	1.375	1.449	1.597
67	Ho	0.223	1.323	1.393	1.540
68	Er	0.216	1.274	1.341	1.487
69	Tm	0.209	1.227	1.291	1.436
70	Yb	0.202	1.183	1.243	1.387
71	Lu	0.196	1.140	1.198	1.341
72	Hf	0.190	1.100	1.155	1.297
73	Ta	0.184	1.062	1.114	1.255
74	W	0.178	1.025	1.074	1.215
75	Re	0.173	0.990	1.037	1.177
76	Os	0.168	0.956	1.001	1.140
77	Ir	0.163	0.924	0.966	1.105
78	Pt	0.158	0.893	0.933	1.071
79	Au	0.153	0.863	0.902	1.039
80	Hg	0.149	0.835	0.872	1.008
81	Tl	0.145	0.807	0.843	0.979
82	Pb	0.141	0.781	0.815	0.950
83	Bi	0.137	0.756	0.788	0.923
84	Po	0.133	0.731	0.762	0.897
85	At	0.129	0.708	0.738	0.872
86	Rn	0.125	0.686	0.715	0.848
87	Fr	0.122	0.665	0.692	0.825
88	Ra	0.119	0.644	0.671	0.803
89	Ac	0.116	0.625	0.650	0.782
90	Th	0.113	0.606	0.629	0.761
91	Pa	0.110	0.587	0.610	0.741
92	U	0.108	0.570	0.592	0.722

TABLE 9.5A. Mass Absorption Coefficients, μ_m (cm^2 g^{-1})(4)

Atomic Number	Element	Wavelength (Å)											
		0.1	0.15	0.2	0.25	0.3	0.4	0.5	0.6	0.7	0.8	0.9	1.0
1	H	0.29	0.32	0.34	0.36	0.37	0.38	0.40	0.42	0.43	0.44	0.44	0.45
2	He	0.114	0.124	0.132	0.140	0.146	0.159	0.173	0.186	0.203	0.222	0.241	0.255
3	Li	0.124	0.132	0.143	0.153	0.163	0.180	0.198	0.223	0.254	0.302	0.358	0.428
4	Be	0.131	0.142	0.153	0.162	0.171	0.185	0.210	0.240	0.292	0.362	0.445	0.57
5	B	0.138	0.152	0.164	0.173	0.182	0.198	0.222	0.277	0.355	0.470	0.61	0.76
6	C	0.142	0.155	0.170	0.186	0.204	0.240	0.305	0.41	0.55	0.75	1.05	1.40
7	N	0.144	0.159	0.175	0.195	0.216	0.288	0.395	0.62	0.89	1.25	1.73	2.20
8	O	0.145	0.162	0.181	0.206	0.236	0.345	0.508	0.87	1.25	1.80	2.40	3.20
9	F	0.147	0.165	0.192	0.228	0.270	0.417	0.675	1.20	1.85	2.60	3.40	4.40
10	Ne	0.149	0.169	0.208	0.256	0.319	0.508	0.865	1.55	2.50	3.50	4.75	6.50
11	Na	0.150	0.175	0.228	0.287	0.380	0.630	1.18	2.05	3.25	4.75	6.70	8.80
12	Mg	0.152	0.190	0.251	0.330	0.445	0.78	1.47	2.70	4.30	6.50	9.10	11.7
13	Al	0.155	0.205	0.277	0.380	0.525	0.97	1.82	3.70	5.75	8.80	11.8	15.2
14	Si	0.159	0.215	0.310	0.442	0.615	1.22	2.25	4.65	7.20	11.1	14.2	18.2
15	P	0.165	0.228	0.346	0.510	0.725	1.48	2.80	5.45	8.30	13.3	16.5	21.7
16	S	0.170	0.241	0.392	0.592	0.855	1.78	3.45	6.40	9.6	15.5	19.3	26.0
17	Cl	0.176	0.260	0.433	0.667	1.02	2.13	4.25	7.45	11.3	18.5	22.6	29.7
18	Ar	0.183	0.278	0.490	0.76	1.21	2.62	5.08	8.60	13.2	21.1	26.4	34.4
19	K	0.191	0.301	0.542	0.86	1.39	3.02	5.75	10.2	15.3	24.0	30.7	40.0
20	Ca	0.200	0.327	0.601	0.98	1.63	3.45	6.50	11.5	18.0	26.5	36.0	48.0
21	Sc	0.210	0.358	0.667	1.12	1.85	3.95	7.4	13.1	20.2	30.0	41.0	55.0
22	Ti	0.221	0.395	0.740	1.26	2.10	4.50	8.4	14.8	22.7	34.0	47.0	62.5
23	V	0.231	0.431	0.830	1.42	2.37	5.15	9.6	16.8	25.5	38.0	54.0	71.5
24	Cr	0.241	0.480	0.925	1.60	2.68	5.88	10.8	18.9	28.7	42.5	60.3	80.5
25	Mn	0.253	0.528	1.03	1.79	3.05	6.65	12.3	21.4	32.2	47.5	66.6	90.0

368

													K
26	Fe	0.265	0.58	1.16	2.02	3.45	7.6	14.1	23.8	36.0	53.2	74.0	100.0
27	Co	0.285	0.64	1.28	2.25	3.80	8.35	15.5	26.6	39.8	58.2	80.5	110.0
28	Ni	0.303	0.71	1.42	2.51	4.15	9.20	17.0	29.6	43.6	64.0	87.5	121.0
29	Cu	0.328	0.785	1.57	2.79	4.55	10.1	18.6	32.4	48.5	70.0	95.5	130.0
30	Zn	0.350	0.818	1.76	3.07	5.00	11.1	20.6	35.3	53.0	76.5	106.0	141.0
31	Ga	0.378	0.94	1.91	3.36	5.50	12.2	22.3	38.7	59.0	82.0	112.0	152.0
32	Ge	0.404	1.02	2.06	3.67	6.15	13.3	24.4	42.1	63.5	88.0	120.0	163.0
33	As	0.427	1.12	2.22	4.01	6.45	14.6	26.3	45.1	68.5	94.0	129.0	175.0
34	Se	0.472	1.21	2.41	4.33	7.00	15.8	28.5	48.4	74.0	101.0	138.0	33.2
35	Br	0.502	1.31	2.60	4.70	7.60	17.0	30.8	52.0	79.5	108.0	149.0	35.2
36	Kr	0.535	1.41	2.81	5.05	8.30	18.3	33.3	55.5	86.0	116.0	26.5	37.5
37	Rb	0.572	1.50	3.03	5.45	9.05	19.7	35.9	59.5	93.0	125.0	28.0	40.2
38	Sr	0.61	1.61	3.28	5.82	9.80	21.3	38.7	63.0	100.0	22.0	29.8	43.0
39	Y	0.65	1.72	3.52	6.20	10.3	22.6	41.3	67.5	108.0	23.2	32.0	46.0
40	Zr	0.69	1.83	3.73	6.60	11.0	24.0	43.5	72.0	16.8	24.5	33.8	49.0
41	Nb	0.74	1.93	4.00	7.02	11.7	25.4	46.0	76.0	17.8	26.0	36.0	51.8
42	Mo	0.79	2.04	4.29	7.50	12.4	26.9	49.0	81.0	19.0	27.7	38.4	54.8
43	Tc	0.84	2.15	4.55	7.95	13.1	28.6	50.5	14.2	20.3	29.5	41.0	58.0
44	Ru	0.89	2.27	4.70	8.4	13.8	30.5	53.0	15.0	21.6	31.4	43.7	61.2
45	Rh	0.94	2.40	5.01	8.9	14.6	32.3	55.5	15.8	23.0	33.5	46.5	65.0
46	Pd	1.00	2.53	5.25	9.4	15.5	34.3	58.2	16.6	24.5	35.5	49.5	69.0
47	Ag	1.05	2.67	5.50	9.9	16.3	36.5	10.0	17.5	26.0	38.0	52.8	72.7
48	Cd	1.10	2.80	5.74	10.3	17.2	38.5	10.7	18.5	27.7	40.2	56.2	76.2
49	In	1.15	2.94	5.98	10.8	17.8	40.6	11.4	19.5	29.5	42.7	60.0	80.0
50	Sn	1.20	3.07	6.22	11.3	18.7	43.1	12.0	20.5	31.0	45.5	63.5	84.0
51	Sb	1.25	3.22	6.48	11.8	19.5	45.5	12.8	21.6	33.1	48.5	67	88.
52	Te	1.30	3.37	6.76	12.3	20.3	7.4	13.6	22.7	35.2	51.5	70.6	93
53	I	1.36	3.52	7.05	12.9	21.2	7.8	14.3	24	37.5	54.5	74.8	97
54	Xe	1.42	3.68	7.33	13.4	21.9	8.2	15.2	25.2	39.8	58	78.5	101
55	Cs	1.48	3.85	7.65	13.9	22.7	8.6	16.2	26.5	42.0	61.3	82.5	106

TABLE 9.5A. (*Continued*)

Atomic Number	Element	Wavelength (Å)											
		0.1	0.15	0.2	0.25	0.3	0.4	0.5	0.6	0.7	0.8	0.9	1.0
56	Ba	1.53	4.03	7.98	14.5	23.1	9.05	17.2	27.9	44.2	64	87	111
57	La	1.60	4.22	8.27	15.0	24.5	9.5	18.1	29.3	46.7	67.2	91	116
58	Ce	1.66	4.37	8.55	15.6	25.2	10.0	19.1	30.9	49.2	70.5	95	122
59	Pr	1.72	4.53	8.88	16.1	4.7	10.5	20.1	32.5	51.6	73.8	100	127
60	Nd	1.80	4.69	9.20	16.7	4.95	11.0	21.1	34.2	54.0	77.2	104	132
61	Pm	1.86	4.86	9.57	17.3	5.20	11.5	22.1	36	56.9	80.7	108	139
62	Sm	1.93	5.03	9.95	17.7	5.45	12.0	23.2	37.7	59.5	84	114	144
63	Eu	2.02	5.20	10.3	18.3	5.71	12.5	24.2	39.8	62.5	87.8	119	150
64	Gd	2.09	5.37	10.5	3.95	6.00	13.2	25.3	41.8	65.5	91.5	125	157
65	Tb	2.18	5.52	10.8	4.15	6.28	13.7	26.6	44.0	69.5	95.5	130	163
66	Dy	2.26	5.70	11.2	4.36	6.57	14.4	27.2	46.2	72	99.5	135	170
67	Ho	2.33	5.87	11.6	4.55	6.90	15.1	28.2	48.6	75	103	141	177
68	Er	2.42	6.03	11.9	4.75	7.2	15.8	30.2	50.8	79	107	147	185
69	Tm	2.50	6.23	12.3	5.02	7.6	16.5	31.4	53.2	82.8	112	152	193
70	Yb	2.58	6.41	12.7	5.27	7.95	17.3	32.7	55.8	87	115	158	201
71	Lu	2.66	6.61	3.07	5.50	8.4	18.2	34.2	58.4	90.5	120	165	210
72	Hf	2.75	6.80	3.22	5.80	8.8	19.0	35.5	61	94.2	124	172	219
73	Ta	2.82	7.02	3.36	6.05	9.2	19.9	37.2	64	99	129	178	229
74	W	2.90	7.24	3.50	6.27	9.7	20.8	38.7	66.5	103	134	185	239 *LI*
75	Re	2.96	7.45	3.63	6.60	10.2	21.8	40.5	69	106	139	192	213
76	Os	3.03	7.65	3.78	6.9	10.6	22.8	42.2	72	110	144	200	222 *LII*
77	Ir	3.10	7.86	3.92	7.3	11.1	23.8	44.0	74.5	113	150	208	157
78	Pt	3.17	8.06	4.08	7.6	11.6	25.0	46.0	77.6	117	155	170	163
79	Au	3.23	8.28	4.22	8.0	12.2	26.2	47.5	80.5	122	160	176	170
80	Hg	3.30	2.17	4.38	8.3	12.7	27.4	49.5	83.7	126	165	135	178

370

		K					LI LII	LIII	LIII			LIII
81	Tl	3.36	2.24	4.53	8.7	13.3	28.6	51.5	86.6	130	170	140	72
82	Pb	3.41	2.32	4.67	9.2	13.9	30.3	53.5	89.2	135	147	145	75
83	Bi	3.45	2.37	4.81	9.6	14.5	31.3	55.5	92	140	109	150	77
84	Po	3.52	2.43	4.88	10.0	15.2	32.8	57.4	94.5	144	112	58.2	79.5
85	At	3.56	2.49	5.02	10.5	15.7	34.0	59.3	97	149	116	60	82
86	Rn	3.61	2.55	5.14	10.9	16.3	35.4	61	100	135	120	62	84
87	Fr	3.66	2.60	5.21	11.2	17.0	36.7	63	102	88	123	64	87
88	Ra	3.70	2.66	5.28	11.7	17.7	38.3	65	105	91	127	66.5	90
89	Ac	3.75	2.71	5.38	12.1	18.6	39.8	67	107	94	131	68	92
90	Th	3.81	2.76	5.45	12.4	19.5	41.2	69	110	97	52	70	95
91	Pa	3.86	2.80	5.53	12.8	20.1	42.6	71	101	100	53.3	72	98
92	U	3.91	2.85	5.61	13.1	20.9	44.0	83.1	79	103	55	74.4	101
93	Np	3.95	2.90	5.68	13.4	21.7	45.2	85.3	82	106	56.8	76.6	104
94	Pu	4.00	2.95	5.75	13.7	22.5	46.7	87.5	85	51	58.7	78.8	107
95	Am	4.05	3.00	5.81	14.1	23.3	47	89.5	87	52	60.5	81	110
96	Cm	2.50	3.04	5.89	14.5	24.2	49.8	91.5	90	53.5	62.3	83.6	113
97	Bk	2.56	3.08	5.97	14.8	25.0	51.2	84.4	92	55	64.4	86.2	117
98	Cf	2.62	3.12	6.04	15.2	25.8	52.8	65	94	57	66.4	89	121
99	Es	2.68	3.16	6.11	15.6	26.6	54.0	67	97	59	68.5	91.6	125
100	Fm	2.74	3.20	6.18	15.9	27.5	55.5	69	46	60	70.5	94.5	128

TABLE 9.5A. (Continued)

Atomic Number	Element	Wavelength (Å)										
		1.4	2.0	2.5	3	4	5	6	7	8	9	10
1	H	0.49	0.52	0.62	0.75	1.25	2.12	3.28	4.85	7.1	10.0	13.7
2	He	0.355	0.715	1.04	1.48	3.55	6.9	11.6	18.1	26.6	37.7	51
3	Li	1.02	2.18	3.98	6.6	15.2	28.8	48.8	76	113	157	213
4	Be	1.55	3.38	6.2	10.3	22.7	43.7	74	118	174	245	333
5	B	2.31	5.05	9.6	15.8	36.0	69	116	187	285	405	560
6	C	4.2	9.7	14.0	32	74	145	250	390	570	810	1100
7	N	6.9	16.0	30.5	52	123	235	400	620	910	1290	1800
8	O	10.5	24.0	45.5	78	180	350	580	920	1350	1900	2600
9	F	14.2	32	62	102	240	450	760	1190	1750	2450	3300
10	Ne	19.8	43	90	145	325	630	1040	1580	2350	3200	4300
11	Na	25.5	53.5	132	250	520	920	1450	2200	3100	4100	5300 *K*
12	Mg	32.5	69	185	280	610	1100	1800	2700	3900	5200	340
13	Al	41.5	87	235	360	780	1400	2250	3300	280	390	520
14	Si	52	112	290	440	950	1700	2800	270	400	550	740
15	P	66	140	330	490	1050	1900	245	380	550	770	1070
16	S	78.5	170	375	590	1250	2200	320	500	730	1040	1400
17	Cl	95	202	430	680	1400	250	410	620	900	1250	1700
18	Ar	112	240	485	780	170	300	490	730	1050	1450	1900
19	K	130	280	540	860	205	370	600	910	1300	1800	2350
20	Ca	150	328	625	950	255	460	730	1100	1550	2100	2800
21	Sc	172	375	720	145	310	550	880	1320	1850	2520	3300
22	Ti	195	425	113	175	360	630	1000	1450	2050	2750	3550
23	V	222	485	125	195	400	710	1120	1620	2250	3000	3700
24	Cr	245	540	136	210	440	760	1200	1720	2350	3200	4100
25	Mn	277	73	148	232	480	820	1260	1850	2550	3400	4350

Z	El.											
26	Fe	312	80	162	255	510	880	1400	2050	2800	3700	4700
27	Co	345	87	177	270	550	960	1500	2200	3000	4000	5200
28	Ni	44	96	193	295	600	1050	1570	2350	3300	4300	5600
29	Cu	48.5	101	212	320	660	1150	1750	2550	3500	4600	6000
30	Zn	53.5	116	227	350	700	1200	1870	2700	3750	4900	6400
31	Ga	59	126	248	385	760	1300	2020	2950	4050	5400	5500
32	Ge	63	138	267	410	820	1400	2150	3100	4300	4700	4000
33	As	69	150	287	440	870	1500	2320	3330	4600	5050	2000
34	Se	74	165	311	470	940	1600	2500	3600	3900	1650	2150
35	Br	80	180	335	510	1020	1720	2700	2180	1250	1750	2300
36	Kr	87	192	359	550	1100	1860	2900	2280	1350	1870	2450
37	Rb	94	207	386	590	1170	2000	3100	950	1440	2000	2620
38	Sr	102	223	412	640	1260	2150	2600	1020	1540	2150	2800
39	Y	109	238	442	690	1380	2320	800	1110	1650	2350	3000
40	Zr	118	255	478	740	1500	1840	850	1200	1770	2450	3200
41	Nb	127	271	507	800	1600	1900	910	1300	1900	2600	3400
42	Mo	136	290	544	860	1700	600	980	1400	2050	2800	3650
43	Tc	147	310	583	920	1850	650	1030	1510	2200	3000	3900
44	Ru	158	342	625	960	1530	700	1090	1650	2350	3200	4150
45	Rh	170	355	670	1050	1050	750	1160	1770	2500	3400	4450
46	Pd	183	380	720	1140	440	800	1230	1900	2700	3650	4750
47	Ag	196	405	770	1250	480	850	1350	2050	2900	3900	5100
48	Cd	210	432	820	1340	510	900	1450	2150	3100	4100	5400
49	In	222	460	880	1120	535	940	1520	2250	3200	4300	5700
50	Sn	236	495	940	775	560	990	1600	2360	3350	4500	5900

Column edge markers: LI, LII, LIII (indicated along the staircase boundary at both top and bottom of the table).

373

TABLE 9.5B Mass Absorption Coefficient, μ_m (cm^2 g^{-1}) (4)

Atomic Number	Element	1.4	2.0	2.5	3	4	5	6	7	8	9	10
51	Sb	250	525	1000	805	590	1050	1660	2480	3500	4700	6200
52	Te	265	557	1070	295	620	1100	1730	2600	3650	4900	6400
53	I	280	590	880	310	650	1160	1810	2700	3700	5100	6700
54	Xe	296	622	710	340	710	1230	1930	2850	4000	5300	7000
55	Cs	312	656	215	360	740	1300	2050	3000	4200	5600	7200 *M* I
56	Ba	330	690	227	390	790	1370	2150	3150	4400	5800	5400
57	La	347	650	240	415	840	1450	2250	3300	4600	6100	5600
58	Ce	365	685	252	440	880	1570	2360	3400	4800	4750	5800
59	Pr	385	540	267	465	920	1600	2470	3650	5000	4950	6000 *M* II
60	Nd	405	179	280	485	980	1670	2580	3750	3900	5100	4500 *M* III
61	Pm	425	187	296	515	1040	1750	2700	3900	3500	5300	3300
62	Sm	446	196	313	535	1080	1830	2810	4100	3600	3900	3400
63	Eu	470	206	330	555	1120	1900	2910	3300	3700	2900	3500 *M* IV
64	Gd	440	215	348	575	1170	1980	3020	3370	2700	3000	2800 *M* V
65	Tb	465	225	368	610	1230	2090	3200	3440	2800	3100	2170
66	Dy	365	236	388	635	1300	2200	3400	3530	2200	3200	2270
67	Ho	390	247	409	665	1350	2280	2500	2800	2260	2500	2380
68	Er	136	258	432	690	1410	2350	2570	2300	2330	1890	2490
69	Tm	142	272	456	720	1480	2500	2650	2390	2400	1970	2500
70	Yb	148	285	482	750	1540	2600	2740	2480	1950	2050	2600
71	Lu	156	298	510	780	1600	1980	2060	2570	1580	2130	2700
72	Hf	162	313	540	825	1660	2050	2100	2650	1660	2210	2800
73	Ta	168	327	570	855	1720	2210	1520	1900	1740	2290	2900
74	W	176	342	600	890	1800	2280	1660	1350	1820	2370	3020
75	Re	184	358	636	925	1900	1750	1700	1400	1900	2450	3130

Column heading: Wavelength (Å)

76	Os	193	376	672	970	1980	1300	1740	1460	1980	2560	3250	
77	Ir	200	392	710	1010	1450	1340	1400	1530	2060	2670	3370	
78	Pt	208	410	750	1070	1500	1380	1160	1590	2150	2780	3490	
79	Au	217	430	795	1130	1550	1430	1200	1650	2250	2900	3600	
80	Hg	225	445	825	1190	1150	1480	1250	1700	2300	2950	3700	
81	Tl	233	460	865	1250	1200	1120	1300	1790	2370	3100	3900	
82	Pb	242	478	900	1310	1250	1150	1360	1880	2440	3270	4100	
83	Bi	252	495	940	1380	920	910	1410	1970	2560	3430	4300	
84	Po	262	507	980	1050	950	960	1460	2060	2680	3590	4500	
85	At	272	532	1020	1080	980	1000	1510	2150	2900	3750	4700	
86	Rn	283	550	1070	1120	1020	1040	1550	2240	3010	3900	4900	
87	Fr	294	572	1110	840	710	1080	1590	2330	3120	4050	5100	
88	Ra	305	592	1160	860	750	1120	1630	2420	3230	4200	5500	*N* I
89	Ac	316	615	890	890	620	1160	1660	2510	3450	4350	4300	
90	Th	328	638	930	920	730	1200	1700	2600	3460	4500	4470	
91	Pa	342	660	725	720	760	1250	1790	2700	3570	4650	4650	
92	U	354	683	755	750	790	1300	1890	2790	3690	3600	4820	*N* II
93	Np	367	709	785	780	820	1350	1980	2890	3800	3700	3500	
94	Pu	382	733	820	810	850	1400	2080	2980	2900	3850	3650	
95	Am	395	760	850	610	880	1450	2170	3070	3000	4000	3800	
96	Cm	412	530	880	640	910	1500	2250	3170	3080	3100	3950	
97	Bk	427	420	570	500	950	1560	2330	3270	3180	3200	4100	*N* III
98	Cf	443	435	595	515	980	1620	2420	2500	3300	3330	3400	
99	Es	461	452	620	530	1020	1690	2520	2580	3400	3450	3550	
100	Fm	480	470	640	550	1050	1730	2600	2600	2600	3570	3700	
			M I *M* II	*M* III	*M* IV *M* V				*N* I	*N* II		*N* III	

TABLE 9.5B Mass Absorption Coefficient, μ_m (cm^2 g^{-1}) (4) (Continued)

Atomic Number	Element	Wavelength (Å)									
		0.0827	0.124	0.155	0.207	0.248	0.310	0.413	0.619	0.827	1.24
1	H	0.265	0.294	0.309	0.326	0.335	0.346	0.357	0.369	0.376	0.385
4	Be	0.119	0.132	0.139	0.147	0.152	0.159	0.171	0.206	0.268	0.536
5	B	0.124	0.138	0.145	0.155	0.161	0.171	0.192	0.266	0.418	1.09
6	C	0.134	0.150	0.158	0.170	0.179	0.193	0.230	0.388	0.722	2.17
7	N	0.134	0.150	0.160	0.174	0.187	0.212	0.276	0.541	1.09	3.57
8	O	0.134	0.152	0.162	0.181	0.199	0.236	0.335	0.754	1.62	5.58
11	Na	0.131	0.152	0.170	0.209	0.254	0.355	0.639	1.88	4.37	15.1
12	Mg	0.136	0.161	0.183	0.236	0.298	0.437	0.839	2.56	5.98	20.3
13	Al	0.134	0.162	0.189	0.255	0.334	0.514	1.03	3.24	7.66	25.8
14	Si	0.140	0.173	0.207	0.292	0.396	0.635	1.31	4.19	9.97	33.6
15	P	0.138	0.175	0.215	0.318	0.444	0.731	1.55	5.10	12.0	40.2
16	S	0.145	0.189	0.238	0.367	0.527	0.891	1.94	6.42	15.2	50.3
18	Ar	0.136	0.189	0.252	0.420	0.630	1.11	2.48	8.27	19.5	63.8
19	K	0.150	0.216	0.296	0.512	0.777	1.39	3.14	10.5	24.6	80.1
20	Ca	0.159	0.237	0.334	0.595	0.925	1.67	3.82	12.6	29.6	95.6
26	Fe	0.184	0.342	0.550	1.13	1.84	3.46	7.88	25.1	55.7	172
29	Cu	0.208	0.427	0.718	1.52	2.50	4.71	10.6	33.0	73.3	223
42	Mo	0.399	1.05	1.92	4.25	6.97	13.0	28.3	11.7	26.8	84.0
50	Sn	0.577	1.60	2.90	6.32	10.4	18.9	40.7	20.2	45.3	139
53	I	0.674	1.91	3.52	7.55	12.3	22.3	7.98	24.7	53.4	158
74	W	1.50	4.29	7.66	3.28	5.40	9.97	21.8	65.1	139	91.2
82	Pb	1.89	5.23	2.07	4.43	7.22	13.1	28.4	83.4	112	128
92	U	2.47	1.71	3.04	6.45	10.4	18.7	39.6	68.5	60.3	173

Edge annotations: *K* edge; *LI,II* edges; *LIII* edge; *K*; *LI,LII,LIII* edges.

376

H_2O	0.149	0.168	0.179	0.197	0.214	0.248	0.338	0.711	1.48	4.99
SiO_2	0.137	0.162	0.183	0.233	0.291	0.422	0.793	2.36	5.52	18.7
NaI	0.590	1.64	3.00	6.42	10.5	18.9	6.86	21.2	45.9	136
Air	0.134	0.151	0.162	0.179	0.196	0.229	0.318	0.691	1.45	4.82
Concrete	0.140	0.170	0.200	0.273	0.361	0.559	1.12	3.45	8.01	26.5
H_2SO_4 (0.8N)	0.149	0.168	0.179	0.199	0.218	0.256	0.358	0.784	1.66	5.57
Polystyrene	0.144	0.161	0.170	0.182	0.191	0.205	0.240	0.386	0.695	2.03
Lucite	0.145	0.162	0.172	0.186	0.198	0.219	0.274	0.503	0.982	3.11
Polyethylene	0.153	0.170	0.180	0.193	0.201	0.215	0.249	0.385	0.672	1.91
Bakelite	0.142	0.158	0.168	0.181	0.191	0.209	0.255	0.448	0.854	2.64
Pyrex Glass	0.136	0.160	0.180	0.225	0.277	0.394	0.724	2.13	4.95	16.7

2. INSTRUMENTATION

Instrumentation for use in X-ray absorption techniques is essentially the same as in X-ray fluorescence methods using a crystal for wavelength dispersion. For this reason, all components have been previously discussed. The instrumental configuration must be changed, however, so that the X-ray beam is focused through the sample just as in molecular or atomic absorption instrumentation.

Care must be taken to ensure that the sample thickness and analyte concentration is not too great, or all the X-ray beam will be absorbed. Also, an absorbing matrix is undesirable. Instrument conditions should be optimized to ensure maximum X-ray intensity, so that the counting statistics will be favorable.

3. QUANTITATIVE ANALYSIS (5)

a. Continuum Absorption Technique

For these techniques, the total output of the X-ray tube (continuum and spectral lines) is focused on the sample, and the total transmitted X-ray signal is detected. The quantitative expression is:

$$A = \log\left(\frac{I_0}{I}\right) = 0.43 l \left(\mu_{m_A} \rho_A + \mu_{m_M} \rho_M \right) \qquad (16)$$

where A is absorbance, I_0 and I are the intensity of the incident and transmitted X-ray beam, respectively, 0.43 is the conversion factor from natural to common logs, l is the sample specimen thickness, μ_m is the mass absorption coefficient for the analyte (subscript A) and the matrix (subscript M), and ρ is the mass (density) concentration (in g cm^{-3}) for the analyte (subscript A) and the matrix (subscript M).

In order to be successful, this method requires matched matrices in both standards and samples, and matched thicknesses in both.

b. Single-Line Absorption Technique

This is the same procedure as the continuum technique, except that the X-ray beam is dispersed such that only one wavelength is transmitted through the sample. This is done by one of the following procedures: using a filter that passes only the major target line or lines using a selective detector, or using a secondary or fluorescence radiation line. The only requirement is that the incident radiation be of a shorter wavelength than that of the analyte species. This procedure is much more selective than the continuum technique, but can be handled by means of equation 16 just as in the continuum case. The selectivity feature means that more than one analyte species can be analyzed in the same sample. As with the continuum technique, as well as with most absorption methods, the analyte concentraction is determined by use of an analytical calibration curve of $\log (I_0/I)$ versus ρ_A (or n_A).

c. Absorption-Edge Spectrometry

A natural extension of the single-line absorption technique is the absorption-edge method, sometimes called differential absorption spectrometry. The essential feature of this technique is that two incident lines are used, as in Figure 9.3, one at a shorter wavelength and one at a longer wavelength than that of the absorption edge. The difference between these two intensities is essentially a signal free of background. The analytical expression in this case is:

$$\log \frac{I_S/I_S^0}{I_L/I_L^0} = 0.43\rho_A l\Delta\mu_m \tag{17}$$

where the subscripts S and L stand for the short- and long-wavelength X-rays, respectively, bracketing the analyte species wavelength, and $\Delta\mu_m$ is

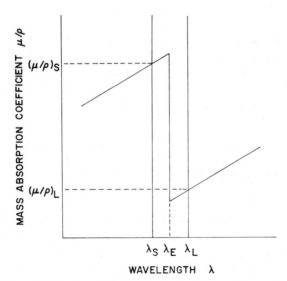

Fig. 9.3 Plot of mass absorption coefficient (linear absorption coefficient divided by density) versus wavelength.

the difference between the mass coefficients at the wavelengths corresponding to S and L. The ideal situation arises when the K_α and K_β or the L_α and L_β lines of the target element bracket the wavelength of the analyte species.

This same measurement can be effected by scanning the continuum radiation across the absorption edge if the intensity level is sufficient for good counting statistics. The absorption technique is not generally as sensitive as the fluorescence methods, because it is more difficult to distinguish between two large signals as compared to two small signals.

D. X-RAY SPECTROMETRY WITH DIRECT ELECTRON EXCITATION

Because the long-wavelength region most effective in exciting K-X-ray lines of light elements is weak in intensity as a result of the X-ray continuum intensity drop-off at long wavelengths and as a result of absorption by the X-ray tube window and by the detector window, because the X-ray fluorescence quantum yield decreases considerably with decreasing atomic number (Auger electron production becoming more important), and because long wavelength X-rays have a small depth of penetration into the sample, X-ray fluorescence spectrometry is a poor means of determining light elements ($Z < 15$). However, to avoid these problems, the sample can be bombarded directly by fast electrons (the sample is essentially the target of the X-ray tube). Disadvantages of this technique include the low pressure requirements for the electron beam ($\sim 10^{-7}$ atom), the extensive heating of the sample causing volatilization, and the X-ray continuum background under the X-ray lines. Therefore for conventional light element determination, this method has little use. However, since electrons can be focused by electric and magnetic fields to a small spot ($\sim 1~\mu m^2$), it can be used for spatial studies (see Section E.).

E. ELECTRON PROBE MICROANALYSIS

There are several microanalysis techniques involving X-rays. The most important of these is the electron probe technique. The electron probe method involves an electron gun with sophisticated focusing such that the electron beam strikes a very small area of the sample target. The electron beam is used to excite X-ray fluorescence. Thus quantitative (or semi-quantitative) analysis can be obtained on a very small area of the surface of the sample. This technique is restricted to solid samples, because of the high vacuum necessary to operate the electron beam with the required degree of flux. Wavelength dispersion can be used by means of a crystal spectrometer, or energy dispersion techniques can be used. The wavelength dispersion technique offers better resolution and sensitivity; however, the energy dispersion technique offers much faster qualitative analysis. The area of excitation is so small that an optical microscope must be used to position the sample.

In Chapter 4, Section B.3, a discussion of the quantitative relationship between X-ray fluorescence irradiance and analyte concentration is given. However, because of the similiarity between X-ray fluorescence spectrometry and electron microprobe analysis, at least as far as the fundamental relationships involving fluorescence irradiance and analyte concentration are concerned, and because of the discussion and expressions given in Chapter 4, Section B.3, no expressions are repeated here.

Because of the very small area of excitation and the shallow penetration, the technique does not offer the sensitivity of other types of X-ray methods. In fact, only in special situations can analysis in the parts-per-million-range be accomplished. The usual limits of detection are in the range 0.1 to 0.01% by weight (about 100 to 1000 ppm).

F. X-RAY DIFFRACTOMETRY

Because X-ray diffractometry has no use in trace analysis, it is not considered in this chapter.

G. ELECTRON SPECTROSCOPY

When an X-ray photon (or an electron) interacts with an atom, other phenomena can also occur if the photon (or electron) is of greater energy than the resultant X-ray fluorescence transition. In Figure 9.4, both photoelectron and Auger electron production are shown schematically. The photoelectron possesses the excess energy of the incident X-ray beam photon:

$$E_{pe} = E_0 - (E_K + \phi) \tag{18}$$

where E_{pe} is the energy of the photoelectron, E_0 is the energy of the incident X-ray photon, E_K is the energy required to just emit the K-shell electron, and ϕ is the work function of the surface of the sample, that is, the energy required to remove the electron from the surface of the sample. The energy of the photoelectrons is characteristic of the sample and thereby of analytical utility. This is also true for Auger electrons (see Figure 9.4). These techniques, like electron probe techniques, are primarily surface techniques and are limited to surface studies. If a vacuum UV light

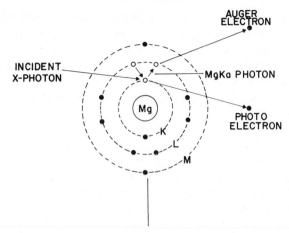

Fig. 9.4   Representation of Auger and photoelectron spectroscopic processes.

source is used, the ejected electrons from molecules in the gaseous state have energies characteristic of the outer electronic levels, as well as characteristic of the vibrational-rotational levels, and so give additional molecular structure information; this method is also potentially useful for trace analysis of molecular species. The detection of electrons requires the use of very low-vacuum chambers; thus the costs of utilizing the techniques are quite high, and little if any use has been made in trace analysis at this time.

REFERENCES

1.   E. P. Bertin, *Principles and Practice of X-Ray Spectrometric Analysis*, Plenum Press, New York, 1970.

2.   L. S. Birks, *X-Ray Spectrochemical Analysis*, Interscience, New York, 1969.

3.   K. A. H. Hooton and M. L. Parsons, *Anal. Chem.*, **45**, 2218 (1973).

4.   J. H. Hubbell, *Photon Cross Sections, Attenuation Coefficient and Energy Absorption Coefficients from 10 keV to 100 GeV*, NSRDS-NBS 29, U. S. Government Printing Office, Washington, D. C., 1969.

5.   L. S. Birks, *X-Ray Spectrochemical Analysis*, Interscience, New York, 1959.

6.   L. S. Birks, *Electron Probe Microanalysis*, John Wiley, New York, 1963.

SPARK SOURCE MASS SPECTROMETRY

R. C. ELSER

Department of Pathology
York Hospital
York, Pennsylvania

A. INTRODUCTION TO INORGANIC ANALYSIS BY MASS SPECTROMETRY

The need to analyze various types of inorganic materials for their trace impurity concentrations has led analysts to search for increasingly sensitive techniques. One such technique is mass spectrometry using high-energy sources to generate elemental ions from the sample. Sources of ionization that have been used are (1) the pulsed RF spark, (2) the thermal ionization source, (3) the laser ion source, and (4) ion sputter sources. Of these, the most commonly used source is the RF spark. Hence the name commonly given to the technique is spark source mass spectrometry (SSMS). The major thrust of the following discussion is directed toward the elucidation of SSMS as an analytical technique. The thermal ionization source is considered only briefly.

The subject of SSMS and its particulars has been recently and thoroughly treated by Ahearn (1); the reader is referred to this excellent treatise for comprehensive analysis of the physical principles underlying high-voltage breakdown between electrodes in a vacuum, the derivation of physical geometries required in spectrometers to achieve focusing, the fundamentals of ion detection by photographic or electrical detectors, and specialized applications of SSMS, such as analysis of radioactive samples, insulators, and thin films.

B. BASIS OF TECHNIQUE

SSMS is *not* a technique that has found widespread usage in analytical laboratories. The reasons for this are severalfold. First, the equipment required is relatively expensive. However, equipment costs can be justified easily if the results obtained by using the equipment effect a savings in manpower expenditures, or if the quality of data is superior to that currently obtained. Second, with spark source mass spectrometers that employ photographic detection, analyses are limited in precision to the homogeneity of a photographic emulsion on a single plate. Inhomogeneities in emulsions between plates and between emulsion lots are of greater magnitude and result in generally quoted precision (RSD) figures for SSMS of 0.3 or poorer. Under research conditions, these figures may be improved but, unless ion-sensitive multiplier phototubes are employed as detectors, relatively poor precision may be expected. With electrical detection, precision rivals that obtainable by any other analytical method, making SSMS much more attractive than it has been in the past.

SSMS involves the generation of elemental ions from a sample by means of a high-voltage spark source which is pulsed at a radio frequency. Ions produced in such a manner have a very broad energy distribution which may range up to several thousand volts per unit charge of the ion. Because of the diverse and broad energy distributions observed for various elements produced in the same spark, spark source mass spectrometers require an energy focusing sector as well as a mass focusing sector.

Double-focusing mass spectrometers may be of the two general configurations depicted in Figure 10.1. The most commonly employed configuration is that illustrated by Figure 10.1a in which the deflection due to the electrostatic energy analyzer and the magnetic mass analyzer are in the opposite sense. This design is attributed to Mattauch and Herzog (2). The advantage of the Mattauch–Herzog geometry is that the ions of all m/e ratios are focused in the same plane. With this arrangement, all elements present in the sample can be recorded simultaneously on a photographic plate located in the focal plane.

With instruments employing Nier–Johnson geometry (3) (Figure 10.1b), the deflections due to the electrostatic analyzer and mass analyzer are in the same sense. In this arrangement, ions of only one m/e ratio are focused on the detector. In order to obtain a mass spectrum, either the magnetic field or the electrostatic field must be varied in order to scan ions of varying m/e ratios past the detector. Nier–Johnson geometry has been employed in the construction of extremely high-resolution instruments.

Interpretation of mass spectra may be accomplished manually with the aid of a densitometer in the case of photographic detection, or by direct readout in the case of electrical detection. Computers have been successfully utilized in the interpretation of both the qualitative and quantative data encoded in the photographic emulsion.

Multielement analysis using electrical detection is accomplished by peak switching. The two approaches used to realize this have been to (1) step the accelerator and electrostatic analyzer voltages proportionately, and (2) step the magnetic field.

(a)

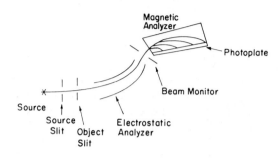

(b)

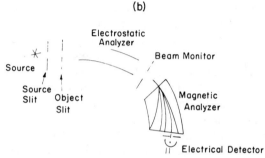

Fig. 10.1 Types of double-focusing mass spectrometers. (*a*) Mattauch–Herzog geometry. (*b*) Nier–Johnson geometry.

SSMS has been used successfully to detect elemental concentrations as low as parts per billion in milligram samples. With photographic detection, it has the capability to detect elements simultaneously over a 36-fold mass range. Some of the kinds of analytical problems to which the application of SSMS is well suited are:

1. Determination of trace components in a bulk matrix.
2. Isotope ratio analysis.
3. Identification of the source of a sample by its isotopic composition.
4. Studies of surface properties.

This chapter has been conceived by the author to fill a need for general information regarding the technique of mass spectrometry in inorganic analysis. Components of systems are discussed from a general utilitarian rather than theoretical standpoint. It is hoped that this approach will provide the reader with an understanding of the technique, its advantages, and its shortcomings.

C. INSTRUMENT

1. SOURCES

a. Thermal Ionization Source

The thermal ionization source is also known as the surface ionization source. Its principle of operation is that, when an atom is evaporated from a surface, it has a probability of being liberated as an ion, the probability being governed by the Langmuir–Saha equation:

$$\frac{n^+}{n^0} = \exp\left[\frac{e(\phi - IP)}{kT}\right] \tag{1}$$

where n^+/n^0 is the ratio of the charged to neutral species, e is the electronic charge (in C), ϕ is the work function of the surface (in eV), IP is the ionization potential of the elemental species (in eV), k is the gas constant (in eV K^{-1}), and T is the absolute temperature (in K). It can be inferred from the equation that surface ionization yields may differ markedly among elements. It is this fact of inherent selectivity that makes thermal ionization useful for studies of isotopic ratios and concentration analysis using the isotope dilution technique. Some advantages and disadvantages of this type of source are listed in Table 10.1.

The thermal ionization source can be employed in either of two forms: (1) single filament, and (2) multiple filament. In the single-filament variation, both evaporation and ionization of the sample occur as a result of heating the filament on which the sample is deposited. In the multiple-filament application, evaporation and ionization temperatures may be

TABLE 10.1.   Criteria of Thermal Ionization Sources

Advantages
No detectable memory effects (with proper design)
Sample sizes necessary for analysis of many elements are 10^2 to 10^4 smaller than for Rf spark sources
The energy spread of the ions is small, thus only a single-focusing mass spectrometer is needed
Selectivity is high
Hydrocarbons are not efficiently ionized by thermal ionization
Disadvantages
Only about half of the elements in the periodic table can be ionized with this source
Sample introduction requires a vacuum lock
Isotopic fractionation may occur

independently controlled. Multiple-filament sources are generally arranged in the form of a three-sided box with the unpaired filament supporting the sample and the opposing paired filaments serving to ionize the evaporated atoms. The filament material is usually tungsten, tungsten oxide, tantalum or platinum.

The chemical form in which a sample is applied to the filament affects the ionization yield. For example, the ionization efficiency for cesium as the chloride is on the order of 10^4 poorer than for the sulfate. The reason for the difference is that CsCl evaporates at low temperatures as CsCl without ionization, while decomposition on the filament occurs for Cs_2SO_4. Thermal ionization sources may also be used to measure gaseous samples by employing the ionization filaments as electron bombarders.

b. RF Spark Source

The RF spark source consists of a pair of electrodes encapsulated in an insulated, evacuated spark housing.The electrodes from part of the secondary circuit of an RF oscillator circuit (Figure 10.2). High voltage from the secondary of the Tesla transformer is fed to the electrodes as a series of pulses varying in duration from 20 to 200 μs. Peak-to-peak voltage may reach 100 kV. Breakdown of the field between the electrodes occurs when the RF voltage reaches a critical value dependent on the width of the gap between the electrodes. After breakdown, the voltage between the electrodes quickly falls to nearly zero and immediately begins to recover, rapidly reaching the breakdown voltage again. Thus, during a long pulse, many breakdowns may occur with the production of a spark during each breakdown. During each spark, an extremely small amount of electrode surface (on the order of 10^{-8} g) is evaporated from the anode. Ion formation occurs within this cloud of evaporated atoms as it expands out

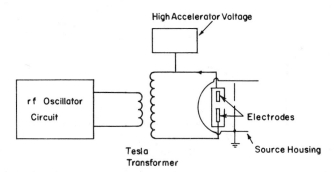

Fig. 10.2   Schematic diagram of RF spark source circuit.

of the volume between the electrodes and collides with high- and low-energy electrons generated by the cathodic electrode.

The spark housing and electrodes are usually electrically connected to the ion acceleration voltage source as in Figure 10.2. The positively charged ions in the cloud are accelerated out of the housing, while neutral and negatively charged particles condense on the electrodes or the housing.

A major consideration in operating spark sources is maintenance of the gap width at some constant value. Magee and Harrison (4) showed that the sensitivity of SSMS may be markedly affected by changes in spark gap width, and that these changes are related to the ionization potentials (*IP*) of the trace element and the matrix (Figure 10.3). In cases in which the *IP* of the impurity element is similar to that of the matrix, little change is noted. However, where large differences in *IP* exist, sensitivity varies greatly with gap width. To minimize these effects, some means of controlling the gap width is required. Several investigators (5, 6) have described such units. These operate by sensing the Rf breakdown voltage and manipulating the electrodes to maintain the voltage at a constant level. Such units are commercially available from all spark source mass spectrometer manufacturers.

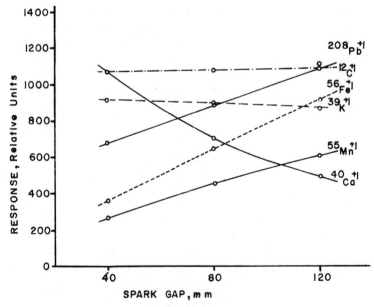

Fig. 10.3 Influence of spark gap width on elemental response in a low-impurity graphite sample.

The Nuclide* Model SGR–1 spark gap regulator controls electrode gap width by maintaining a constant RF voltage. The SGR–1 includes a sample-and-hold amplifier which "remembers" the DC voltage during the relatively long interval between pulses, thus making it possible to have a calibration that is very nearly independent of the pulse repetition frequency. The AEI* Autospark maintains the spark by controlling the mean gap width. This device rotates one electrode with respect to the other, maintaining a constant mean gap width and creating a more uniform current. One of the effects is to average out local inhomogeneities in electrode surfaces. The JEOL* automatic sparking system includes both an automatic spark gap control unit which monitors breakdown voltage and a sample electrode vibrator.

2. MASS SPECTROMETERS

a. General

The beam of ions emanating from the source housing has a relatively broad energy distribution because of the nature of ion production in the spark. As the ion beam passes between the exit aperture of the source and the entrance slit of the mass spectrometer, the beam diverges and the ion flux density decreases. Since the analytical signal of a mass spectrometer depends on the absolute number of ions striking the detector, the sensitivity of the measurement depends in part on how wide a beam it can accept and focus.

The mass analyzer is nothing more than a stable, controllable magnetic field which causes ions to be deflected along curved paths according to their mass-to-charge (m/e) ratio. The radius of curvature of the path a deflected ion follows is proportional to its momentum and inversely proportional to the strength of the magnetic field. The resolution of such a magnetic mass analyzer is given by:

$$R = \frac{r_m}{w_m} \qquad (2)$$

where r_m is the radius of curvature for a given m/e species, and w_m is the width of the entrance slit. If it is remembered that the ions impinging on the slit have a broad energy distribution, it becomes apparent that r_m may take a range of values for a given m/e since r_m is proportional to momentum or, in other words, apparently broaden the slit width with a resultant degradation of resolution.

*No endorsement of commercial equipment is intended.

The so-called *single focusing* mass spectrometers have only a magnetic mass analyzer section. Because of the resolution limitations imposed by equation 2, they are unsatisfactory for use with sources producing ion beams having a broad energy distribution, for example, RF spark sources. However, they are eminently useful in applications in which the source is more selective, for example, thermal ionization sources.

In SSMS, extremely high resolution is required. To obtain that resolution, an improvement must be made in the system to reduce the energy spread of the ions entering the mass analyzer. This is accomplished by using energy or velocity focusing of the ion beam after it emerges from the source and before it enters the mass analyzer. The path of a charged particle in an electrostatic field is dependent on the ion energy and inversely dependent on the field strength. Thus an electrostatic field focuses ions of like energy without regard to mass. The combination of an electrostatic field and a magnetic field is referred to as a *double-focusing* mass spectrometer.

b. Electrostatic Deflection Fields

An electrostatic deflection field is incorporated into a double-focusing mass spectrometer between the source housing and the mass analyzer. The field is produced by a pair of parallel metal plates which can be held at some appropriate potential with respect to one another.

Focusing is accomplished by acceleration or deceleration of ions as they enter the electrostatic field. Positive ions traveling more closely to the positive plate are slowed down, while those traveling more closely to the negative plate are accelerated. The resultant effect is to focus ions having equal velocity.

Many double-focusing mass spectrometers utilize cylindrical electrostatic fields. These are generated by plates whose surfaces are parts of coaxial cylinders. They provide a field whose strength is proportional to $1/r$. The voltages $\pm V/2$ supplied to the plates are related to the ion energy eU by equation 3, where U is the potential associated with the ion, r_1 and r_2 are the radii of the outer and inner plates, respectively, and V is the voltage applied to the plates:

$$V = 2U \ln\left(\frac{r_1}{r_2}\right) \approx 2U \frac{r_1 - r_2}{r_2} \tag{3}$$

For the center orbit of the field, U becomes identical to the acceleration potential from the source, and the radius of curvature of a deflected particle in the electric field r_e is given by:

$$r_e = \frac{2Ur_1(r_1 - r_2)^{1/2}}{V} \tag{4}$$

Here it can be seen that, for the center orbit condition to be fulfilled, the acceleration voltage and electrostatic field voltage must be maintained at a constant ratio.

The principles of geometric optics have been applied to cylindrical fields to establish focal points and image magnification (7). The condition employed in Mattauch–Herzog type mass spectrometers is to place the object, that is, the source exit aperture, at the first focal point, and thereby the image at infinity, resulting in a collimated beam of ions.

A focusing effect along the axial direction may also occur, that is, z-focusing, if the cylinders are not at least four times longer than the distance $r_2 - r_1$. In instruments employing photographic detection, z-focusing is generally not beneficial, in as much as sensitivity changes occur with shortening or lengthening of the image of the resolving slit.

Spherical fields have higher energy dispersion than cylindrical fields and have been used in applications in which high sensitivity is important. The spherical field acts as a spherical lens, focusing a point source to a point, in contrast to a cylindrical lens which focuses a point to a line. The voltage applied to a spherical field is related to the acceleration voltage by:

$$V = U\left(\frac{r_1}{r_2} - \frac{r_2}{r_1}\right) \tag{5}$$

where the symbols are identical in meaning as in equation 4.

c. Magnetic Deflection Fields

Magnetic deflection fields may be either homogeneous or nonhomogeneous. Both types have been used by investigators in the construction of mass spectrometers. Discussion is limited here to homogeneous fields, that is, the condition in which the gap between the pole pieces of the magnet is small compared to their size. Under this condition, the radius of curvature of a charged particle may be described by:

$$r_m = 143\left(\frac{m}{e}\right)^{1/2} U^{1/2} H^{-1} \tag{6}$$

where U is the acceleration potential, r_m is the radius of curvature in the magnetic field, and H is the magnetic field strength.

Fringe magnetic fields at the boundaries of the pole pieces may exert focusing effects in the z-direction. Depending on the magnitude and direction of z-focusing, these may or may not be negligible. Negative z-focusing or divergence in the z-direction may cause a loss in sensitivity in electrical detection. The condition of negative z-focusing occurs in the Mattauch–Herzog geometry in which the angle between the pole piece boundaries is larger than the deflection angle. As long as the ion detector is located close to the pole piece boundary, the effect is small. As the detector is moved away from the boundary, the signal density decreases and the sensitivity falls.

The resolving power, R of a double-focusing mass spectrometer is given by:

$$R = \frac{M}{\Delta M} = \frac{r_e}{2w_e}\left(1 - \frac{x_m}{x_e}\right) \tag{7}$$

where r_e is the radius of curvature in the electric field, w_e is the width of the entrance slit of the electric sector, x_m is the image size at the entrance of the magnetic sector, and x_e is the image size at the entrance of the electric sector. The interesting point illuminated by this equation is that, for a double-focusing mass spectrometer employing an electric field followed by a homogeneous magnetic field, the resolution exhibits no dependence on any parameters of the magnetic field. High resolution requires that the entrance slit width be small compared with the radius of curvature in the electric field and that the image of the entrance slit be demagnified by the electric field.

Dispersion by the magnetic field is affected by the magnitude of the field strength. However, the size of the image is affected proportionately, with the result that the resolution remains constant. The advantage in using a larger magnetic field is to increase the size of the ion beam that can be focused and to reduce image defects due to field effects.

3. DETECTORS

a. Photographic Devices

Until approximately a decade ago, ion-sensitive photoplates provided the only means of recording mass spectra obtained from spark sources. Even with the advent of ion multipliers, the photoplate has remained a very useful means of detection. However, precision limitations of ion-sensitive photoplates place constraints on the technique of SSMS in applications requiring high accuracy and precision. Generally speaking, results may be considered accurate only to within a factor of ≈ 3. Even so, the advantages to be realized by using photoplates as detectors are not small. In fact, any application utilizing SSMS requires that the instrumentation have the capability of photoplate recording, as well as some other detection system. The primary advantage of the photoplate is its capability to record information concerning all elements (within a 36-fold mass range) in the sample specimen simultaneously. A lesser advantage concerns accurate calibration of the detection system with respect to mass. Since the radius of curvature of an ion in a magnetic field varies with the square of the magnetic field strength, the mass spectrum is nonlinear with respect to mass. Accurate mass calibration for the photoplate can be achieved simply by obtaining the mass spectrum of a sample of a high-molecular-weight paraffin, which will yield lines at virtually every integral mass unit throughout the entire mass spectrum. Direct comparison of photoplates obtained from unknown samples with the calibration plate allows mass assignments to be easily made.

Characteristics of ion-sensitive plates are enumerated in Table 10.2. In practice, not all these characteristics can be realized on a single type of plate. An important consideration in the usage of plates involves the concept of *saturation blackening*. The degree of blackening can be expressed as percent absorbance, where total blackness is arbitrarily assigned 100% absorption. At long exposure good plates should yield areas that approach 98% absorption. Even though quantitative measurements cannot be made using areas as black as 98% absorption, this figure represents an estimate of the quality and sensitivity of the photoplate.

The sensitivity of the photoplates is a function of both ion energy and ion mass. The sensitivity can be expressed by:

$$S = k' E^x m^y \tag{8}$$

TABLE 10.2. Characteristics of ion-sensitive photoplates

High sensitivity to ions but preferably not to light
Good spatial resolution
Good spatial homogeneity of response
Low background and noise level
Adherence to the law of reciprocity
Low water and gas content
Good storage characteristics
Large dynamic range
Good contrast for quantitative accuracy
Adequate surface conductivity
Adequate silver bromide density in the top layer, sufficient
  to produce saturation blackening near unity

where k' is a constant relating photoplate sensitivity to ion mass and energy, E is the ion energy, m is the mass of the ion, x is the exponent for energy dependence, and y is the exponent for mass dependence; x has experimentally been determined to have a value of roughly 2.0, while y appears to have a value of roughly -0.5.

As ions strike the photoplate, *secondary positive* ions may be sputtered from the plate itself and deflected by the magnetic field to the other locations on the plate. The result of secondary ion interaction with the emulsion on the plate is to produce a low-level fogging of the plate on the higher-mass side of the primary line. Since the secondary ions have much less energy than the primary ions, their penetration into the gelatin is much less than that of the primary ions and, as a result, fogging is restricted to a relatively superficial layer of the emulsion. Secondary blackening effects can be minimized during processing of the plates by first bleaching the emulsion to remove the top layer and then allowing internal development to take place.

Spatial resolution of the emulsion is a function of homogeneity of size and distribution of silver halide granules in the gelatin. If homogeneity is not adequate, both resolution and response may vary with location. In quantitative work, this is unacceptable. Even good emulsions probably vary as much as 5% within individual plates, and more among plates. Various techniques, such as vacuum deposition of silver halide to produce a gelatin-free plate and the use of special emulsion layers which can be stripped from the silver halide grain, have been tried in an attempt to improve homogeneity of response from plate to plate, but with only moderate or little success. Low-gelatin or Schumann-type plates, such as the Ilford Q2, remain the only commercially available useful detector photoplates.

Background and noise level of plates is generated primarily by trauma to the emulsion during handling. *Background fog* levels may vary by several percent absorption. *Noise* is a result of dust particles and scratches.

When the ion-sensitive photoplate is to be used as a quantitative detector, some means must be available to estimate the amount of exposure the plate receives. Since the blackening of the emulsion is proportional to the number of ions striking it, exposures can be controlled by allowing some predetermined number of ions to strike it. An ion beam monitor, located between the electrostatic and magnetic sectors, allows the analyst to measure some fraction of the total ion beam passing from the electrostatic analyzer and thus to estimate the relative number of ions that reach the photoplate. The ion beam monitor is constructed of a series of grounded plates which limit the size of the beam emerging from the electrostatic analyzer and generally pass about as many ions as they intercept. The total charge accumulated by the monitor is measured using a vibrating reed electrometer and may range from 1×10^{-7} to 1×10^{-14} C for a given exposure.

If all constituents of a sample are to be measured, a graded series of exposures must be made, since the concentration range for a sample from matrix elements to trace impurities may span over nine orders of magnitude. A photographic emulsion, however, is generally only capable of recording a concentration range of 10^3. In order to record all elemental constituents, a series of exposures which differ from one another by a factor of 3 are made. A typical exposure schedule might be 1, 3, 10, 30, 100, 300, and 1000 pC as measured by the ion beam monitor. Corresponding exposure times might range from a fraction of a second to several minutes. If these charge accumulations are to bear any meaningful relationships to actual exposure, the fraction of the total beam collected by the monitor must remain constant. The major factor influencing the proportion of the total beam being intercepted is the spark location. Of course, very short exposures are difficult to control with much accuracy because of drift and amplifier noise in the monitor measuring circuit. One method of circumventing problems associated with making short exposures has been to employ ion beam gating techniques. By interrupting the beam through electrical deflection, the source can be maintained at the same conditions used for longer exposures. However, gating techniques that are not synchronous with spark production can cause errors in interpretation of results. The production of ions during a spark is variable, more volatile elements being ionized during the initial breakdown while the temperature is still low, and less volatile elements ionizing later as the temperature increases. In order to obtain representative ion populations for measurement, both the breakdown and cessation portions of the spark must be excluded from the measurement.

There are several practical drawbacks associated with the use of photoplate detection systems. Each photoplate can accommodate a limited number of exposures (15 to 20), after which it must be removed and replaced by an unexposed plate. Changes necessitate a vacuum break in the detector section with subsequent reestablishment of vacuum conditions before the next analysis can be made. Plates, on introduction, must be positioned within 20 μm of the focal plane and must be adjustable through a racking mechanism to allow successive exposures to be made. Results of an analysis are not immediately available, but must await development, calibration, and interpretation of the plate.

b. Faraday Cup Collector Device (see Figure 10.4)

The Faraday cup collector provides a simple and effective means of monitoring ion current in the focal plane. It consists of a cup with suitable suppressor electrodes and guard electrodes (Figure 10.4). A vibrating reed

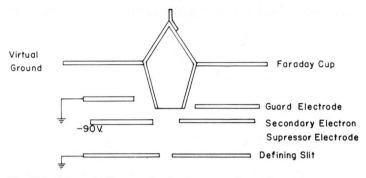

Fig. 10.4 Schematic diagram of a faraday cup collector [from Herzog (7)].

electrometer is used to measure the current arising from the detector, which may be as low as 10^{-15} A.

c. Ion Multiplier Devices

Ion multiplier tubes used as detectors in mass spectrometers have several advantages and disadvantages which are listed in Table 10.3.

Typical ion multiplier tubes used in mass spectrometers have between 15 and 20 copper–beryllium dynodes arranged in either venetian blind or box-and-grid fashion. Operating anode-to-first-dynode voltages are between 3 and 5 kV, and current gain ranges from 10^5 to 10^7. Table 10.4 lists some commonly used electron multipliers.

TABLE 10.3.   Criteria of Ion Multiplier Tubes

Advantages
    Immediate accessibility of data
    Good detection sensitivity
    More linear response than photoplates
    Good precision
    Requirement for absolute number of ions should be less as a result of
        multiplicative effect of detector
    Data amenable to computer handling
    Counting or integration may be employed
Disadvantages
    Only one measurement can be made at one time
    Saturation of multiplier may occur at intense lines
    Multiplier characteristics may be variable with time
    Resolution is poorer than for photoplates
    Electrostatic peak switching limits mass range to M to $2M$

TABLE 10.4. Typical Characteristics of Commercial Ion Multipliers

Manufacturer	Dynode Material	Number	Typical Gain	Operating Voltage (kV)
EMI 970913	Cu–Be	17	6×10^5	2.85
Hammamatsu P515	Cu–BeO	19	4×10^7	5.0
RCA 31019B	Cu–Be	14	5×10^5	4.2

All commercially available mass spectrometer systems have a capability for incorporating electrical detectors in their focal planes. Electrical detection of the mass-analyzed ions allows operation of the system in one of several modes: (1) scanning to obtain a conventional mass spectrum, (2) peak switching to obtain information on a small number of selected nuclides, and (3) nonscanning to achieve low limits of detection by long-term integration of the signal.

Scanning the mass spectrum across an electrical detector is an extremely inefficient way to collect information, since the detector views gaps in the spectrum for a much greater proportion of the time than it gathers information. Consequently, this mode is seldom used. Peak switching, however, allows the detector to view selected regions of the spectrum, bypassing those that contain no information or unwanted information (a slewing method—see Appendix A.1).

In either of these modes, the spectrum must be moved with respect to the fixed position of the detector. Scanning can be accomplished either by varying the voltage applied to the electrostatic analyzer (and proportionally varying the acceleration voltage) or by varying the magnetic field. Either alternative has drawbacks. If the electrostatic scanning approach is used, the scan is limited to the mass range M to $2M$ for a given magnetic field setting. At the end of each scan range, the magnetic field must be changed and a new scan initiated. Scanning the magnetic field allows the entire mass range of M to $36M$ to be viewed. Stepping the electrostatic analyzer voltage (and acceleration voltage proportionately) is the most common approach to peak switching. However, the mass range is limited to M to $2M$, just as in the scanning mode. The major reason for choosing to employ electrostatic peak switching has been the necessity for obtaining highly reproducible field settings which can be achieved in the electrical sector. With the advent of Hall probes, the magnetic field can now be reset reproducibly and the full mass range of the instrument can be utilized. Resolution generally suffers in peak switching modes because the slits must be widened to allow for slight inaccuracies in field settings while still viewing a portion of the mass spectrum that includes the mass number of interest.

4. VACUUM SYSTEMS

Standard pumping systems include differentially pumped source and analyzer sections. Because of the necessity of breaking vacuum to change samples, the source pumping system is at a higher pump rate, usually at least twice the analyzer rate. To attain the 10^{-8} torr analyzer operating pressure,* typical pump rates might be on the order of 80 to 100 l s^{-1} in the source and 50 l s^{-1} in the analyzer. Additional pumping may be required in the photoplate magazine section to minimize pressure increases there as a result of outgassing of photographic emulsions.

Since the only communication between the source section and analyzer section is the final source slit, pressure can be maintained by differential pumping. This allows outgassing of sample with rises in source pressure of up to 10^{-4} torr without affecting analyzer pressure.

Systems consist of either ion pumps or oil diffusion pumps backed by rotary vacuum pumps. Rotary pumps are sufficient to lower pressures to the neighborhood of 10^{-3} torr, beyond which ion pumping, oil diffusion pumping, or mercury vapor pumping is required. The addition of cold traps minimizes pumping time.

If low source pressures, for example, $\lesssim 10^{-7}$ torr, are to be maintained, careful attention must be paid to the source pumping system. High-speed pumps capable of pumping at 200 to 400 l s^{-1} should be used as well as large-diameter, short-length pumping tubes. Conductance in a pumping system is proportional to the radius of the pump tubes and inversely proportional to their length. Therefore the pumps should be located as close to the source section as possible.

Bakeout of source and analyzer sections can rid the system of residual gases and allow systems with minimal pumping capacity to attain respectable vacuum in a nonoperating, that is, nonsparking, mode. The proof of the quality of any system is the operating pressure attainable while a sample is being sparked.

5. COMMERCIAL SSMS SYSTEMS

a. General Classes

Instrumental systems, for purposes of discussion here, may be divided into two groups: those whose sources produce ions having a narrow energy spread, that is, thermal ionization sources, and those whose sources generate ions of widely disparate energy, that is, RF sources.

*1 torr = 133 N m^{-2}.

b. Thermal Ionization

A mass spectrometer equipped with a thermal ionization source is exemplified by the Nuclide Corporation Model 12-90-SU2. This is a single-focusing instrument designed primarily for isotopic abundance studies and for concentration analysis using isotope dilution.

Resolution of thermionic source instruments is generally set such that it is no greater than required by the experiment. Since sensitivity varies inversely with resolution, the resolving slits are set to maximize sensitivity.

The figure of merit to assess performance of this type of instrument is abundance sensitivity. *Abundance sensitivity* is defined as the ratio of the intensity of a signal due to an isotope of mass M to the signal at $M + 1$ or $M - 1$, that is, the ratio of the peak to its tail at $\Delta M = 1$. Good values for abundance sensitivity, which should routinely be obtained, are in the range of 5×10^4 to 5×10^5. With careful technique and decreased background pressure, these can be increased to 10^8 or greater.

Detection in such systems is accomplished using 16 to 20 stage electron multipliers. The ion current can be measured with conventional (analog) high-impedance electrometers, or alternatively pulse counting (digital) may be employed. Computer acquisition of this data is possible, as well as computer control of peak stepping devices. Nuclide's entry into this field is the DA/CS-III which employs a minicomputer to execute instructions from the operator, store data as taken, calculate isotope ratios, and perform statistical analysis of the collected data. Peak stepping is accomplished by magnetic field switching. A temperature-regulated Hall probe eliminates problems associated with the magnet's hysteresis loop.

c. RF Ionization

RF spark source mass spectrometers are typified by the AEI MS702. These instruments must be *double-focusing*, because of the nature of the source. They have extremely high sensitivity with limits of detection on the order of 1 ppb.* Normal operating resolution is on the order of 3000 although higher resolving power is attainable.

Samples may be made directly into electrodes if they consist of a conductor or semiconductor, although powders are usually mixed with a conductive material and compressed into electrodes.

The ions produced by the high-voltage discharge possess a broad distribution of kinetic energy. RF sources are generally operated at voltages near to or exceeding 30 kV. The resulting ions may have a kinetic energy in the neighborhood of 20 keV. Energy focusing of the ions emanating from the

*Concentration of trace element in sample.

source is accomplished by the electrostatic analyzer prior to their entrance into the mass analyzer. Mass separation occurs with ions of differing m/e ratios traversing paths of differing radii and coming to focus on a photographic plate or electron multiplier in the focal plane of the magnetic sector.

Ions having low m/e ratios are deflected the most, that is, have the smallest radii of curvature. In instruments having Mattauch–Herzog geometry, the positional distribution of ion impacts along the focal plane is proportional to $(m/e)^{1/2}$.

The resolution of double-focusing mass spectrometers may be selected according to the analytical requirements. As the resolution increases, however, sensitivity suffers a proportional decrease. Thus resolution is never held at a higher level than it need be. An example of a resolution requirement calculation is seen in the analysis of tin in nickel alloys (9). The major isotope of tin appears at m/e 120, with possible interference by a dimer of 60Ni isotope. Due to the mass defect (Figure 10.5), the actual absolute mass of the tin isotope is 119.902, and that of the nickel dimer is 119.860. The minimum resolving power necessary to separate these isotopes is equal to $M/\Delta M$, where M is the nominal mass and ΔM is the difference in isotopic masses.

$$Resolving\ power = \frac{120}{119.902 - 119.860} = 2857$$

A resolving power greater than 2857 may be necessary to separate these isotopes adequately, depending on their relative proportions.

Instruments that are to be used for both comprehensive survey and trace impurity analysis ought to have the capability of both photographic and electrical detection. Photographic detection allows simultaneous detection of elements spanning a 36-fold mass range, while electrical detection permits reduction in limits of detection as a result of the capability for employing long measurement times at a specific m/e.

Reduction of data by computers speeds analysis time, allowing higher throughput. A discussion of computers or in-depth computer applications is beyond the scope of this chapter. The mass spectroscopist with time on his hands can develop his own computer data and instrument handling system, or he can purchase commercial systems dedicated to the mass spectrometer with software developed by the mass spectrometer manufacturer.

The AEI system utilizes a PDP8/I to acquire data from an electrical detection system. An interpretation algorithm computes and assigns masses to observed peaks with an accuracy of 0.1 amu. This is accom-

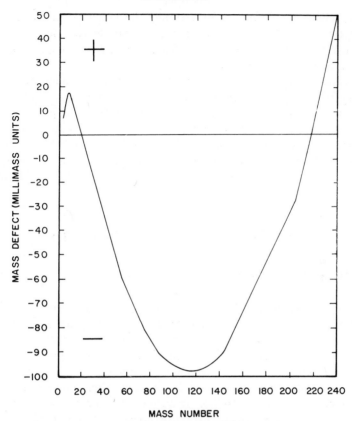

Fig. 10.5 Mass defect curve [from Morrison (26)].

plished by initiating an exponential decay of the magnetic field and converting signals from the time domain to the m/e domain. Element identification is made using a unique isotope, when possible, for each element. A peak due to the doubly charged ion at half-mass provides confirmation of the presence of an isotope. Concentration is calculated relative to an internal standard and its relative abundance.

The disadvantage of scanning systems is that most of the measurement time is spent viewing portions of the spectrum devoid of analytical information. The technique of peak switching obviates this difficulty. In the peak switching mode, a mass spectrometer detector spends virtually all of its time acquiring analytical data, since empty or analytically uninteresting portions of the mass spectrum are bypassed. Peak switching can be accomplished in either of two ways: by stepping the electrostatic analyzer and accelerator voltages, or by stepping the magnetic field. For a given magnetic field strength, a mass range of twofold can be covered using electrostatic peak switching. For isotopes beyond the limits of this range, the magnetic field must be reset.

Nuclide Corporation has a computer-controlled magnetic field stepping system available, which permits monitoring of more than 100 field positions. In this application, a temperature-regulated Hall probe is used to monitor the magnetic field strength and provide feedback control for stabilization. Magnetic regulation allows the full M to $36M$ mass range to be observed for a given electrostatic analyzer voltage, thus speeding analytical throughput by obviating the need for resetting fields multiple times for each analysis.

D. QUALITATIVE AND QUANTITATIVE ASPECTS

1. QUALITATIVE ANALYSIS

As a qualitative tool for elements present in trace quantity, the mass spectrometer with photographic detection is virtually unsurpassed by any other technique. All elements in the periodic table from ^7Li to ^{238}U, inclusive, may be identified in a single analysis. Multiple-exposure techniques permit not only a qualitative estimate but a semiquantitative estimate of all elements present in concentration ranging from matrix constituents to parts per billion impurities. Interpretation of photoplates requires some experience and expertise, however. The presence of multiply charged species which produce lines at fractions of the normal m/e positions, polymeric species, and molecular ions combine to make the interpretation of photoplate spectra a demanding task. An excellent discussion of photoplate interpretation has been given by Kennicott (10).

In making semiquantitative estimates of impurity concentration levels, it is generally assumed that the sensitivity factors, that is, the signals due to some known quantity of an element, are identical for all elements. In reality, this is not so, but is useful since most sensitivity coefficients H_R are needed for accurate quantitative estimations. A relative sensitivity coefficient is a ratio that relates the concentration of one element x to a reference element y. H_R for element x in a matrix z, $H_R(x/y)_z$, is defined as the ratio of the concentration of elements x and y in the matrix material z determined by mass spectrometry, divided by the true ratio of the concentrations of elements x and y (1). These factors may then be used to calculate true values of concentration from measured values.

2. QUANTITATIVE ANALYSIS

a. General

The nonreproducible nature of the Rf spark severely limits the precision attainable in quantitative SSMS. The several approaches that enable good quantitative data to be obtained rely either on some part of the measurement process being reproducible or on simultaneous measurement of a standard and the unknown.

The former technique is employed when electrical detection is used. The assumption is made that a constant fraction of total ion beam as monitored in the field-free region between the electrostatic analyzer and mass analyzer sectors can be measured for separate loadings of the same or different samples. The ion current at the focal plane is measured as a ratio of the total beam monitor current for a series of standards. If consistent multiplier-to-monitor ratios can be obtained, quantitative analyses of good precision can be made. Evans, Guidoboni, and Leipziger showed that, for the assay of silver in a copper matrix, a precision (RSD) of 0.027 was realized.

Referencing* to the signal generated by an element contained within the sample provides a modality of minimizing error due to effects of instrumental parameters which may change between analyses. Either the major matrix element or an impurity element doped to a known concentration may be used to provide the reference signal. Both of these techniques involve referencing collector signals to a common signal, the monitor signal. In all cases, elemental sensitivities must be ascertained if maximum accuracy is to be achieved. The major disadvantage of matrix referencing is the requirement for a large dynamic range of measurement and corrections for multiplier gain changes.

b. Isotope Dilution

Isotope dilution is a substantially different approach to the determination of impurity concentrations. For an element to be determined, a known amount of tracer or "spike," whose isotopic composition is different than that of the sample, is mixed intimately with the sample to form a homogeneous admixture. The altered isotopic ratios are then measured. The technique is limited to the 85% of the elements having two or more naturally occurring or long-lived isotopes. It is, however, relatively free

*Reference standard method—see Chapter 2.

from interferences due to matrix or other effects and is very sensitive and accurate.

Spikes of stable isotopes are available in 80 to 99% enrichment from Oak Ridge National Laboratory or from Harwell, England. A spike isotope having an abundance markedly different from that occurring in the sample is usually selected. The spike is added to the unknown as early in the sample preparation process as practicable, to minimize differences in manipulative losses between unknown and spike. Preconcentration of the element of interest is desirable, but limits of detection are set by background contaminants. The equation used to calculate concentration has been given by Paulsen, Alvarez, and Mueller (13):

$$c = q_a K \left(A_{sp} - B_{sp}\rho \right) q_s^{-1} (B\rho - A)^{-1} \tag{9}$$

where c is the concentration (in wt ppm), q_a is the weight of the isotopically enriched material added (in μg), q_s is the weight of the sample (in g), A and B are the natural abundances of the sample isotopes a and b, A_{sp} and B_{sp} are the isotopic abundances of a and b in the spike, ρ is the measured altered ratio of isotope a to isotope b, and K is the ratio of the natural atomic weight of the element to the atomic weight of the spike.

Isotope dilution may be used with either thermal ionization sources or Rf sources. The sample requirements are fewer for thermal ionization, but sensitivity is dependent on the element of interest. SSMS affords nearly equal sensitivity for all elements. Because of the sample manipulative steps, the technique is time-consuming; it has been estimated (14) that, in addition to reagent preparation time, it would require 10 to 15 man-days to analyze five samples for six elements.

c. Limits of Detection

Limits of detection of elements measured by SSMS are dependent on several factors, including sample matrix, detection mode, instrumental variables, and sample preparation. Werner (15) has described a theoretical approach to estimate detection limits of SSMS. The ion current of an element M^+ at the ion collector for an element M of mass m is defined as the sum of the currents of all isotopes of element M and is given by:

$$i_{M^+} = i_p Q_{M^+} \eta_{M^+} \frac{c_M}{10^6} \tag{10}$$

where i_p is the primary ion current at the target (in A), c_M is the relative atomic concentration of element M (in ppm), Q_{M^+} is the absolute yield of secondary positive ions of element M, and η_{M^+} is the fraction of secondary current arriving at the detector.* In terms of the limit of detection $c_{M,lim}$ equation 10 becomes:

$$c_{M,\,lim} = \frac{10^6 i_{M+,\,lim}}{i_p Q_{M^+} \eta_{M^+}}$$ (11)

Now if the abundance of isotope M_i is α_{M_i} and if the ion current is $i_{M_i^+}$, equation 11 becomes:

$$c_{M,\,lim} = \frac{10^6 i_{M_i^+,\,lim}}{\alpha_{M_i} i_p Q_{M^+} \eta_{M^+}}$$ (12)

Actually, in real analyses (15–17), absolute current measurements are not used, but rather relative current measurements with an internal standard (see Chapter 2), and so Q_{rel} is defined as:

$$Q_{rel} = \frac{Q_{M^+}}{Q_{R^+}} = \frac{i_{M^+} \eta_{R^+} c_R}{i_{R^+} \eta_{M^+} c_M}$$ (13)

for any concentration ratio of element M and internal standard R, and the Q's and c's are the respective absolute yields and concentrations. Some typical values of Q_{rel} are given (15) in Table 10.5. In terms of Q_{rel}, $c_{M,\,lim}$ is given (15) by:

$$c_{M,\,lim} = \frac{10^6 i_{M_i^+,\,lim}}{\alpha_{M_i} Q_{R^+} Q_{rel} \eta_{M^+} i_p}$$ (14)

TABLE 10.5. Relative Yield Values Q_{rel} of Some Elements for the Formation of Positive Ions M^+

Element	Q_{rel}
Al	15.4
Cr	9
Ti	8
Mo	3.2
W	0.5
Re	2.5
Fe	1
Si	3.6
Ta	0.2
Au	0.02

*η_{M^+} is a function of the element mass m, the geometry of the mass analyzer, and the acceleration voltage. Q_{M^+} is independent of c_M but may depend on the matrix of the sample.

and, substituting in terms of $i_{R_j^+}$, the ion current for isotope j of the internal reference at the ion collector:

$$c_{M,\,\text{lim}} = \frac{10^6 i_{M^+,\,\text{lim}}\alpha_{R_j}}{\alpha_{M_i^+} i_{R_j^+} Q_{\text{rel}}} \tag{15}$$

where $i_{R_j^+}$ is the ion current of the jth isotope of the reference element R, and α_{R_j} is the fractional absorbance of the jth isotope of the reference element R_j.

The limiting detectable ion current $i_{M^+,\,\text{lim}}$ when using an electron multiplier detector, an ion-to-electron transducer, is limited for practical purposes to $\sim 10^{-19}$ A and is independent of i_p, which is often of the order of 10^{-6} A. In Table 10.6 is a listing of calculated (15) limits of detection of seven elements in an iron matrix assuming $i_p = 10^{-6}$ A, $i_{M^+,\,\text{lim}} = 10^{-19}$ A, and $i_{R^+} = 10^{-12}$ A, where R is ^{56}Fe.

The limit of detection $c_{M^+,\,\text{lim}}$ is also dependent on the volume zA sputtered (consumed) away per unit time. According to Werner (15), the relationship is:

$$c_{M^+,\,\text{lim}} = \frac{10^6 i_{M^+,\,\text{lim}}\alpha_{R_j^+}}{\alpha_{M_i^+} zAKQ_{\text{rel}}} \tag{16}$$

where z is the layer thickness sputtered away per unit time (in cm s^{-1}), A is the target area (in cm^2) hit by the primary ion beam, and K is defined as:

$$K = \frac{i_{R_j^+}}{\alpha_{R_j^+} zA} \tag{17}$$

TABLE 10.6. Calculated Limits of Detection of Several Elements in an Ion Matrix[a]

Element	Q_{rel}	$c_{M,\,\text{lim}}$(ppm)
^{27}Al	15.4	0.005
^{48}Ti	8	0.01
^{28}Si	3.6	0.02
^{43}Ca	0.35	0.3
^{181}Ta	0.2	0.4
^{208}Pb	0.04	3
^{197}Au	0.02	4

[a]Conditions: $i_p = 10^{-6}$ A; $i_{M^+,\,\text{lim}} = 10^{-19}$ A; $i_{Fe^+} = 10^{-12}$ A; ^{56}Fe matrix isotope.

and is a constant for a given matrix element. The sputtering rate z is defined as:

$$z = \frac{3.6 \times 10^{-4} j_p N_p \rho}{M} \tag{18}$$

where j_p is the primary ion current density (in A cm^{-2}) impinging on the target of area A, N_p is the number of sputtered matrix atoms per primary ion (usually ~ 2 to 10), M is the molecular weight of the matrix atom, and ρ is the density of the matrix (in g cm^{-3}). For example, for a primary current (15) of 10^{-6} A, $z \sim 8.6 \times 10^{-2}$ μm hr^{-1} or 2.4×10^{-9} cm s^{-1}. If A is 0.15 cm^2 and i_{Fe+} measured at the ion collector is $\sim 10^{-12}$ A, $K \sim 4 \times 10^{-3}$ C cm^{-3}. In Table 10.7, calculated limits of detection of several elements in an iron matrix are given. It should be noted that even though $c_{M^+, lim}$ decreases with sputtering rate z, the amount of sputtered material increases.

Finally, it should be stressed that the detection limits estimated via equations 12, 14, and 16 are based primarily on the signal level, and the noise level is not considered directly except as it affects the minimum detectable current level $i_{M^+, lim}$. Therefore the detection limits may actually be larger in certain cases than those estimated by the above theoretical approach.

In Table 10.8, the number of ions found per mass spectrometric peak are given (16). The minimum number of ions required for a peak to be positively established is about six with an electron multiplier. Therefore the

TABLE 10.7. Calculated Limits of Detection $c_{M^+, lim}$ of Some Elements in an Iron Matrix[a]

Element				Matrix		
	Q_{rel}	i_p (μA)	$c_{M^+, lim}$ (ppm)	$\dot{z}$ (Å hr^{-1})	$\dot{V}$ (mm^3 hr^{-1})	$\dot{M}$ (μg hr^{-1})
^{48}Ti	8	1	0.01	—	—	—
^{181}Ta	0.2	1	0.4	860	0.0013	10
^{197}Au	0.02	1	4	—	—	—
^{48}Ti	8	10	0.001	—	—	—
^{181}Ta	0.2	10	0.04	8600	0.013	100
^{197}Au	0.02	10	0.4	—	—	—
^{48}Ti	8	100	0.0001	—	—	—
^{181}Ta	0.2	100	0.004	86000	0.13	1000
^{197}Au	0.02	100	0.04	—	—	—

[a]Conditions: $A = 0.15$ cm^2; ^{56}Fe matrix element; $\dot{z} = dz/dt$; $\dot{V} = dv/dt$; $\dot{M} = dm/dt$.

TABLE 10.8. Number of Ions Observed per Mass Spectrometric Peak

Mode	Monitor Signal, Integrated Charge (nC)	Ions Collected at Monitor, Total	Number of Ions Observed per Peak at					
			100 ppm	10 ppm	1 ppm	0.1 ppm	0.01 ppm	0.001 ppm
Peak switching[a] or	1	6×10^9	6×10^5	6×10^4	6×10^3	6×10^2	6×10^1	6×10^0
Photographic[c]	0.1	6×10^8	6×10^4	6×10^3	6×10^2	6×10^1	6×10^0	—

Mode	Monitor Signals Current (A)	Ions Collected at Monitor, Ions per second	Numbers of Ions Observed per Peak at					
			100 ppm	10 ppm	1 ppm	0.1 ppm	0.01 ppm	0.001 ppm
Scanning[b]	10^{-9} (high)	6×10^9	6×10^4	6×10^3	6×10^2	6×10^1	6×10^0	—
	10^{-10} (typical)	6×10^8	6×10^3	6×10^2	6×10^1	6×10^0	—	—
	10^{-11} (low)	6×10^7	6×10^2	6×10^1	6×10^0	—	—	—

[a] Peak switching is only useful over a twofold mass range without changing the magnetic field.
[b] A 3 min scan over a mass range of 7 to 238.
[c] It is assumed that the equivalent of 1 nC or 0.1 nC of total charge (ions) is available at the photographic plate.

TABLE 10.9. **"Experimentally" Determined Detection Limits**[a]

Absolute (ng)[b]			
Ar	0.03	Mn	0.05
Ag	0.2	Mo	0.3
Al	0.02	N	0.01
As	0.06	Na	0.02
Au	0.2	Ne	0.02
B	0.01	Nb	0.08
Ba	0.2	Nd	0.4
Be	0.008	Ni	0.07
Bi	0.2	O	0.01
Br	0.1	Os	0.4
C	0.01	P	0.03
Ca	0.03	Pb	0.3
Cd	0.3	Pd	0.3
Ce	0.1	Pr	0.1
Cl	0.04	Pt	0.5
Co	0.05	Rb	0.1
Cr	0.05	Re	0.2
Cs	0.1	Rh	0.09
Cu	0.08	Ru	0.03
Dy	0.5	S	0.03
Er	0.5	Sb	0.2
Eu	0.2	Sc	0.04
F	0.02	Se	0.1
Fe	0.05	Si	0.03
Ga	0.09	Sm	0.5
Gd	0.5	Sn	0.3
Ge	0.2	Sr	0.09
H	0.0008	Ta	0.2
Ho	0.003	Tb	0.1
Hf	0.4	Te	0.2
Hg	0.6	Ti	0.2
He	0.1	Tl	0.1
I	0.1	Tm	0.2
Ir	0.3	V	0.04
K	0.03	W	0.5
Kr	0.1	Xe	0.4
La	0.1	Y	0.07
Li	0.006	Yb	0.5
Lu	0.1	Zn	0.1
Mg	0.03	Zr	0.1

[a] According to AEI Scientific Apparatus Ltd., Manchester, England, detection limits are below 0.003 ppm (by weight) for over half the elements and below 0.1 ppm (by weight) for essentially all the elements. The lowest detected detection limit reported by R. Brown of AEI is 0.0003 ppm (by weight). According to Nuclide Corporation, State College, Pa., detection limits for most elements are of the order of 0.01 to 0.001 ppm (by weight). According to JEOL, Tokyo, Japan, the detection limit of most elements is 0.0002 ppm. Unfortunately, there has been no tabulation of experimental concentrational or absolute detection limits of the elements in selected materials (as far as we are aware).

[b] Values estimated by G. H. Morrison, *Trace Analysis*, Interscience, New York, 1965, p. 10–14.

concentrational limit of detection is represented by the concentrations corresponding to the six ions in Table 10 corresponding to about 10^{-3} ppm for peak switching and about 10^{-2} ppm for scanning methods, the latter being applicable to a much wider range of masses without changing experimental conditions. However, with photographic emulsions about 6×10^3 ions are needed to produce a just visible line on the photographic plate; and so the concentrational detection limit is about 10 ppm. If it is assumed that 6×10^9 total ions are needed for the peak switching system and the photographic methods, and about 6×10^9 ions s^{-1} for the scanning method, the sample consumption is $\sim 1 \mu g$ for the former case and $\sim 1 \mu g$ s^{-1} for the latter case.

From Tables 10.5 to 10.7, the detection limits (absolute and concentrational values) can be estimated for any element in any matrix (assuming the appropriate parameters such as Q_{M^+}, n_{M^+}, and n_{R^+} are known). Table 10.8 also allows estimation of detection limits depending upon the mode of measurement of the ions or ion currents.

In Table 10.9, detection limits estimated by Morrision (17) for a large number of elements are given assuming the use of a photographic plate.

It should be indicated that it is also possible to measure negative ions as well as positive ions but no improvement of detection limits generally occurs because the reference (internal standard) current of R$^-$ is about 10^2 times less than R$^+$; this is despite the improvement in negative ion yield over positive ion yield. Nevertheless, negative ion measurements allow estimation of low concentrations of electronegative elements such as fluorine, chlorine, and oxygen.

The precision (RSD) for the scanning mass spectrometric method is ~ 0.35 for a 3 min scan of isotopes having a mass range of 7 to 238 and a concentration of more than 1 ppm; the RSD can be improved to ~ 0.10 with averaging techniques. The precision (and systematic errors) of the peak switching mass spectrometric method is $\cong 0.01$ to 0.03 for all components greater than 1 ppm. The precision (RSD) of the photographic plate is of the same order as that of the scanning method, or perhaps a little better, for example, 0.03 to 0.3.

E. APPLICATIONS

1. GENERAL

The applications of SSMS to analytical problems are essentially bounded only by the imagination of the analyst. Problems involving studies of surface composition, migration of elements in alloys, trace impurities in alloys and other materials, isotopic abundance determinations, geological dating, environmental pollution, and trace elements in biological media may all be approached using this technique.

2. INDUSTRIAL ANALYSES

Brown and Jacobs (9) have reported on the analysis of "tramp" elements in heavy-duty alloys. The presence of lead at concentrations greater than 10 ppm has been shown to cause cracking failures of turbine blades used in jet engines. Other elements, such as bismuth, tin, antimony, cadmium, silver, selenium, arsenic, and gallium, are also known to be deleterious in their effect on the operating properties of nickel and cobalt alloys; they analyzed turbine blades that had been in service in aircraft engines for 7000 hr. One of the blades exhibited significant areas of damage. Table 10.10 lists the concentration of various elements in samples taken from various locations on the blade. It can be seen that there is an increase in lead in the burned area of the leading edge of the blade. If the increase is due to migration of lead from other areas with accumulation at the leading edge, reduction in the bulk lead concentration in the alloy could increase the useful lifetime of the blade. Assay of production melts of the alloy by SSMS could lead to improved specifications for lead concentrations,

TABLE 10.10. Concentrations of elements in a high-duty alloy Turbine Blade

Element[a]	Location[b]			Undamaged Blade[c]
	A	B	C	
Pb	3.6	9.0	12.	4.0
Ba	0.12	2.4	3.0	0.30
Sb	0.70	1.7	0.60	1.0
Sn	8.8	7.0	4.0	5.0
Cd	0.4	0.4	0.4	0.4
Ag	0.1	0.1	0.1	0.1
Se	0.67	0.36	0.18	1.3
As	1.6	1.3	0.65	2.2
Ga	4.5	4.5	1.9	4.5
Zn	18.	18.	4.5	1.0
Ca	2.3	17.	51.	1.2
K	1.7	17.	32.	1.0

[a]All concentrations are in parts per million.
[b]Location A was undamaged trailing edge, location B was damaged leading edge, and location C was detached oxide coat over damaged leading edge.
[c]A blade from a different aircraft which showed no damage.

allowing rejection of certain melts for use in fabrication of turbine blades.

Low-loss glass fibers are required in the production of light wave quides for use in laser communication systems. In order to manufacture glass of the quality required, its trace element composition must be precisely known. Among the problems arising in the analysis of glass by SSMS are (1) the fact that it is nonconductive and (2) multinuclear complex spectral lines arise from combinations of oxygen and silicon in the glass and carbon in the graphite used to pelletize the sample into a conductive form. Ikeda and Umayahara (18) showed that, by pelletizing a pulverized glass sample with graphite powder, trace concentrations of iron, cobalt, nickel, copper, and chromium could be analyzed at levels of less than 1 ppm with an RSD of better than 0.02. They compared their results with those from atomic absorption spectrometry where applicable (Table 10.11).

On-stream analysis of liquid sodium has been discussed by Hickam and Berkey (19). The possible future use of breeder reactors using liquid sodium as a heat transfer medium makes important the analysis of the coolant for radioactive by-products. The advantage of SSMS over emission spectrographic analysis is the ability to analyze for several additional elements of importance, namely, carbon, nitrogen, oxygen, and chlorine.

Carter and Sites (20) employed SSMS in the analysis of radioactive transuranium elements. In the production of higher-weight elements in high neutron flux reactors, the purity of the initial target material is important. Quality control of target materials of plutonium, americium, and curium is possible only by SSMS. High-purity plutonium targets as well as irradiated targets containing "sister" elements have been analyzed. Samples containing chemically separated fission products such as ^{111}In, [129]I, [137]La, [147]Pm, and [170]Tm have also been analyzed. Since sample contamination of the source assumes special importance because of the radioactive nature of the samples, demountable source housings are a necessity.

TABLE 10.11.   Concentrations of Impurities in Glass

Element	Glass No. 1		Glass No. 2		Glass No. 3	
	SSMS	AA	SSMS	AA	SSMS	AA
Cr	0.23	[a]	0.039	[a]	0.015	[a]
Fe	2.0	1.8	2.5	3.6	0.5	[a]
Co	0.0075	0.00	0.0075	0.003	0.0075	[a]
Ni	0.089	0.12	0.029	0.07	0.015	[a]
Cu	0.23	0.17	0.17	0.08	0.08	[a]

[a]Cannot be detected by atomic absorption spectrophotometry.

3. GEOCHEMICAL ANALYSES

Herzog (21) has reported on the use of mass spectrometry in geo-chemistry, including the use of Rb–Sr dating of geological samples by measurement of the abundance of ^{87}Sr produced as a daughter of rubidium. The development of sulfur bacteria can be similarly dated using ^{32}S/^{34}S isotope ratio changes. The time scale for differentiation of the earth into crust, mantle, and core may be established by studies of the increase in daughter isotopes of ^{87}Sr, ^{206}Pb, ^{207}Pb, and ^{208}Pb in sedimentary materials.

Other geochemical applications included analyses of meteoritic im-pactglass (22), the Smithsonia meteorite (23), and Lanthanides in basalt (24) have been reviewed by Deines (25).

Morrison (26) has reviewed the analytical techniques used in the assay of lunar materials collected by the *Apollo 11, 12,* and *1*14 flights. SSMS played an important role in multielement analysis. A group at Cornell University was able to estimate quantitatively the concentrations of 56 elements in the basalts, breccias, and soils they analyzed.

4. ENVIRONMENTAL POLLUTION ANALYSES

Ball, Barber, and Vossen (27) have used SSMS for the determination of mercury in fish as part of an environmental surveillance program. Their findings showed that fish taken in the Morecambe Bay area near the western coast of Great Britain contained much higher than normal con-centrations of mercury—averaging about four times the concentration in fish taken elsewhere. Sample preparation was accomplished by combustion and extraction of mercury as the dithizonate with chloroform. Electrodes were prepared by mixing graphite with the extract, allowing the solvent to evaporate, and compressing the graphite-sample into the desired form. Mercury was recovered with an efficiency of 94.7% using this technique.

Brown (28) has reported on the use of an MS702 R in the analysis of air and water samples for inorganic pollutants. Samples of airborne pollutants were collected using standard sampling techniques. Either glass fiber, nitrocellulose, or cellulose acetate filters were used. After completion of sampling (usually 24 hours), the filters were crushed (in the case of glass fibers) or dissolved (in the case of nitrocellulose or cellulose acetate), mixed with graphite, and compressed into electrodes in a molding die. Typical results for air pollution particulates in the atmosphere over New York City are shown in Table 10.12. The high concentrations of lead and vanadium are due to automobile exhaust emission.

Water samples were simply collected at the site and mixed directly with

TABLE 10.12.   Trace Elements in Air Pollutant Particulates

Element	Concentration ($\mu g\ m^{-3}$)
Pb	4.3
Ba	0.02
Sn	0.07
Mo	0.01
Zr	0.004
Sr	0.05
Br	0.12
As	0.005
Zn	1.7
Cu	0.26
Ni	0.32
Co	0.007
Fe	2.4
Mn	0.07
Cr	0.30
V	1.9
Ti	0.23
K	40
Ca	2.8
Cl	0.28
S	2.3
P	1.1
Si	63
Al	1.1
Mg	11
Na	>5.5
F	0.11
B	0.004

graphite using ethyl alcohol as a wetting agent. Evaporation was accomplished by gentle heating under an IR lamp. Analysis was performed using ^{107}Ag as an internal standard. Table 10.13 lists the results for samples collected by the Environmental Protection Agency. The samples from Lake Erie and Lake Alamagordo represent clean water, while the sample from Sweeney's Pond is polluted. The levels of lead, cadmium, zinc, and antimony are on the order of 10 to 1000 times higher than those observed in natural waters. This must be attributed to contamination by industrial effluents. It is probable that SSMS could be used to locate sources of pollution.

5.   BIOLOGICAL ANALYSES

Trace elemental studies of biological material were, until recently, made using emission spectrographic techniques. Several groups have begun to

TABLE 10.13. Trace Elements in Water Pollution Samples

Element	Concentrations ($\mu g \ ml^{-1}$)		
	Sweeney's Pond	Lake Alamogordo	Lake Erie
Pb	34	<0.04	<0.04
Ba	0.09	0.70	1.3
Sb	0.30	<0.04	<0.04
Cd	2.0	<0.04	<0.04
Sr	0.10	0.70	0.72
Rb	0.04	<0.04	<0.04
Zn	6.0	0.40	0.60
Cu	0.24	0.10	0.15
Co[a]	0.35	0.70	0.90
Fe	7.0	7.3	9.0
Mn	1.0	0.35	0.38
Cr	0.36	0.32	0.90
V	<0.02	<0.02	0.12
Ti	0.40	1.3	0.70
K	30	29	33
Ca	50	470	360
Cl	8.0	9.5	6.6
S	20	22	36
P	0.1	1.1	1.4
Si	170	120	64
Al	50	10	9.0
Na	>10	>10	>10
Mg	50	110	200
B	2.3	2.0	2.4

[a]Possible interference from CaF^+ ion. Sample mixed in PTFE.

investigate tissue levels of trace elements using SSMS. Brown, Jacobs, and Taylor (29) analyzed the composition of particulate matter collected in personal samplers carried by coal miners. They have also analyzed samples of lung tissue taken from miners who have spent most of their working years in coal mines. Figure 10.6 shows the relative concentration of elements in coal dust and dry lung. A remarkable correlation exists for most elements.

Fitchett, Buck, and Mushak (30) have recently reported on the direct determination of heavy elements in biological media by SSMS. They prepared samples by lyophilizing serum or homogenized liver. Electrodes were prepared by mixing lyophilate with graphite and molding the mixture under pressure in an electrode molding die. Individual electrodes were broken in the middle, and the halves sparked against each other. Uranium was employed as an internal standard and the elements bismuth, gold, lead, thallium, and platinum were measured. These investigators concluded

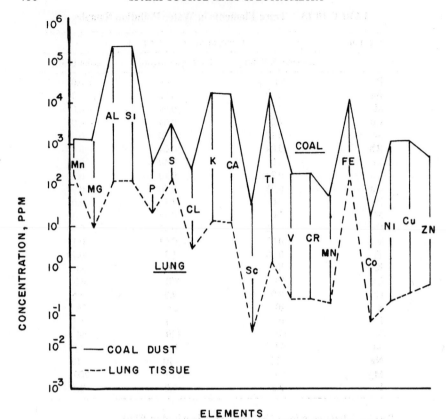

Fig. 10.6   Relative concentrations of elements in coal dust and in dry lung (26).

that the procedure was a valuable multielement screening tool for biological media and should find application in regional surveys and other screening efforts.

REFERENCES

1. A. J. Ahearn, Ed., *Trace Analysis by Mass Spectrometry*, Academic Press, New York, 1972

2. J. Mattauch, and R. F. Herzog, *Z. Phys.*, **89**, 786 (1934).

3. E. G. Johnson, and A. O. Nier, *Phys. Rev.*, **91**, 10 (1953).

4. C. W. Magee, and W. W. Harrison, *Anal. Chem.*, **45**, 852 (1973).

5. C. W. Magee, and W. W. Harrison, *Anal. Chem.*, **45**, 220 (1973).

6. B. N. Colby, and G. H. Morrison, *Anal. Chem.*, **44**, 1263 (1972).

7. R. F. Herzog, in *Trace Analysis by Mass Spectrometry*, The transmission of

Ions through Double Focusing Mass Spectrometers, A. J. Ahearn, Ed., Academic Press, New York, 1972, pp. 58.

8. R. E. Honig, in *Trace Analysis by Mass Spectrometry*, Detection and Measurement of Ions by Ion-Sensitive Plates, A. J. Ahearn, Ed., Academic Press, New York, 1972, pp. 102.

9. R. Brown, and M. L. Jacobs, Presented at the 142nd National Meeting of the Electrochemical Society, October 1972.

10. P. R. Kennicott, in *Trace Analysis by Mass Spectrometry*, Interpretation of Mass Spectrograph Plates, A. J. Ahearn, Ed., Academic Press, New York, 1972, pp. 179.

11. H. Farrar, in *Trace Analysis by Mass Spectrometry*, Relating the Mass Spectrum to the Solid Sample Composition, A. J. Ahearn, Ed., Academic Press, New York, 1972, pp. 240.

12. C. A. Evans, R. J. Guidoboni, F. D. Leipziger, *Appl. Spectrosc.*, **24**, 85 (1970).

13. P. J. Paulsen, R. Alvarez, and C. W. Mueller, *Anal. Chem.*, **42**, 673 (1970).

14. P. J. Paulsen, R. Alvarez, and D. E. Kelleher, *Spectrochim. Acta*, **24B**, 535 (1969).

15. H. W. Werner, in *Developments in Applied Spectroscopy*, Investigations of Solids by Means of an Ion Bombardment Mass Spectrometer, Vol. 7A, E. L. Grove and A. J. Perkins, Ed., Plenum Press, New York, 1969, pp. 239.

16. R. A. Bingham and R. M. Elliott, *Anal. Chem.*, **43**, 43 (1971).

17. G. H. Morrison, Ed., *Trace Analysis Physical Methods*, Interscience, New York, 1965.

18. Y. Ikeda and A. Umayahara, *JEOL News*, **11a**, 22 (1973).

19. W. M. Hickam and E. Berkey, in *Trace Analysis by Mass Spectrometry*, The Analysis of Low-Melting and Reactive Samples, A. J. Ahearn, Ed., Academic Press, New York, 1972, pp. 323.

20. J. A. Carter and J. R. Sites, in *Trace Analysis by Mass Spectrometry*, The Analysis of Radioactive Samples by Spark-Source Mass Spectrometry, Academic Press, New York, 1972, pp. 347.

21. L. F. Herzog, *Int. J. Mass Spectrom. Ion Phys.*, **4**, 253 (1970).

22. S. R. Taylor, *Geochim. Cosmochim. Acta*, **30**, 1121 (1966).

23. E. Berkey, and G. H. Morrison, Paper 25 presented at the 17th Annual Conference on Mass Spectrometry and Allied Topics, Dallas, 1969.

24. A. L. Graham, G. D. Nicholls, *Geochim. Cosmochim. Acta*, **33**, 555 (1969).

25. P. Deines, Earth and Mineral Sciences Experiment Station Circular 78, Pennsylvania State University, University Park, Pa., 1970.

26. G. H. Morrison, *Anal. Chem.*, **43**, 22A (7), (1971).

27. D. F. Ball, M. Barber, and P. G. T. Vossen, *Sci. Total Environ.*, **2**, 101 (1973).

28. R. Brown, *AEI Publ. TP*, **36**, 1971.

29. R. Brown, M. L. Jacobs, and H. E. Taylor, *Am. Lab.*, **27**, 29 (1972).

30. A. W. Fitchett, R. P. Buck, and P. Mushak, *Anal. Chem.*, **46**, 710 (1974).

11

COMPARISON OF SPECTROSCOPIC METHODS

J. D. WINEFORDNER

Department of Chemistry
University of Florida
Gainesville, Florida

The choice of an analytical method is quite complex. Several factors must be taken into account: (1) the nature and composition of the sample, which affect the sample treatment prior to measurement; (2) the approximate concentration of the analyte present in the measured sample and the limit of detection by the analytical procedure to be used; (3) the precision and accuracy required in the final analysis; (4) the speed required for obtaining a result; (5) the amount of sample available; (6) the number of samples to be measured; (7) the variability of the sample matrix from sample to sample; (8) the cost per analysis; (9) the instrumentation and methodology available. In Table 11.1, a comparison is given of the various spectroscopic methods discussed in this book, as well as some others of limited use which are not discussed in this book. Table 11.1 should give the novice a useful review of the types of spectroscopy of interest to one who is doing trace analysis and also should be useful to anyone wishing to obtain a rapid survey of spectroscopic methods. The comparison is taken after a similar tabulation by deGalan (1). It should be stressed that the analyst's choices are not always limited to just spectroscopic methods, and other methods (see Chapter 1) may also have to be considered; however, the vast majority of trace analyses are performed with spectroscopic methods.

**TABLE 11.1. Comparison of Spectroscopic Methods
of Analysis of Elements**

I. Atomic emission spark spectrometry
   A. General aspects
      1. Cost of instrument
         a. Minimal: $20,000
         b. Normal: $35,000
         c. Specialized: $70,000 (small dedicated computer)
      2. Experience required
         a. Instrument operation: ∼1 month

TABLE 11.1. (*Continued*)

  b. Setting up instrument: $\sim$1 yr

  c. Need for specialist for maintenance: sometimes

 3. Sample characteristics

  a. Physical state: solid or solution

  b. Chemical state: conductive (native or mixed with suitable powder to form briquet)

  c. Amount normally required for analysis: a few milligrams*

  d. Minimal amount required for analysis: a few milligrams*

   *Actually, the sample size may be many orders of magnitude greater than this, but only a few milligrams are volatilized.

 4. Elements analyzable: essentially all elements, especially metals—some difficulties (experimental) with nonmetals

B. Qualitative analysis—not normally done

C. Quantitative analysis

 1. Limits of detection

  a. Relative (μg g^{-1}): $10^1 - 10^3$ for most elements

  b. Absolute (ng): 10^1–10^3 for most elements

 2. Interferences and accuracy

  a. Systematic errors—gross (making analysis impossible): seldom present

   Systematic errors—minor: often present

 3. Precision of analysis: 0.05–0.10

 4. Time required

  a. Setting-up procedure: several months

  b. Sample pretreatment: several minutes

  c. Measurement: several minutes

 5. Multielement capability: yes (multichannel spectrometers)

D. Uses

 1. Trace to major elements in metals and alloys

 2. Metallic impurities in bones, caustic liquids, alkali and alkaline earth salts, and so on

 3. Trace metals in minerals, corrosion products, and other solid materials

 4. Trace and major elements in plastics, resins, and other organic materials

 5. Trace elements in biological fluids, tissues, bones, teeth, and so on

 6. Trace metals in petroleum products

II. Atomic emission arc spectrometry

A. General aspects

 1. Cost of instrument

  a. Minimal: $7500

  b. Normal: $17,000

  c. Specialized: $35,000 (plasma jet, dedicated computer)

 2. Experience required

  a. Instrument operation: $\sim$1 month

  b. Setting up instrument: $\sim$1 yr

  c. Need for specialist for maintenance: sometimes

 3. Sample Characteristics

  a. Physical state: solids

  b. Chemical state: inorganic mainly

  c. Amount normally required for analysis: 1 mg

TABLE 11.1. (*Continued*)

  d. Minimal amount required for analysis: 1 mg*
   *Depends on concentration of analyte.
 4. Elements analyzable: essentially all elements, especially metals—some difficulties (experimental) with nonmetals
B. Qualitative analysis: commonly done—usually with photographic plate and comparison of sample plate with reference plate
C. Quantitative analysis
 1. Limits of Detection
  a. Relative (μg g^{-1}): $10^0 - 10^2$
  b. Absolute (ng): 10^0–10^2
 2. Interferences and accuracy
  a. Systematic errors—gross (making analysis impossible): seldom present
   Systematic errors—minor: often large ones
  b. Minimization of minor systematic errors: buffering of sample (and discharge) with a large concentration of some easily ionizable material, such as a lithium matrix, to supply electrons to discharge
 3. Precision of analysis: 0.1–0.2
 4. Time Required
  a. Setting-up procedure: $\sim$ a few weeks
  b. Sample pretreatment: $\sim$5 minutes
  c. Measurement: $\sim$2 hr
 5. Multielement Capability: yes (multichannel spectrometers)
D. Uses
 1. Trace elements in metals and alloys
 2. Metallic impurities in minerals, corrosion products, geochemical samples, and metallurgical materials
 3. Trace metals in biological samples
 4. Trace metals in petroleum products

III. Atomic emission RF (or microwave) plasma spectrometry
 A. General aspects
  1. Cost of instrument
   a. Minimal: $\sim$\$12,500
   b. Normal: $\sim$\$25,000
   c. Specialized: $\sim$\$40,000
  2. Experience required
   a. Instrument operation: $\sim$1 wk
   b. Setting up instrument: $\sim$3 months
   c. Need for specialist for maintenance: possible but not requirement
  3. Sample characteristics
   a. Physical state: solutions (solids can be done by special sputtering or furnace methods)
   b. Chemical state: inorganic mainly
   c. Amount normally required for analysis: $\sim$5 ml
   d. Minimal amount required for analysis: $\sim$2 ml
  4. Elements analyzable: primarily metals and a few of nonmetals, for example, boron and phosphorus
 B. Qualitative analysis: not normally done, but a wavelength scan of the emission spectra can be used or a photographic plate can be used

TABLE 11.1. (*Continued*)

C. Quantitative analysis
1. Limits of detection
    a. Relative ($\mu g\ g^{-1}$): $10^{-6} - 10^{1}$
    b. Absolute (ng): $10^{-3} - 10^{4}$
        For solids, multiply limits of detection by $\sim 10^{2}$.
2. Interferences and accuracy
    a. Systematic errors—gross (making analysis impossible): seldom present
        Systematic errors—minor: ionization
    b. Minimization of minor systematic errors: adjustment of plasma conditions via electrical parameters and via ionization buffer
3. Precision of analysis: 0.01–0.05
4. Time required
    a. Setting-up procedure: days to weeks
    b. Sample pretreatment: a few minutes to a few hours
    c. Measurement: a few minutes
5. Multielement capability: yes (multichannel spectrometers)

D. Uses
1. Trace metals in metallurgical and mineralogical samples
2. Trace metals in brines and salts
3. Trace metals in air and water pollution samples
4. Trace metals in biological materials

IV. Atomic emission flame spectrometry
A. General Aspects
1. Cost of Instrument
    a. Minimal: $1500
    b. Normal: $5000
    c. Specialized: $12,500
2. Experience required
    a. Instrument operation: few days to 1 wk
    b. Setting up instrument: 1–3 months
    c. Need for specialist for maintenance: no
3. Sample characteristics
    a. Physical state: solutions (solids can be analyzed only by special pulsed evaporation methods)
    b. Chemical state: inorganic mainly
    c. Amount normally required for analysis: ~ 1–5 ml
    d. Minimal amount required for analysis: a few microliters (by means of special pulsed sampling methods)
4. Elements analyzable: only metals via atomic emission; some nonmetals via molecular band emission

B. Qualitative analysis: not normally done, but a wavelength scan or a spectrograph can be used
C. Quantitative analysis
1. Limits of detection
    a. Relative* ($\mu g\ g^{-1}$): $10^{-3} - 10^{3\dagger}$
    b. Absolute (ng): $10^{0} - 10^{6}$
        *For solids, multiply limits of detection by $\sim 10^{2}$.
        †For elements with resonance lines above ~ 350 nm, the detection limits

TABLE 11.1. (*Continued*)

are at the lower end and, for elements with resonance lines below ~350 nm, the detection limits are at the upper end of the detection range.

2. Interferences and accuracy
   a. Systematic errors—gross (making analysis impossible): seldom present
      Systematic errors—minor: a few (physical, chemical)
   b. Minimization of minor systematic errors: preparation of simulated standards, addition of buffers, and separations
3. Precision of analysis: 0.005–0.05
4. Time required
   a. Setting-up procedure: days to weeks
   b. Sample pretreatment: a few minutes to a few hours
   c. Measurement: a few minutes
5. Multielement capability: yes (multichannel spectrometers)

D. Uses
   1. Trace metals in metallurgical and mineralogical samples
   2. Trace metals in brines and salts
   3. Trace metals in air and water pollution samples
   4. Trace metals in biological materials

V. Atomic absorption spectrometry
   A. General aspects
      1. Cost of instrument
         a. Minimal: $3000
         b. Normal: $7000
         c. Specialized: $15,000
      2. Experience required
         a. Instrument operation: a few days
         b. Setting up instrument: a few weeks
         c. Need for specialist for maintenance: no
      3. Sample characteristics
         a. Physical state: solutions for flames, solutions of solids for furnaces
         b. Chemical state: inorganic mainly
         c. Amount normally required for analysis: 1–5 ml for flame, 1–100 μl solutions for furnace, 1–10 mg solids for furnace
         d. Minimal amount required for analysis: see low end of ranges in part c above; actually, the smallest sample size measureable is limited by the analyte concentration and by the atomization process
      4. Elements analyzable: only metals; nonmetals can sometimes be determined by indirect methods and in a few instances directly by furnace atomization
   B. Qualitative analysis: virtually never used for qualitative analyses
   C. Quantitative analysis
      1. Limits of detection
         a. Relative (μg g^{-1})*: flames $10^{-4}-10^2$ furnaces $10^{-5}-10^1$
         b. Absolute (ng): flames $10^{-1}-10^5$, furnaces $10^{-5}-10^1$
            *For solids with flames and furnaces, multiply by $\sim10^2$.
      2. Interferences and accuracy
         a. Systematic errors—gross: (making analysis impossible): seldom, few with flames
            Systematic errors—minor: often with furnaces

TABLE 11.1. (*Continued*)

   b. Minimization of minor systematic errors: with flames, it is best to prepare simulated standards, add buffers, and use separations; with furnaces, it is best to prepare simulated standards, use separations, and make measurements at the proper temperature, at the best height (if possible) above furnace, and in the proper atmosphere
3. Precision of analysis: flame 0.005–0.02, furnace 0.02–0.10
4. Time required
   a. Setting-up procedure: a few days to a few weeks
   b. Sample pretreatment: a few minutes to a few hours
   c. Measurement: a few minutes (extremely simple)
5. Multielement capability: not generally done—possible via photodiode array detector

D. Uses
   1. Trace metals in metallurgical and mineralogical samples
   2. Trace metals in brines and salts
   3. Trace metals in air and water pollution samples
   4. Trace metals in biological materials

VI. Atomic fluorescence spectrometry
   A. General aspects
      1. Cost of instrument: $3000–$30,000*
         *No commercial instruments are available, and so cost is estimated for a single element (~$3000) and a multielement (~$30,000) instrument.
      2. Experience required
         a. Instrument operation: a few days[†]
         b. Setting up instrument: a few weeks[†]
            [†]See comment in item 1.
         c. Need for specialist for maintenance: no
      3. Sample characteristics
         a. Physical state: solution for flames, solutions or solids for furnaces
         b. Chemical state: inorganic mainly
         c. Amount normally required for analysis: 1–5 ml for flames, 1–100 μl solutions for furnaces, 1–10 mg solids for furnaces
         d. Minimal amount required for analysis: see low end of above ranges in item 3.c; actually, the smallest sample size measurement is limited by the analyte concentration and by the atomization process.
      4. Elements analyzable: only metals; nonmetals can be determined indirectly or directly in some cases with furnace atomization
   B. Qualitative analysis; yes (rapid or slew scan spectrometers)
   C. Quantitative analysis:
      1. Limits of detection
         a. Relative (μg g^{-1})*: flames $10^{-5} - 10^2$, furnaces $10^{-6} - 10^1$
         b. Absolute (ng): flames $10^{-2} - 10^4$, furnaces $10^{-6} - 10^1$
            *For solids by flames and furnaces multiply by ~10^2.
      2. Interferences and accuracy
         a. Systematic errors—gross (making analysis impossible): seldom present
            Systematic errors—minor: few with flames, often with furnaces
         b. Minimization of minor systematic errors: same as item V. C.2.b
      3. Precision of analysis: flame 0.005–0.02, furnaces 0.02–0.10

TABLE 11.1. (*Continued*)

4. Time required
    a. Setting-up Procedure: a few days to 2 weeks
    b. Sample pretreatment: a few minutes to a few hours
    c. Measurement: a few minutes
5. Mulitelement capability: excellent—simple spectra

D. Uses
1. Trace metals in metallurgical and mineralogical samples
2. Trace metals in brines and salts
3. Trace metals in air and water pollution samples
4. Trace metals in biological materials

VII Neutron activation analysis
A. General aspects
1. Cost of instrument
    a. Minimal: not possible
    b. Normal: $20,000 plus reactor ($\sim$\$1,000,000)
    c. Specialized: \$45,000 plus reactors ($\sim$\$1,000,000)
2. Experience required
    a. Instrument operation: $\sim$1 month
    b. Setting up instrument: $\sim$1 yr
    c. Need for specialist for maintenance: yes
3. Sample characteristics
    a. Physical state: solid or liquid
    b. Chemical state: organic or inorganic
    c. Amount normally required for analysis: 1 g
    d. Minimal amount required for analysis: depends on analyte concentration
4. Elements analyzable: most elements
5. Qualitative analysis: seldom used
B. Qualitative analysis: via γ-ray spectra
C. Quantitative analysis
1. Limits of detection
    a. Relative (μg g^{-1}): $10^{-5}-10^{-1}$
    b. Absolute (ng): $10^{-2}-10^2$
2. Interferences and accuracy
    a. Systematic errors—gross (making analysis impossible): often present
       Systematic errors—minor: often present
    b. Minimization of minor systematic errors: separation necessary
3. Precision of analysis: $\sim$0.02–0.20
4. Time required
    a. Setting-up procedure: weeks
    b. Sample pretreatment: minutes to days including irradiation time
    c. Measurement: minutes to 1 hr
5. Multielement capability: yes, but interferences common
D. Uses
1. Trace elements in foods
2. Trace elements in petroleum products
3. Trace elements in air and water pollution samples
4. Nonmetals in polymers, metals, and other organic and metallurgical samples

TABLE 11.1. (*Continued*)

VIII. X-ray fluorescence spectrometry
- A. General aspects
  - 1. Cost of instrument
    - a. Minimal: $25,000
    - b. Normal: $37,500
    - c. Specialized: $75,000 (dedicated computer)
  - 2. Experience required
    - a. Instrument operation: weeks to months
    - b. Setting up instrument: months to 1 yr
    - c. Need for specialist for maintenance: yes
  - 3. Sample characteristics
    - a. Physical state: solids or solutions
    - b. Chemical state: inorganic
    - c. Amount normally required for analysis: $\sim$1 g
    - d. Minimal amount required for analysis: depends on analyte concentration
  - 4. Elements analyzable: all elements with atomic numbers $Z \gtrsim 14$. Actually, if concentration levels are $\gtrsim 0.1\%$, then can go down to $Z \sim 11$
- B. Qualitative analysis: for all elements with $Z \gtrsim 14$ (see item A.4 above)
- C. Quantitative analysis
  - 1. Limits of detection
    - a. Relative (μg g^{-1}): 1–100
    - b. Absolute (ng): 10^3–10^5
  - 2. Interferences and accuracy
    - a. Systematic errors—gross (making analysis impossible): never present
      Systematic errors—minor: often
    - b. Minimization of minor systematic errors: simulation of unknown via standards; buffering of interfering species
  - 3. Precision of analysis: 0.001–0.10
  - 4. Time required
    - a. Setting-up procedure: months
    - b. Sample pretreatment: 1–15 min
    - c. Measurement: 5 min to 1 hr
  - 5. Multielement capability: yes
- D. Uses
  - 1. Determination of nonmetals in polymers
  - 2. Determination of rare earths (minor and trace) in solids and solution
  - 3. Determinations of metals (minor and trace) in metallurgical materials
  - 4. Determination of nonmetals in foods

IX Spark source* mass spectrometry
- A. General aspects
  - 1. Cost of instruments
    - a. Minimal: $25,000
    - b. Normal: $60,000
    - c. Specialized: $\gtrsim$$100,000
  - 2. Experience required
    - a. Instrument operation: months
    - b. Setting up instruments: $\gtrsim$1 yr
    - c. Need for specialist for maintenance: yes

TABLE 11.1. (*Continued*)

3. Sample characteristics
   a. Physical state: solid, liquid, gas (sample must be volatilized prior to analysis)
   b. Chemical state: organic and inorganic
   c. Amount normally required for analysis: $\gtrsim 1$ mg (~ 1 g) (amount is for organic sample ionized with nonspark method; with Rf spark method, much greater amounts are needed even though much less sample is actually volatilized
   d. Minimal amount required for analysis: $\sim 10^{\sim 2}$ μg
4. Elements analyzable: all elements (ions) via spark source; all molecules (ions) via other ionizing source, such as electron impact, chemical ionization, laser, and field ionization
   *Thermal ionization also allows sensitive measurements of about 40 elements, and electron impact ionization allows sensitive measurement of gaseous compounds of elements as nitrogen, oxygen, carbon and silicon.

B. Qualitative analysis: all elements, all organic substances
C. Quantitative analysis:
   1. Limits of detection
      a. Relative (μg g^{-1}): $10^{-3} - 10^{-1}$
      b. Absolute (ng): $10^{-2} - 10^{2}$
   2. Interferences and accuracy
      a. Systematic errors—gross: (making analysis impossible): never present
         Systematic errors—minor: often large ones
      b. Minimization of minor systematic errors: simulation of unknowns via standards
   3. Precision of analysis: isotope dilution $\lesssim 0.01$, organic analysis 0.01–0.05, elemental 0.25
   4. Time required
      a. Setting-up procedure: weeks to months
      b. Sample pretreatment: 5 min to 1 hr
      c. Measurements: 2 hr
   5. Multielement capability: yes
D. Uses
   1. Trace analysis of elements in small amounts of solids (conducting or semiconducting)
   2. Trace analysis of elements in biological materials

X. Molecular UV-Visible* Absorption Spectrometry
A. General aspects
   1. Cost of instrument
      a. Minimal: $400 (colorimeter)
      b. Normal: $4000
      c. Specialized: $12,500–$25,000 (double-beam, double wavelength)
   2. Experience required for
      a. Instrument operation: hours to days
      b. Setting up instrument: days to weeks
      c. Need for specialist for maintenance: no
   3. Sample characteristics
      a. Physical state: solution

TABLE 11.1. (*Continued*)

      b. Chemical state: organic and inorganic

      c. Amount normally required for analysis: 10 ml

      d. Minimal amount required for analysis: 100 μl

  4. Elements analyzable: all elements that absorb strongly (native absorption) or can be transformed chemically (chelate formation) to absorb strongly

      *IR and microwave absorption spectrometry are not methods for elemental analysis and are not compared. Similarly, Raman (resonance and non-resonance) spectrometry is not an elemental analysis method, except for isolated air pollution studies involving small-molecule analysis.

B. Qualitative analysis: often used to give gross structural information

C. Quantitative analysis

  1. Limit of detection

      a. Relative (μg g^{-1})*: $10^{-3} - 10^1$

      b. Absolute (ng): $10-10^4$

      *Multiply by 100 for solid samples.

  2. Interferences and accuracy

      a. Systematic errors—gross (making analysis impossible): often present

        Systematic errors—minor: often present

      b. Minimization of minor systematic errors: use masking agents and separations

  3. Precision of analysis: 0.005–0.10

  4. Time required

      a. Setting-up procedure: days

      b. Sample pretreatment: minutes (liquid samples) to hours (solid samples)

      c. Measurement: 1 min

  5. Multielement capability: no

D. Uses

  1. Determination (structural) of functional groups following reactions with specific reagents

  2. Determinations (quantitative) of transition metals in complexes

  3. Determination (concentration) of organic molecules in organic materials

  4. Determination (quantitative) of organic molecules in biological materials and in pollution samples

XI. Molecular fluorescence* spectrometry

A. General aspects

  1. Cost of instrument

      a. Minimal: $1000 (filter fluorimeter)

      b. Normal: $6000 (excitation and emission monochromator)

      c. Specialized: $25,000 (correction of spectra)

  2. Experience required

      a. Instrument operation: hours to days

      b. Setting up instrument: days to weeks

      c. Need for specialist for maintenance: no

  3. Sample characteristics

      a. Physical state: solution

      b. Chemical state: organic and inorganic

      c. Amount normally required for analysis: 10 ml

      d. Minimal amount required for analysis: 100 μl

  4. Elements analyzable: all elements (ions) that can be transformed (generally

TABLE 11.1. (*Continued*)

via chelation) into a strongly absorbing, fluorescing complex. Some rare earths, soft metals, a few transition metals work well
*Molecular phosphorescence and light scattering spectrometry are not normally used elemental methods and are not discussed here.
B. Qualitative analysis: often used to give gross structural information
C. Quantitative analysis
   1. Limits of detection
      a. Relative (μg g^{-1})†: $10^{-4}-10^{-1}$
      b. Absolute (ng): $10^{0}-10^{3}$
      †For solid samples, multiply by 100.
   2. Interferences and accuracy
      a. Systematic errors—gross (making analysis impossible): often present
      Systematic errors—minor: often present
      b. Minimization of minor systematic errors: use masking agents and separations
   3. Precision of analysis: 0.01–0.20
   4. Time required
      a. Setting-up procedure: days
      b. Sample pretreatment: minutes (liquid samples) to hours (solid samples)
      c. Measurement: 1 min
D. Uses
   1. Determination of trace organic pollutants
   2. Determination of orgainc compounds in biological materials
   3. Determination of trace elements via complexes.
XII. Microprobe analysis*
A. General aspects
   1. Cost of instrument
      a. Laser microprobe (LMP): $50,000
      b. Electron microprobe (EMP): $125,000
      c. Ion microprobe (IMP): $250,000
   2. Experience required
      a. Instrument operation: months
      b. Setting up instrument: months to years
      c. Need for specialist for maintenance: yes
   3. Sample characteristics
      a. Physical: Solids for all three microprobes—all microprobe methods are surface methods (LMP has $\sim$10–250 μm lateral resolution; EMP has $\sim$0.2–2 μm lateral resolution; IMP has $\sim$1–5 μm lateral resolution, as well as the ability to study a few atomic layers)
      b. Chemical: inorganic (LMP has also been used for biological samples, as tissues)
      c. Amount normally required for analysis: $\sim$10^{0}–10^{2} pg per exposure, but entire sample is many, many orders larger than this, for example, grams to kilograms
   4. Elements analyzable
      a. LMP: all metals and many nonmetals
      b. EMP: elements with atomic numbers $\gtrsim$5
      c. IMP: all elements
*LMP utilizes a laser (pulse) to volatilize, atomize, and excite (an auxilary

TABLE 11.1. (*Continued*)

spark can be used) the atomic vapor in a crater 10 to 250 μm in diameter, and the atomic emission is measured. IMP utilizes an accelerated focused beam of ions on a surface to produce secondary ions of elements in a 1 to 5 μm diameter, and the resulting secondary ions are analyzed in a mass spectrometer.

   B. Qualitative analysis: yes, however, microprobe methods are most often used to obtain spatial profiles of elemental concentrations of surfaces

   C. Quantitative analysis

      1. Limits of detection

         a. Relative (μg g^{-1}): LMP $10^{-1} - 10^2$, EMP $10^1 - 10^3$, IMP $10^{-2} - 10^1$

         b. Absolute (ng): LMP 10^{-12}–10^{-15}, EMP 10^{-15}–10^{-17}, IMP 10^{-16}–10^{-19}

      2. Interferences and accuracy

         a. Systematic errors—gross (making analysis impossible): seldom in EMP, often in IMP

            Systematic errors—minor: present in all microprobe methods

         b. Minimization of minor systematic errors: use simulated standards (difficult to obtain for surface studies) and internal standard method

      3. Precision of analysis: LMP 0.05–0.25, IMP 0.02–0.05, IMP (semiquantitative) 0.50–0.75

      4. Time required

         a. Setting-up procedure: months for all three

         b. Sample pretreatment: minutes to days

         c. Measurement time: minutes to 1–2 hr

      5. Multielement capability: yes

   D. Uses

      1. Characterization of residues left on metals by reagents

      2. Distribution of elements in biological materials

      3. Partitioning of elements in various phases

      4. Determination of elements in individual plant and animal cells

      5. Identification of elements associated with mechanical failures of welds, pipes, and so on

XIII. Summary of other lesser used elemental methods

   A. Electron spin resonance (ESR)

      1. General: This method is primarily of use for structural analysis of inorganic, organic, and biological materials containing one or more unpaired electrons.

      2. Sample: The sample (1–500 mg) can be gas, liquid, or solid. Sample preparation can be simple or complex, depending on sample material and information desired.

      3. Quantitative analysis: The precision (RSD) is 0.10; systematic errors are seldom less than $\pm 50\%$. The limit of detection is ~ 1 μg of analyte with spin.

      4. Time required and cost: Instrument time ~ 5–30 min. The time required to become acquainted with the instrument and the procedure may be of the order of 1 yr. The cost of the instrumentation is high ($\gtrsim \$30,000$).

      5. Uses mainly concerned with the structure of free radicals resulting from electron irradiation of polymers, photolysis of organic materials, reaction intermediates, conformation of triplet-state species, conformation of biological materials, and exchange rates of radical anions of aromatic molecules.

   B. Microwave spectrometry

      1. General: This method is used mainly for the identification and analysis of

TABLE 11.1. (*Continued*)

organic and inorganic species having a permanent dipole movement and a vapor pressure $\gtrsim 10$ mtorr at room temperature.

2. Sample: The sample can be solid, liquid or gas, as long as it has a vapor pressure $\gtrsim 10$ mtorr at room temperature.

3. Quantitative analysis: The precision (RSD) is ~ 0.10; the accuracy is good for small molecules with permanent dipole movement ($\pm 20\%$ systematic errors or better). The limit of detection is in the ppm to ppb range for gases.

4. Time required and cost: The time required to obtain a spectrum is 1 hr. The time required to become acquainted with the instrument is ~ 1 year. The cost of the instrument is $\sim\$40,000-\$100,000$.

5. Uses: It is used primarily for identification of small organic molecules.

C. Mossbauer Spectroscopy

1. General: The Mossbauer effect can be obtained only for a few dozen nuclear isotopes, of which ^{57}Fe and ^{119}Sn have been the most studied.

2. Sample: The samples must be solids or very viscous liquids (sometimes at 77 K) and must contain > 1 mg of the element being studied. Samples must be thin (~ 0.02 mm foil to about 20 mm for solutions) to allow transmission of radiation.

3. Quantitative analysis: No real uses.

4. Time required and cost: The measurement time is 10 min to 200 hr (most samples require 3–6 hr). The cost per analysis is 3 man-hours. The instrument cost is about $\$25,000-\$50,000$.

5. Uses: Mainly concerned with ^{119}Sn and ^{57}Fe amounts in materials, and with catalytic activity of iron catalysts, lattice vibrational characteristics of crystals, and the oxidation state of iron.

D. Electron spectroscopy of chemical analysis (ESCA)

1. General: There are four basic types of ESCA: (a) electron impact spectroscopy in which a free electron collides with a molecule or atom, transferring energy and resulting in a less energetic electron; (b) X-ray-induced photoelectron spectroscopy in which X-ray excitation causes ejection of electrons from inner electronic shells of an atom, (c) vacuum UV-induced photoelectron spectroscopy in which a high-energy photon (helium—21.2 or 40.8 eV) causes a valence shell electron to be ejected; and (-d-) Auger spectroscopy in which an X-ray (or electron) causes ionization of an atom via inner-shell electron ejection and, during the rearrangement of inner-shell electrons, a photon (X-ray fluorescence) or an electron may be emitted. The emission of electrons is called Auger spectrometry, and the kinetic energy of the electrons is independent of the ionization process.

2. Sample: In electron impact spectrometry and vacuum UV photoelectron spectroscopy, gaseous samples are used. In X-ray photoelectron spectroscopy and Auger spectroscopy only solid surfaces (10–100 Å) are studied.

3. Quantitative analysis: ESCA methods have been used primarily for structural and surface related studies. Vacuum UV photoelectron spectroscopy has been linked to gas chromatography and used quantitatively in a few cases.

4. Time required and cost: The measurement time is of the order of hours. The instrumentaion is costly ($\$40,000-\$150,000$) and requires much expense (several years training) for "routine" use.

5. Uses: The major uses of the four types of electron spectroscopy are indicated in item 1.

TABLE 11.1. (*Continued*)

Footnotes to Table 11.1.

A. General aspects
   1. Cost of instrument (as of early 1974)
      a. Minimal: the lowest cost instrument available, for example, filters instead of spectrometers
      b. Normal: the price of commonly used instruments
      c. Specialized: the price of commonly used instruments including special features—a digital computer, photon counting, double-wavelength spectrometers, and so on; such instruments are often only of research use
   2. Experience required
      a. Instrument operation: the time required for a technician to learn how to obtain good spectra and good quantitative results once the method is set up
      b. Setting up instrument: the time required for a researcher to determine the optimum instrumental conditions
      c. Need for a specialist for maintenance: complex instruments require a specialized service engineer
   3. Sample characteristics
      a. Physical state: the physical state of the sample—solid, liquid, or gas
      b. Chemical state: the chemical state of the sample—organic or inorganic
      c. Amount normally required for analysis: this amount of sample generally gives results with no problems
      d. Minimal amount required for analysis: the smallest amount of sample that will give reliable results
   4. Elements analyzable: the number and kinds of elements analyzable with the instrumental method
B. Qualitative analysis: the possibility of qualitative analysis—identification of elements
C. Quantitative analysis
   1. Limits of detection
      a. Relative detection limits in terms of micrograms per gram of measured sample. If the sample is dissolved in a solution, the concentrational detection limits must be increased by a factor of $\sim 10^2$; Concentrations as well as amounts of analyte are determined in almost all cases through the use of a calibration curve prepared from standards (often simulated to have the same matrix as the unknowns)
      b. Absolute detection limits in terms of nanograms, that is, the smallest amount of an analyte that can be measured with a certain confidence.
   2. Interferences and accuracy
      a–c. Systematic errors: two types of systematic errors are considered; gross errors make the analysis completely impossible, whereas minor interferences can cause systematic errors but can be accounted for or minimized by certain methods, including simulation of all standards to be similar to the samples, the reference standard method, physical separations of the analytes, and so on
   3. Precision of analysis: the relative standard deviation (standard deviation per mean) of a determination for optimum conditions, good instruments, and concentrations well above the limits of detection
   4. Time required: the time required here involves setting up the analytical procedure, preparing the sample, and measuring the analyte signal

5. Multielement capability: this refers to the potential of the method to measure more than one spectral element simultaneously or rapidly sequentially

REFERENCE

1. L. deGalan, *Analytical Spectrometry*, Adam Hilger, London, 1971.

QUANTITATION IN ANALYSIS

J. D. WINEFORDNER

Department of Chemistry
University of Florida
Gainesville, Florida

In 1970, Kaiser (1) published an excellent article on quantitation in elemental analysis, which grew out of a talk at the 22nd Annual Summer Symposium of the Analytical Division of the American Chemical Society. In addition, primarily as a result of Kaiser, IUPAC (2) prepared a booklet on data interpretation. In this appendix, some of the most important concepts from these two reports, as well as from a translation of two of Kaiser's articles on limits of detection by Menzies (3), are summarized in order to aid the analyst in efficiently optimizing his experimental system and analytical procedure, in efficiently condensing his data, and in clearly presenting his data and interpreting his results.

A.1. THE COMPLETE ANALYTICAL PROCEDURE

Analytical figures of merit refer only to a definite, complete analytical procedure, which is specified in every detail, that is, the analytical task, the apparatus, the external conditions, the experimental conditions, the experimental procedure, the measurement process, the means of calibration, and the evaluation of data.

A.2. DEFINITION OF MEASUREMENT PARAMETERS

The measure x is the measured value for the quantity in question, that is, the analyte signal. An error is the difference between the measured value and the true value. Chemical (quantitative) analyses are performed to determine either the relative concentration c (wt %, ppm, mol l^{-1}, etc.) or the absolute amount q (g, ng, etc.). Chemical contents *cannot* be measured directly but only indirectly through an appropriate measure, such as weight, volume, voltage, and intensity. Measures x are dependent not only on the chemical content, but also the parameters r which influence the relationships between x and c (or q). Parameters r can be measured.

A.3. CALIBATION OF ANALYTICAL PROCEDURE

The relationship between the measure x_i and the chemical content c_i for species i is described by a analytical calibration function:

$$x_i = g_i(c_i) \qquad \text{(A-1)}$$

The inverse of this simple *calibration function* is the *analytical* evaluation *function*:

$$c_i = f_i(x_i) \qquad \text{(A-2)}$$

If x_i is determined for a series of standards of known c_i, the results establish an *analytical calibration curve*. After the calibration curve is plotted, it is then called the *analytical evaluation curve* and is used for quantitative analysis. Analytical evaluation curves, or at least analytical evaluation functions, are generally necessary for quantitative analysis studies, since there are *no* absolute analytical procedures. Calibration curves are generally determined with synthetic standard samples (prepared from pure substances to simulate the unknown samples), with analytical standard samples (often used in the metallurgical industries), or with standard additions of known amounts of the analyte to the sample to be measured (this is used in cases in which synthetic standards are difficult or impossible to prepare and no series of analyzed samples is available). Other calibation procedures exist but are primarily for special cases.

A.4. OPTIMIZATION OF ANALYTICAL PROCEDURES

The optimum analytical procedure for a given task is based on such practical concepts as cost and time for analysis, availability of equipment, and such fundamental concepts as the analytical figures of merit for the methods, the capabilities (informing power) of various available methods, and the topology of the procedure, that is, the logical interrelation of operations and decisions which in the analysis must lead from the sample to the result. Kaiser (4) has discussed four basic types of topology with respect to analytical procedures: the *tree structure*, in which several branched operations and reactions are utilized, for example, after an extraction step, the upper layer is analyzed by several separate steps, as is the lower layer; the *bundle structure*, in which a single analytical principle operates, as in a multichannel spectroscopic method based on a direct reader for a DC arc analysis; the *chain structure*, in which a series of sequential steps is performed; and the *network structure*, the most complex process, in which many interrelated (often reversible) steps are performed, such as in the determination of the structure of an organic molecule.

A.5.   CAPABILITY (INFORMING POWER) OF ANALYTICAL METHODS

From information (communication) theory, P_{inf} is defined by:

$$P_{inf} = \left(\sum_{i=1}^{i=n} \log_2 S_i \right) \text{ bits} \qquad \text{(A-3)}$$

where S_i is the number of distinguishable steps in each of the n quantities (measures). If each measure has $\bar{S}$ steps,

$$P_{inf} = \left(n \log_2 \bar{S} \right) \text{ bits} \qquad \text{(A-4)}$$

Note that the unit for P_{inf} is binary bits. Note that in equation A-3, if the number of parameters doubles, P_{inf} doubles, but if S_i doubles, P_{inf} increases by only 1 bit.

In many types of science, for example, spectroscopy, electrochemistry, and separations, for example, the n different measures belong to a variable parameter, for example, frequency v, magnetic field H, mass m, and voltage V. The ratio of the variable r to the smallest distinguishable difference in r, dr, is called the *resolving power R* of the method, that is,

$$R(r) = \frac{r}{dr} \qquad \text{(A-5)}$$

and

$$R_0(r) = \frac{r}{d_0 r} \qquad \text{(A-6)}$$

for the theoretical resolving power, where $d_0 r$ is the theoretical limiting* value of dr, and R_0 is the theoretical resolving power (no units). Thus there are $\Delta r / dr$ different measureable parameters (each having a certain number S_i of distinguishable steps); Δr is the range of the variable, for example, in spectroscopy, Δr is from an upper frequency v_u to a lower frequency v_l. The number of distinguishable steps in the variable r is therefore given by:

$$\frac{\Delta r}{\delta r} = R(r) \frac{\Delta r}{r} \qquad \text{(A-7)}$$

---

*If $d_0 r$ is the smallest distinguishable frequency in spectroscopy, it is related to the effective duration of the signal Δt by a form of the Heisenberg uncertainty principle:

$$d_0 v \, \Delta t = 1 \qquad \text{or} \qquad d_0 v = \frac{1}{\Delta t}$$

and so the general expression for P_{inf} is given by:

$$P_{inf} = \int_{r_1}^{r_u} R(r) \log_2 S(r) \frac{dr}{r} \qquad (A-8)$$

and if $R(r)$ and $S(r)$ are essentially constants over the range r_l to r_u,

$$P_{inf} = \overline{R} \log_2 \overline{S} \ln\left(\frac{r_u}{r_l}\right) \qquad (A-9)$$

If another variable, say t, independent of r is present in the method (e.g., time in time resolution spectroscopy),

$$P_{inf} = \int_{r_l}^{r_u} \int_{\tau_l}^{t_u} R(r) R(t) \log_2 S(r,t) \frac{dr}{r} \frac{dt}{t} \qquad (A-10)$$

and so assuming $R(r)$ and $R(t)$ are constants ($\overline{R}_r$ and $\overline{R}_t$) over the ranges r_u to r_l and t_u to t_l, respectively,

$$P_{inf} = \left[\overline{R}_r \overline{R}_t \log_2 \overline{S} \ln\left(\frac{r_u}{r_l}\right) \ln\left(\frac{t_u}{t_l}\right) \right] \text{ bits} \qquad (A-11)$$

In Table A.1, the values of P_{inf} for several instrumental methods are given. Some comments should be made concerning Table A.1:

1.   To obtain a high P_{inf}, it is necessary to have a large $\overline{R}_r$ and/or $\overline{R}_t$, as well as a reasonable ratio r_u/r_l (and/or t_u/t_l). A large $\overline{S}$ has little effect on the number of binary bits.

2.   A large resolving power direct reader in emission spectroscopy has a rather low P_{inf}, despite the great selectivity of the instrumental system, that is, it wastes most of the spectral information.

3.   A large resolving power only is not sufficient if r_u/r_l correspondingly decreases, as for the Fabry–Perot emission spectrometric system.

4.   Single-channel systems have extremely a small P_{inf} but can be highly selective, that is, there are few interferences and they are simple to operate.

The amount of information M_{inf} needed for a logical analysis must be exceeded by P_{inf} if the analysis is to be possible. In addition, sometimes much of the informing power of the procedure is wasted; for example, CN bands in arc emission spectrometry "wipe out" a large spectral region; and so, to be on the safe side, $P_{inf} \gtrsim 10\, M_{inf}$. For example, the amount of information needed to allow 10^2 elements to be determined over a concentration range of 10^{-8} to $10^2\%$ in 100 steps per decade is:

$$M_{inf} = 10^2 \log_2 10^3 = 10^3 \text{ bits}$$

As can be seen from Table A.1, there are several possible (universal) methods that could be used. Of course, in addition to $P_{inf} \gtrsim 10\, M_{inf}$, for an analytical procedure to be applied, the analytical figures of merit must also be adequate (see Section A.6); for example, the limit of detection must be sufficiently low to allow the analysis to be performed.

Kaiser has also indicated the desirability of categorizing analytical problems and analytical procedures by a set of independent parameters (noted by numbers) including:

1. Elements and/or compounds to be detected and determined *and* their respective concentration ranges.
2. Type, composition, homogeneity, and variety of samples to be analyzed.
3. Quantity of sample available.
4. Spectral and practical problems, such as local analysis, microanalysis, production control, and *in vivo* studies.
5. General analytical figures of merit, such as precision, accuracy, limits of detection, and selectivity.
6. Restriction with respect to cost of analysis, laboratory time, instruments available, skill needed, and so on.

A multidimensional space of problems and procedures could then be set up (via a computer). Any given analytical problem would be represented by a single point in the space of analytical problems, and each analytical procedure would also be presented by single points in the space of analytical procedures.

A.6 FIGURES OF MERIT DESCRIBING THE ANALYTICAL PROCEDURES

Functional Figures of Merit for a Simple Procedure (No Interferences). These include *accuracy*, *sensitivity*, *selectivity*, and *specificity*. The analytical calibration function and the evaluation functions were previously given (see equations A-2 and A-1. These functions are the inverse of each other. The *inversion* is possible only within the range of concentrations (or amounts) where the function $g(c_i)$ exists and where it is > 0. The *sensitivity of a measurement procedure is defined as the slope S of the analytical calibration curve when plotted as c versus x, that is,*

$$S_i = \left(\frac{\partial c}{\partial x} \right)_{c_i} \qquad (A\text{-}12)$$

A sensitive method is one that has a small S_i, that is, a large variation in x results from a small variation in c. The sensitivity has nothing to do with

TABLE A.1. Informing Powers of Several Analytical Measurement Methods

Method[a]	$\bar{R}_r$[b]	$\dfrac{r_u}{r_l}$	$\bar{R}_t$	$\dfrac{t_u}{t_l}$	$\bar{S}$	P_{inf} (bits)
Molecular absorption UV-visible spectrometry	10	5	—	—	10^2	1.1×10^2
Molecular absorption IR spectrometry	100	10	—	—	10^2	1.6×10^3
Molecular fluorescence UV-visible spectrometry	$10(10)^c$	$5(5)^c$	—	—	10^2	1.8×10^3
Molecular time-resolved fluorescence UV-visible spectrometry	$10(10)^c$	$5(5)^c$	10^2	10^2	10^2	1.3×10^6
Atomic emission spectrometry, small monochromator	10^4	5	—	—	10^2	1.1×10^5
Atomic emission spectrometry, large spectrograph	10^5	5	—	—	10^2	1.1×10^6
Atomic emission spectrometry, Fabry–Perot	10^7	1.0002	—	—	10^2	1.4×10^4
Atomic emission spectrometry, direct reader (50 channels)	50^d	—	—	—	10^2	3.5×10^2
Atomic fluorescence spectrometry, small monochromator (time resolution)	10^{4e}	5	10^2	10^2	10^2	7.7×10^7
Atomic fluorescence spectrometry, nondispersive (one channel)	1^f	—	—	—	10^2	7.0×10^0

Atomic fluorescence spectrometry, nondispersive photon counting (one channel)	1^f	—	—	—	10^{4h}	1.4×10^1
X-ray fluorescence spectrometry, counting detector	10^2	5	—	—	10^{4h}	2.2×10^3
Spark source mass spectrometry	10_4	35	—	—	10^2	2.5×10^5
Polarography (conventional)	10	10	—	—	10^2	1.6×10^3
Gravimetry	1^g	—	—	—	10^4	1.4×10^1

[a] The methods are described only generally. For most atomic emission cases, atomic fluorescence is also possible.

[b] The resolving powers were estimated from the width of the molecular bands for the molecular spectrometric methods.

[c] Two separate monochromators are used: an excitation monochromator to select the exciting radiation and an emission monochromator to select the emission (fluorescence) radiation.

[d] A direct reader consists of m channels; here $m = 50$.

[e] The same instrumentation is used for the atomic emission spectrometry. Small monochromator case is assumed except that time resolution is possible via a gated detector and or a pulse source.

[f] A single-channel system ($m = 1$) is assumed.

[g] In gravimetry, only one channel ($m = 1$) is possible for a given analyte.

[h] All detection systems are assumed to be analog, except for the two designated counting systems.

the limit of detection and should not be used as such.

The accuracy of an analytical procedure is determined by the presence of systematic errors and can always be increased by a critical investigation of the analytical technique to reduce such errors. If Δc is the interval describing the upper and lower limits of the systematic errors, the accuracy is given by Kaiser:

$$A_c = \text{Min} \frac{c}{\Delta c} \tag{A-13}$$

As Δc is reduced, A_c increases.

Functional Figure of Merit for Complex Analytical Procedures (*Presence of Interferences*). Assuming the need for p independently measurable quantities (concentrations, for example, of p different elements), the measure x is replaced by the measure set $x_1, x_2, \ldots, x_{p-1}, x_p$, which consists of the coordinates of a point in a space of p dimensions. Correspondingly, the contents $c_1, c_2, \ldots, c_{p-1}, c_p$ are the coordinates of a content space. The calibration of an analytical procedure gives the relationship between the measure space and the content space, leading to a system of calibration functions, that is, the measure x of element 1 may depend not only on c but also on $c_2, c_3, \ldots, c_p$, and the measure of x_2 of element 2 depends not only on c_2 but also on $c_1, c_3, c_4, \ldots, c_p$, and so on; thus the analytical *calibration functions* are:

$$x_1 = g_1(c_1, c_2, \ldots, c_p)$$

$$x_2 = g_2(c_1, c_2, \ldots, c_p)$$

$$x_p = g_p(c_1, c_2, \ldots, c_p) \tag{A-14}$$

and the inverse system, the *analytical evaluation functions* are

$$c_1 = f_1(x_1, x_2, \ldots, x_p)$$

$$c_2 = f_2(x_1, x_2, \ldots, x_p)$$

$$c_p = f_p(x_1, x_2, \ldots, x_p) \tag{A-15}$$

In other words, the concentration of any element p may depend on the concentration of other elements. The first-order functional approximation of an element j on an element i can be expressed as a constant numerical multiplier α_{ij}. This approximation can be used only over a small range for each of the c values (5 to 10% of the nominal value of c). Assuming the first-order functional approximation, a set of linear simultaneous equations can be written:

$$c_1 = \alpha_{11}x_1 + \alpha_{12}\alpha_2 + \alpha_{13}x_3 + \cdots + \alpha_{1p}x_p$$

$$c_2 = \alpha_{21}x_1 + \alpha_{22}x_2 + \alpha_{23}x_3 + \cdots + \alpha_{2p}x_p$$

$$c_p = \alpha_{p1}x_1 + \alpha_{p2}x_2 + \alpha_{p3}x_3 + \cdots + \alpha_{pp}x_p \qquad \text{(A-16)}$$

The numerical values for the coefficients α_{ij} can be determined by measuring multicomponent standards (p standards) and solving the simultaneous equations via algebraic methods. The coefficients reflect the sensitivity of the concentration of an interferent j on the analyte i. If there is no interference, $\alpha_{ii} = 0$. Of course, α_{ii} is simply the sensitivity (slope) of the analyte described previously. The coefficients α_{ij} are defined by:

$$\alpha_{ij} = \frac{\partial g_i}{\partial c_j} \qquad \text{(A-17)}$$

which is the slope of the calibration curve for the analyte i measured as a function of the concentration c_j of a potential interferent.

From equations A-16, a calibration matrix can be set up:

$$\begin{matrix} \alpha_{11}\alpha_{12}\cdots\alpha_{1p} \\ \alpha_{21}\alpha_{22}\cdots\alpha_{2p} \\ \alpha_{p1}\alpha_{p2}\cdots\alpha_{pp} \end{matrix} \qquad \text{(A-18)}$$

The calibration matrix must not be zero in the range of application. Many analytical procedures follow a linear relationship in the range of application (i.e., valid locally).

A better format for regression equations with numerical coefficients is:

$$\frac{c_1}{x_1} = \beta_{10} + \beta_{11}c_1 + \beta_{12}c_2 + \cdots + \beta_{1p}c_p$$

$$\frac{c_2}{x_2} = \beta_{20} + \beta_{21}c_1 + \beta_{22}c_2 + \cdots + \beta_{2p}c_p$$

$$\frac{c_p}{x_p} = \beta_{p0} + \beta_{p1}c_1 + \beta_{p2}c_2 + \cdots + \beta_{pp}c_p \qquad \text{(A-19)}$$

The above equations are nonlinear in x_i, which is often a better approximation of the analytical curve in real analytical procedures than equation A-16; they do not require the curved line for any local c_i to pass through the origin, are simple to evaluate with standards and mathematically, and are linear equations because the x's are measured and the c's can be determined by algebraic methods.

In some types of spectroscopy and nonspectroscopy, higher-order terms, for examples, $c_j c_k$, c_j^2, $c_j^2 c_k$, and so on, must be included. However, the solution of this higher-order equation is much more complicated mathematically and is not treated here. Nevertheless such cases can be handled by suitable mathematical "know-how."

Selectivity. A selective method is one in which there is no interference from other species, that is, $\alpha_{ij} = 0$ (or $\beta_{ij} = 0$) for all values of $j \neq i$. If $\alpha_{ij} > 0$ (or $\beta_{ij} > 0$), an interference is said to exist and the method is said to lack selectivity. No formal expressions for selectivity are given, since no generally accepted form has been approved by IUPAC or even utilized sufficiently by workers to allow us to recommend one form.

Specificity. A specific method is one in which the signal is representative of only one species, the analyte, that is, the informing power is low but the method is usable for selective, specific determination of one species only. In this case, only *one sensitivity factor* α_{ii} (or β_{ii}) is nonzero, and all the rest are zero for a truly specific method. Again, no IUPAC recommendation for specificity has been given and no generally accepted expression has been used by workers to allow us to recommend one form.

Statistical (Descriptive) Figures of Merit. These figures are useful in compressing and reducing experimental data and to interpret the results in terms of predictions, risks, and so on. One way to describe data is via a histogram, a plot of occurrence versus numerical values. To compress the data even more and to give more quantitative information, the moments of different order are calculated. The kth moment for individual results $x_i(i-1$ to $n)$ is given by:

$$m_k = \frac{1}{n} \sum_{i=1}^{n} x_i^k \tag{A-20}$$

However, the first moment m_1 is simply the average $\bar{x}$ of the series and is:

$$\bar{x} = m_1 = \frac{1}{n} \sum_{i=1}^{n} x_i \tag{A-21}$$

The higher moments are generally taken with respect to the average, $\bar{x}$, and so:

$$m_{k \geqslant 2} = \frac{1}{n} \sum_{i=1}^{n} (x_i - \bar{x})^{k \geqslant 2} \tag{A-22}$$

The second moment m_2 is the square of the standard deviation s and is a measure of scatter of the results. However the square of the standard deviation s^2 is generally defined slightly differently from the format of equation A-22 to account for the use of the set mean $\bar{x}$ rather than population mean $\bar{X}$; and so:

$$s^2 = m_2 = \frac{1}{n-1} \sum_{i=1}^{n} (x_i - \bar{x})^2 \qquad \text{(A-23)}$$

The third moment m_3 accounts for the *skewdness of the data* and is:

$$m_3 = \frac{1}{n} \sum_{i=1}^{n} (x_i - \bar{x})^3 \qquad \text{(A-24)}$$

and the fourth moment is a measure of the *peakedness of the data* and is:

$$m_4 = \frac{1}{n} \sum_{i=1}^{n} (x_i - \bar{x})^4 \qquad \text{(A-25)}$$

Of the above figures of merit, only the mean $\bar{x}$ and the standard deviation s are used to any great extent in quantitative analysis. A combination of s and $\bar{x}$, called the relative standard deviation s_r, is also used:

$$s_r = \frac{s}{\bar{x}} \qquad \text{(A-26)}$$

To avoid confusion with concentrations expressed as weight percents, *do not* multiply by 100 and call it relative standard deviation.

Statistical (Prognostic) Figures of Merit. In order to appraise an analytical procedure for the smallest content detectable, it is necessary to determine whether an observed measure is "real," that is, due to the analyte or due to an uncontrollable fluctuation. Uncertainty is limited but *not* removed via a statistical approach. A decision with calculable risk can be made using a criterion agreed on by convention, namely, that chance disturbing influences are operative when an analytical procedure is applied to a sufficiently large number (at least 20) of blank analyses and then treating the blank measures statistically to obtain $\bar{x}_{bl}$ and s_{bl}. Chance fluctuations are due to several reasons (see Table A.2). It should of course be kept in mind that the size of the blank is not critical in determining the detection limit but rather the chance fluctuations in it.

A criterion used to decide whether the observed value of the measure x is real or must be rejected (see Figure A.1) is given by:

$$x_L = \bar{x}_{bl} + k s_{bl} \qquad \text{(A-27)}$$

where x_L is the limiting detectable measure, and k is a protection factor. Since $c = f(x)$, the limits of detection in concentration or in amount of

TABLE A.2. Fluctuational and Disturbing Factors Operating on Blank (and Sample) Measures

I. Random errors
    A. Chemical impurities
        1. Residual quantities of analyte in reagents, solvents, and so on
        2. Foreign substances acting like analyte
        3. Wandering residuals from air, water, and vessels used
        4. Memory effect
    B. Losses during analysis
        1. Incomplete reaction
        2. Vaporization
        3. Walls of vessels and precipitation
        4. Transfer
    C. Variation of parameters
        1. Temperature and pressure during weighing and volumetric procedures
        2. Stray light
        3. Wavelength of spectrometer
    D. Meter reading errors and other blunders
II. Noises
    A. Signal-limited noises
        1. Fluctuation $(1/f)$
        2. Photon
    B. Background-limited noises
        1. Fluctuation $(1/f)$
        2. Photon
    C. Detector-limited noises
        1. Shot
        2. Johnson
        3. Semiconductor
    D. Electronic measurement noise
    E. Statistical fluctuations of pulse counts
    F. Grain fluctuations of emulsions

---

analyte can also be found for the given analytical procedure. Generally k is chosen as 3. However, $k = 3$ does not give a 99.86% (one-sided) confidence limit but rather a practical confidence limit of $\sim$95% in most cases, because the normal gaussian distribution does not strictly apply [the Tshebyscheff theorem which gives a minimal value for the confidence level gives $(1/k^2) \times 100$, or $\sim$89%] and because $\bar{x}_{bl}$ and s_{bl} are only estimates of $\bar{x}_{bl}$ and σ_{bl}. Therefore,

$$x_L = \bar{x}_{bl} + 3s_{bl} \qquad \text{and} \qquad c_L = f(x_L) \tag{A-28}$$

The uncertainty in c_L is $\sim$0.33 $(\sigma(c)/c = s_{bl}/x \approx 0.33)$, and so detection limits c_L should never be written with more than *one* significant figure.

One must never guarantee the purity of a sample based on the limit of

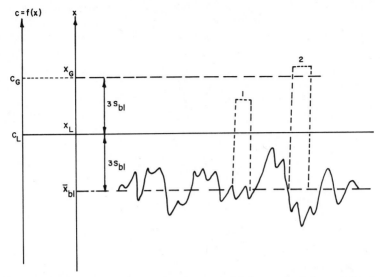

Fig. A.1 Representation of a fluctuating measure (signal) x and of the variation in concentration of analyte c corresponding to the measure. Two special cases are shown: Case 1: the measure exceeds x_L and so the analyte is just detectable, case 2, the measure exceeds x_G and so the analyte concentration is $\sim c_G$. If case 2 is not obtained, the sample can be guaranteed to have a concentration of analyte less than c_G [from Kaiser (3)].

detection for the analytical procedure. The uncertainty is too great. The limit of guarantee of purity c_G is expressed by:

$$x_G = \bar{x}_{bl} + 6s_{bl} \qquad \text{and} \qquad c_G = f(x_G) \qquad \text{(A-29)}$$

or if the sample is inhomogeneous,

$$x_G = x_{bl} + 3s_{bl} + 3\left(s_s^2 + s_{bl}^2\right) \qquad \text{(A-30)}$$

where s_s is the standard deviation describing sample heterogeneity. Thus it is then "safe" to say that the limit of guarantee of purity of an analyte in a given sample is c_G. Note that

$$c_G \sim 2c_L \qquad \text{(A-31)}$$

REFERENCES

1. H. Kaiser, *Anal. Chem.*, **42**, (21, 24A (1970)); **42**, (41, 26A (1970)).

2. IUPAC Comission V-4, Part II, Data Interpretation.

3. H. Kaiser, *Two Papers on the Limit of Detection of a Complete Analytical Procedure*, A. C. Menzies, transl., Hafner, New York, 1969.

4. H. Kaiser, *Foundations for Critical Discussion of Analytical Methods*, in "Methodicum Chimicum," Academic Press, in press.

SEQUENTIAL VERSUS MULTICHANNEL VERSUS MULTIPLEX MEASUREMENTS METHODS

J. D. WINEFORDNER

Department of Chemistry
University of Florida
Gainesville, Florida

In optical spectroscopy, there are basically four measurements systems: the sequential linear scan (SLS), the sequential slew scan (SSS), the multichannel (MC), and the multiplex (MX). All these systems can be used to measure spectral components. However, it is not readily apparent which method(s) is(are) better for rapidly measuring more than one spectral component. Most of this section has been taken from an article by Winefordner et al. (1).

The SLS method generally involves a single-slit dispersive spectrometer with a grating or a prism disperser which scans the wavelength region of interest at a uniform rate. In such a system, much time is lost scanning spectral regions where no spectral information is present. All normal and rapid scan spectrometers are SLS devices.

The SSS method involves a system similar to the one used for SLS, except that essentially no time is lost between spectral components because the spectrometer is programmed to remain on each spectral component for a certain time or to achieve a certain signal level and then rapidly changes to a new spectral component. A nondispersive system utilizing rotating filters is also an SSS system.

The MC method generally involves a spectrometer with a single entrance slit and multiple exit slit-detector combinations at the exit focal plane, each combination being preset for certain desired spectral components. Other MC methods utilize a series of interference filter-detector combinations or a series of resonance monochromators. In the MC system, *all desirable spectral components are measured all the time*. Image devices, such as photodiode arrays and television camera tubes, used with conventional dispersive spectrometers are considered MC systems.

MX methods (1–5) are quite different from the previous systems generally based on dispersion of radiation and an exit slit–detector for each spectral component. In MX methods, all the spectral components are measured all or most (e.g., 50%) of the time by a *single* detector. As a result, the spectral information must be encoded by an appropriate means, and the resulting complex signals must later be decoded (a computer is usually needed to reproduce the spectral information (spectrum) of interest. The major encoding systems for MX methods consist of either a cyclic multislit (Hadamard) mask at the exit focal plane [double encoding (double cyclic Hadamard masks) with a similar mask at the entrance focal plane is also possible] of a dispersive spectrometer, or a movable mirror in a Michelson interferometer.

In the Hadamard case, the multislit cyclic mask is stepped sequentially, one slot position at a time, to allow certain spectral components either to reach or not to reach the detector; if sufficient positions are available, and if the detector receives a series of signals dependent on these positions, it is possible by means of a fast Hadamard transform (the cyclic mask is constructed according to Hadamard matrices) to decode the resulting complex array of signals and to retrieve the original spectrum. This approach is called Hadamard transform spectrometry (HTS).

The other major MX system is Fourier transform spectrometry (FTS), in which a Michelson interferometer (or a modification of it) is used and the radiational signals are encoded via a moving mirror (very high mechanical tolerances are required for the mirrors) in one of the arms of the interferometer. If the mirror has a velocity of v (in cm s^{-1}), the encoded signal for a spectral line with a wavelength λ has an audio frequency f (in s^{-1}) of $2v/\lambda$. Other types of FTS include modulation of source radiation at different frequencies to produce signals at different audio frequencies at a single detector (a monochromator is not needed) and similar systems utilizing one detector for more than one spectral component. MX systems either require a digital computer to perform the fast Hadamard transform (FHT) or the fast Fourier transform (FFT), or an analog device (in the case of FTS) called a wave or frequency analyzer. The FHT is much simpler to perform on a digital computer than the FFT, and the HTS has less stringent mechanical restrictions.

In order to compare the four measurement methods for UV-visible-IR spectra, the signal-to-noise ratios for several hypothetical spectra are calculated and the results compared (1).

The three major noise types to be considered are detector, photon (shot), and fluctuation. No other noise types are significant for UV-visible-IR spectroscopy, that is, electronic and digitizing noises can always be made negligible compared to the above three. Also, for simplicity all signals and noises are given in digital form (counts) rather than analog form (coulombs).

The *detector noise* (see Chapter 2) in terms of the dark current rate R_d (in s^{-1}) is given by:

$$N_d = \sqrt{R_d t_c} \tag{B-1}$$

where t_c is the counting time (averaging time).

The *photon noise* in terms of the photon flux R_p (in s^{-1}) incident on the cathode (assuming no losses due to cathode or dynode inefficiencies is given by:

$$N_p = \sqrt{R_p t_c} \tag{B-2}$$

The *fluctuation noise* (see Chapter 2) in terms of the photon flux R_p is given by:

$$N_{fl} = \xi_f R_p t_c \tag{B-3}$$

where ξ_f is the measurement frequency-dependent fluctuation fraction, that is, the ratio of fluctuation noise power to the spectral radiant power being studied.

The total noise (on counts) is given by:

$$N_T = \sqrt{N_d^2 + N_p^2 + N_{fl}^2} \tag{B-4}$$

assuming the independency of the three noise types.

According to Winefordner et al. (1), the signal-to-noise ratios (SNR) for spectral element i for the four measurement systems are given by:

$$(\text{SNR})_{SLS_i} = \frac{R_{L_i}}{\sqrt{\frac{m}{t}\left[R_d + (R_{L_i} + R_{B_i})\right] + \left[(R_{Li}\xi_{Lf})^2 + (R_{Bi}\xi_{Bf})^2\right]}} \tag{B-5}$$

assuming 1 s per spectral interval and m spectral intervals, and by:

$$(\text{SNR})_{SSS_i} = \frac{\frac{t}{r} R_{Li}}{\sqrt{R_d \cdot \frac{t}{r} + \frac{t}{r} \cdot (R_{L_i} + R_{B_i}) + \frac{t^2}{r^2}\left[(R_{Li}\xi_{fl})^2 + (R_{Bi}\xi_{Bf})^2\right]}} \tag{B-6}$$

assuming m spectral intervals, r spectral components, and equal time, t/r, per spectral interval with no time for slewing, and by:

$$(\text{SNR})_{MC_i} = \frac{tR_{L_i}}{\sqrt{R_d t + t(R_{L_i} + R_B) + t^2\left[(R_{L_i}\xi_{Lp})^2 + (R_{B_j}\xi_{Bf})^2\right]}}$$

(B-7)

assuming m spectral components and t, s per spectral element and

$$(\text{SNR})_{MX_i}$$

$$= \frac{\frac{1}{2}tR_{L_i}}{\sqrt{R_d t + \frac{1}{2}t\sum_j (R_{L_j} + R_{B_j}) + \frac{t^2}{4}\left[\left(\sum_{j=1}^{m} R_{L_j}\xi_{Lf}\right)^2 + \left(\sum_{j=1}^{m} R_{B_j}\xi_{B_j}\right)^2\right]}}$$

(B-8)

where the signal level is effectively measured $t/2$, s in either the HTS or the FTS systems. In the above equations,

R_d = detector dark count rate (in s^{-1})
R_{L_i} = photometric count rate of line of analyte of interest (in s^{-1})
R_{B_i} = photometric background count rate at wavelentgh λ_i (in s^{-1})
R_{L_j} = photometric count rate of line at any wavelength λ_j (in s^{-1})
R_{B_j} = photometric count rate of background at any wavelength λ_j (in s^{-1})
m = number of spectral intervals, that is, $\Delta\lambda = m\Delta\lambda_s$ (no units)
ξ_{Lf} = fluctuation factor for line (assumed to be independent of λ_j) (no units)
ξ_{Bf} = fluctuation factor for background (assumed to be independent of λ_j) (no units)
λ_i = wavelength of interest (in nm)
λ_j = any wavelength (in nm)
$\Delta\lambda_s$ = spectral bandwidth (resolution) of system (in nm)

In the above expressions, it has been assumed that all noises are independent. If dependent noises exist, the appropriate noise cross-products must be considered. It has also been assumed that ξ_{Lf} is the same for all lines; however, this may not be so for lines of different elements, and so it may be necessary to evaluate ξ_{Lf} at each wavelength λ_j. Similarly ξ_{Bf} is assumed the same for all background components but may also be dependent on λ_j.

The *multiplex advantage* (the *Fellgett advantage*) is often discussed in the literature, especially literature concerning the IR region which is said to be detector noise–limited. If the detector noise limitation applies, that is, noise that is *not* dependent on the signal level, in MX spectroscopy $(SNR)_{MX_i}$ becomes:

$$(SNR)_{MX_i} = \frac{\frac{1}{2}tR_{L_i}}{\sqrt{R_d t}} = \frac{1}{2}R_{L_i}\sqrt{\frac{t}{R_d}} \qquad (B\text{-}9)$$

assuming a duty factor of unity whereas the signal-to-noise ratio for a sequential linear scan system for the same ith spectral component is:

$$(SNR)_{SLS_i} = R_{L_i}\sqrt{\frac{t}{R_d m}} \qquad (B\text{-}10)$$

and so the multiplex advantage is:

$$\frac{(SNR)_{MX_i}}{(SNR)_{SLS_i}} = \frac{\sqrt{m}}{2} \geqslant 1 \qquad (B\text{-}11)$$

as long as $m \geqslant 4$. Generally in the literature, the factor of 2 is omitted, but for a fair comparison it must be retained. If one wishes to accept the same signal-to-noise ratio for both methods, the analysis time for the MX_i method is reduced by $2t_{anal}/m$, where t_{anal} is the time for the SLS method. Of course, it is well-known that the detector noise limitation applies only in long-wavelength spectral regions, such as the IR, but not in the UV–visible–near-IR where good detectors and sources are available.

The *multiplex disadvantage* for MX methods occurs in spectral regions that are photon and/or fluctuation noise–limited and results in $(SNR)_{MX_i}$ being less than $(SNR)_{LS_i}$. The multiplex disadvantage can be considered as follows. When $(SNR)_{MXi}$ becomes just smaller than $(SNR)_{SLS_i}$, any further increase in the overall light level further decreases $(SNR)_{MX_i}$ with respect to $(SNR)_{SLS_i}$. The multiplex disadvantage is therefore given by:

$$\frac{(SNR)_{MX_i}}{(SNR)_{SLS_i}} = \frac{\frac{m}{2}\sqrt{\frac{t}{m}(R_{L_i}+R_{B_i}) + \frac{t^2}{m^2}\left[(R_{L_i}\xi_{Lf})^2 + (R_B\xi_{Bf})^2\right]}}{\sqrt{\frac{t}{2}\left(\sum_j R_{L_j} + \sum_j R_{B_j}\right) + \frac{t^2}{4}\left[\left(\sum_j R_{L_j}\xi_{Lf}\right)^2 + \left(\sum_j R_{Bj}\xi_{Bf}\right)^2\right]}}$$

$$(B\text{-}12)$$

assuming a duty factor of unity. For the *worst possible case* for MX methods, namely, where *background flicker limits the total MX and SLS noises* at the ith component, and assuming R_{B_j} is constant over all spectral intervals, the worst possible multiplex disadvantage is given by:

$$\left[\frac{(\text{SNR})_{MX_i}}{(\text{SNR})_{SLS_i}} \right]_{\text{worst}} = \frac{1}{m} \tag{B-13}$$

if *photon noise in the background is the limiting noise*, there is a slight but constant multiplex disadvantage because:

$$\left[\frac{(\text{SNR})_{MX_i}}{(\text{SNR})_{SLS_i}} \right]_{\text{photon noise}} = \frac{1}{\sqrt{2}} \tag{B-14}$$

assuming a duty factor of unity and the background R_{B_j} is the same for all spectral intervals ($j = 1$ to m) and is designated R_B. If we had neglected the factor of $\frac{1}{2}$, as in most previous papers, we would have observed neither a multiplex advantage nor a disadvantage in the photon noise case.

Finally, the *influence of noises in MX methods is not necessarily the same as for SLS, SSS, and MC methods.* For example, detector noise occurs evenly distributed in all channels (at all wavelengths) for both SLS and MX methods. However, photon and fluctuation noises are unevenly distributed in the SLS, MC, and SSS methods, that is, such noises are greatest where the signals are greatest and essentially nonexistent where there are no signals. However, for MX methods, these noises occur evenly throughout the spectral range (channels), and so weak signals are obviously hidden by the noise (multiplex disadvantage), whereas strong signals are enhanced by the multiplexing process (multiplex advantage); these statements will be clearly indicated by the theoretical comparison to follow.

The above signal-to-noise ratio expressions apply to all of optical spectroscopy (UV-visible-IR) and apply to any other type of spectroscopy in which there are independent detector, photon (or photonlike), and fluctuation noises. For example, it is clear that these expressions apply to absorption, emission, luminescence, and scattering (elastic Rayleigh and inelastic Raman) spectrometry. However, these signal-to-noise ratio expressions with a few changes may also apply to electron spectroscopy, X-ray spectroscopy, and mass spectrometry.

By means of the signal-to-noise ratio expressions above, signal-to-noise values for the four basic measurement methods and for nine hypothetical spectra are estimated in order to evaluate the measurement system.

The nine hypothetical spectra are listed in Table B.1. All other assumed parameters are also given in Table B.1. Calculated signal-to-noise ratios for two spectral lines measured on nine hypothetical spectra are given in Table B.2. Several conclusions are evident from the calculations given in Table B.2.

TABLE B.1.   Hypothetical Spectra for Comparison of Spectral Measurement Systems[a,b]

Spectrum Type	Lines		Background	
	A	B	C	D
I	x	—	—	—
II	—	x	—	—
III	x	—	x	—
IV	x	—	—	x
V	—	x	x	—
VI	—	x	—	x
VII	x	x	—	—
VIII	x	x	x	—
IX	x	x	—	x

[a]Line A: $R_{L_A} = 10^2 \text{su}^{-1}$; line B: $R_{L_B} = 10^4 \text{ s}^{-1}$; background C: $R_{B_C} = 10^1 \text{s}^{-1}$ (constant over all channels); background D: $R_{B_D} = 10^3 \text{s}^{-1}$ (constant over all channels).

[b]Other assumed parameters: (1) The spectral range in all four systems is $m\delta\lambda_s$. (2) The total measurement time for each spectrum by each system is m (in s): (a) SLS: 1 s per spectral component (interval); (b) SSS: 500 per 1 s per spectral component if one line (A or B) is measured, and 500 per 2 s per spectral component if both lines (A or B) are measured; (c) MC: 500 s per spectral component; (d) MX: 500 s per spectral component (the factor of $\frac{1}{2}$ is accounted for separately for the HTS and FTS systems). (3) Detectors: EMI 6256, $R_d \sim 10^2$ s^{-1}; EMI 9558, $R_d \sim 10^4 \text{ s}^{-1}$. (4) Resolving power times luminosity product for the four systems (SLS, SSS, MC, and MX) is approximately the same. (5) Fluctuation factors: $\xi_{L_f} = 0.01$ and $\xi_{B_f} = 0.01$.

1.   As long as the spectroscopic systems are equivalent in all ways, that is, they have the same resolving power–luminosity product (see Appendix C) for the spectroscopic method, the same detector, and the same source of radiation with the same optical arrangement for transferring source radiation to the spectroscopic system, *the MC method is better than or at least equal to the other methods in terms of the signal-to-noise ratio.* This result indicates the importance of improving state-of-the-art image detectors for example, television detectors and photodiode arrays. Presently, such detectors have poor sensitivity and limited dynamic range as compared to

TABLE B.2. Signal-to-Noise Ratios for Hypothetical Spectra (see Table A.1) Measured by the Four Major Spectral Measurement Systems

Measurement System	Spectrum Type	Signal-to-Noise Ratio			
		Line A		Line B	
		9558 Detector	6256 Detector	9558 Detector	6256 Detector
SLS	I	0.5	4.1	—	—
	II	—	—	38	57
	III	0.5	4.0	—	—
	IV	0.5	1.9	—	—
	V	—	—	38	52
	VI	—	—	37	56
	VII	0.5	4.1	38	57
	VIII	0.5	4.0	38	57
	IX	0.5	1.9	37	57
SSS	I	11	67	—	—
	II	—	—	99	100
	III	11	67	—	—
	IV	7.3	9.7	—	—
	V	—	—	99	100
	VI	—	—	99	100
	VII	8.0	54	99	100
	VIII	8.0	54	99	100
	IX	6.1	9.5	99	99
MC	I	11	67	—	—
	II	—	—	99	100
	III	11	67	—	—
	IV	7.3	10	—	—
	V	—	—	99	100
	VI	—	—	99	99
	VII	11	67	99	100
	VIII	11	67	99	100
	IX	7.3	10	99	99
MX	I	5.6	45	—	—
	II	—	—	98	100
	III	1.9	1.9	—	—
	IV	0.02	0.02	—	—
	V	—	—	88	89
	VI	—	—	2.0	2.0
	VII	1.0	1.0	97	99
	VIII	0.9	0.9	87	88
	IX	0.02	0.02	2.0	2.0

photomultiplier tubes and so do not compete favorably with the MC or SSS methods utilizing photomultiplier tube detectors.

2.   For the same conditions as in item 1, *the SSS system approaches the MC system as the spectrum becomes simpler*, that is, if only one spectral line (spectral component) is measured, it is identical to the MC system in signal-to-noise ratio. However, as the spectrum becomes more complex and approaches m spectral lines (components) per spectral range (i.e., m components in m intervals), the SSS system approaches the SLS method in terms of signal-to-noise ratio.

3.   For the same conditions as in 1, *the MX method is equivalent to the MC method for strong (intense) spectral lines in the absence of strong background*. When the background noise term becomes dominant, the MX method results in smaller signal-to-noise ratios than the MC method. However, for weak spectral lines in the presence of background, the MC method is considerably better in signal-to-noise ratio than the MX method, because of the presence of the multiplex disadvantage in the latter method.

4.   As the detector noise increases, for example, EMI 9558 compared to EMI 6256 photomultipliers, *the signal-to-noise ratio for weak signals in the presence of low background becomes detector noise–limited*. This is of course the case in absorption and emission IR spectrometry but is true only in UV–visible–near-IR emission, luminescence, and scattering (not absorption) spectrometry and for very low cell backgrounds and for low-intensity lines (spectral components of interest) in a low-density spectrum (few lines per m spectral intervals). The last-mentioned situation could arise in atomic fluorescence spectrometry with furnace cells and in Raman spectrometry if the source scatter can be eliminated. However, it is unlikely to be true for UV-visible molecular emission, UV-visible molecular absorption, UV-visible molecular luminescence, and UV-visible scattering spectrometry, because of the presence of appreciable background signals over wide spectral regions in all these methods.

5.   *For relatively simple UV-visible line spectra, as in AFS, the SSS method is undoubtedly the least expensive, most versatile multielement method of measurement*. As the line spectra become more complex, as in Rf induction-coupled or microwave plasmas, DC arcs, or plasma jets, the need for higher resolution, lower luminosity spectroscopic methods arise, and the MC method becomes more ideal. MX methods have some possible use in (AFS) and possibly in (AAS), but probably will never be useful in (AES). The SLS method probably is (or will be) of no general analytical (quantitative analysis) use for any type of spectrometry, but is of course of qualitative use in order to determine the presence of spectral components and background over a wide spectral region. The importance of pulsing light sources in AFS is obvious from the present study, that is, by pulsing

the source and gating the detector, considerable increases in the signal-to-noise ratios are possible.

6. *For broad band molecular UV–visible–near-IR spectra, the best technique for quantitative analysis* (with conditions stated in item 1) *is MC but, for molecular spectra, a MC method would be unwieldy and cumbersome and result in little improvement over the SSS method, and so the SSS method* (with provision for SLS to determine the spectral features) *is recommended.* Of course, if the spectra is in the IR region, the MX method is recommended. However, as more intense sources, for example, the tunable semiconductive laser and better detectors for the IR become available, the SSS method will find more use and the MX methods less use in the IR region.

REFERENCE

1. J. D. Winefordner, R. Avni, T. L. Chester, L. P. Hart, J. L. Fitzgerald, D. J. Johnson, and F. W. Plankey, *Spectrochim, Acta B*, in press.

MULTIPLEX METHODS—FOURIER TRANSFORM SPECTROMETRY (FTS) AND HADAMARD TRANSFORM SPECTROMETRY (HTS)

J. D. WINEFORDNER

Department of Chemistry
University of Florida
Gainesville, Florida

In FTS, the various spectral components are coded with a cosine function, such as $2\pi f_i t$, where f_i is the encoded frequency of a spectral component of wavelength λ_i. And so, for a Michelson interferometer (two-beam), f_i is given by:

$$f_i = \frac{2v}{\lambda_i} \tag{C-1}$$

where v is the drive speed of the movable interferometer mirror. In an ideal symmetric interferometer, the interferogram $F(x)$, that is, the detector signal as a function of the optical path difference between the movable mirror and the stationary mirror, is an even function of the path difference x, and the spectrum $S_p(\lambda)$ is given by a cosine transform:

$$S_p(\lambda_i) = 2\int_0^\infty F(x)\cos\left(\frac{2\pi x}{\lambda_i}\right) dx \tag{C-2}$$

In practice, the interferometer reaches some finite value X_{max} of the path difference which limits the resolution to:

$$\Delta\lambda = \frac{\lambda_i^2}{X_{max}} \tag{C-3}$$

In HTS, the various spectral components are separated with a conventional dispersive spectrometer and singly encoded via a multislit cyclic mask at the focal plane, which allows various spectral components to be transmitted to the detector at each mask position. In the case of a mask with m open and closed slits (slots), with each slit being equivalent to $\delta\lambda_s$

in the SSS or SLS methods, and $2m+1$ total slits in the cyclic mask so that the mask can achieve m different orientations by moving sequentially one step at a time, the signal for the ith spectral component is:

$$R_{L_i} = \sum_{j=1}^{j=m} b_{ij} H_j \qquad\qquad (C-4)$$

where

$$H_j = \sum_{i=1}^{i=m} a_{ji} R_{L_i} \qquad\qquad (C-5)$$

where the a_{ji}'s ($i=1$ to m) are either 0 (if the light is blocked at spectral component R_{L_i}) or 1 (if the light is allowed to pass at spectral component R_{L_i}), and the b_{ij}'s are the Hadamard transform matrix elements of the a_{ij}'s.

Multiplexed and nonmultiplexed spectroscopic systems may differ greatly in their RL product and, as a result, more analysis time may be needed for a spectral analyses by the nonmultiplex (even with MC methods) method than by the multiplexed methods. For example, the count rate of the spectral component of interest is directly proportional to the luminosity L, and so instrument A which has a luminosity 10-fold less than that of instrument B requires an analysis time 10 times longer than instrument B for the same spectrum, assuming the same signal levels are desired in both cases. Similarly, instrument C with a resolving power 10-fold less than that of instrument D can result in unresolved spectral components in a complex spectrum.

The luminosity (also called throughput and *étendue*) in general for a given optical instrument determines the amount of light that can be transferred by the system. It is defined:

$$L = A\Omega_\tau \qquad\qquad (C-6)$$

where A is the area of the collimator, Ω is the solid angle subtended by the slit at the collimator, and τ is the overall transmittance of the optical system. For a Michelson interferometer, it is known (1–3) that:

$$\Delta\sigma_{min} = \frac{1}{2X_{max}} \qquad\qquad (C-7)$$

$$R_{Mich} = \frac{\sigma}{\Delta\sigma_{min}} \qquad\qquad (C-8)$$

$$\Omega_{Mich} = \frac{2\pi}{X_{max}\sigma_{max}} = \frac{2\pi}{R_{Mich}} \qquad\qquad (C-9)$$

where σ_{max} is the wave number (in cm^{-1}) of the spectral interval corresponding to a path difference of $X = X_{max}$. Combining the above equations,

$$L_{Mich} = \frac{2\pi A\tau}{R_{Mich}} \quad \text{or} \quad L_{Mich} R_{Mich} = 2\pi A\tau \qquad (C-10)$$

For a single-slit dispersive spectrometer,

$$\Delta_{Disp} = \frac{wh}{f_{col}^2} \qquad (C-11)$$

where w is the slit width (in cm), h is the slit height (in cm), and f_{col} is the collimator focal length (in cm). For a resolution of $\Delta\sigma$ (in cm^{-1}) and an angular dispersion of $d\theta/d\sigma$, the exit slit width w is:

$$w = f_{col} \frac{d\theta}{d\sigma} \Delta\sigma = f \cdot \frac{d\theta}{d\sigma} \cdot \frac{\bar{\sigma}}{R_{Disp}} \qquad (C-12)$$

and so

$$\Omega_{Disp} = \frac{h}{f_{col}} \cdot \frac{1}{R_{Disp}} \cdot \frac{d\theta}{d\theta} \cdot \bar{\sigma} \qquad (C-13)$$

However, it can easily be shown that:

$$\frac{d\theta}{d\sigma} \bar{\sigma} = \tan\theta \cong 1 \qquad (C-14)$$

assuming θ is not grossly different from 45°, that is, not close to 0 or 90°, and so the luminosity, L_{disp}, is given (1–3) by

$$L_{Disp} = \frac{h}{f_{col}} \cdot \frac{t}{R_{Disp}} \cdot A \quad \text{or} \quad L_{Disp} R_{Disp} = \frac{h}{f_{col}} tA \qquad (C-15)$$

The luminosity for a singly encoded (one-mask) Hadamard spectrometer is given by the same expression as for the single-slit dispersive spectrometer. For a doubly encoded (entrance and exit masks—not discussed here) Hadamard spectrometer, the solid angle Ω is increased greatly, that is, it is determined not by just one slit (entrance or exit) as in the single slit case but by the total active mask width.

Other types of MX techniques, including the Mock interferometer, the lamellar grating interferometer, and the field-widened Michelson interferometer, have certain advantages and disadvantages compared to the conventional Michelson interferometer, but are of little use in UV-visible-IR analytical spectroscopy. The reader can refer to references 1–4 for thorough discussions of these devices. It should be pointed out that all interferometric systems have need of considerable mechanical tolerance, particularly in the UV-visible region, and that all MX systems require the use of a computer for transformation.

It should also be pointed out that the product LR for dispersive spectroscopic system generally cannot be varied greatly except by complete reconstruction of the optical system. Also, it should be noted that, even for a high-aperture grating spectrometer, the ratio h/f_{col} is seldom greater than $\frac{1}{30}$, and since

$$\frac{L_{Mich}}{L_{Disp}} = \frac{\dfrac{2\pi\tau A}{R_{Mich}}}{\dfrac{h\tau A}{f_{col}R_{Disp}}} \cong \frac{2\pi f_{col}}{h}\frac{R_{Disp}}{R_{Mich}} \tag{C-16}$$

then (assuming $h/f \cong \frac{1}{30}$, τA is the same for both and $R_{Disp} \cong R_{Mich}$),

$$\frac{L_{Mich}}{L_{Disp}} \cong 200 \tag{C-17}$$

And so the Michelson interferometer, assuming equal resolving powers, transfers about 200 times as much radiation to the detector per unit time as the single-slit dispersive spectrometer. The gain in signal-to-noise ratio for a given spectral component in the HTS system—singly encoded as compared with the SLS—is a result of time per spectral component rather than increased throughput. Finally, it should be noted that, if we accept the same luminosities for the Michelson interferometer as for the dispersion spectrometer, the resolving power of the Michelson is 200 times greater than that of the dispersive system. As a result, the comparison (see Appendix B) of the Michelson systems (FTS) with the dispersive systems (SLS, SSS, MC, and HTS) is complicated by the inequality of the LR product. However, because of the multiplex disadvantage for weak spectral components, the FTS system is *not* recommended, despite the LR advantage for UV–visible–near-IR spectroscopy in which detector noise does *not* normally apply. In the IR region, where detector noise does apply, the FTS system is certainly the method of choice for spectra determination and is equivalent or better than the MC method in terms of signal-to-noise ratio. However, as narrow-line, tunable lasers for the IR become available, SSS methods will find more use and possibly replace FTS methods for intense emission spectra.

Of course, as shown in Appendix B.2, if the MX system is background fluctuation noise–limited, an increase in luminosity can actually result in a decrease in the signal-to-noise ratio as compared to a single-slit grating spectrometer operated in the SLS mode.

REFERENCES

1. G. A. Vanesse, Paper given at Spring Meeting of the Optical Society of America, New York, 1972.
2. H. Sakai, Paper given at Aspen Conference on Source Spectroscopy, Aspen, Colorado, 1972.
3. A. Girard and P. Jacquinot, "Instrumental Methods," in *Advanced Optical Techniques*, A. C. S. Van Heel, Ed., North Holland, Amsterdam, 1965.
4. L. Mertz, *Transformations in Optics*, John Wiley, New York, 1965.

INFLUENCE OF SPECTROMETRIC SYSTEM ON THE MEASURED FLUX

J. D. WINEFORDNER

Department of Chemistry
University of Florida
Gainesville, Florida

The instrumental system (the optical and electrical systems) converts radiation emitted, absorbed, or luminesced by the sample (and blank) into an instrumental signal. The arrangement of a typical optical-electrical measurement system is shown schematically in Figure D.1. No attempt is made here to show the exact optical (arrangement of lenses, mirrors, fibers, etc.) or electrical (arrangement of active and passive filters, amplifiers, etc.) components. However, it is assumed that (1) the radiation emitted, transmitted, or luminesced by the analyte in the sample is $B_{s\lambda}$ (in other words, no attempt is made to indicate the factors affecting $B_{s\lambda}$); (2) the solid angle of radiation collected by the entrance optics Ω_E exceeds the solid angle of the spectrometer Ω_S; (3) the size of the emitted, transmitted, or luminesced beam (active sample cross-sectional area) is larger than the spectrometer entrance slit width w and height h (or aperture width and height if a filter or interferometer is used) that is, no magnification of images is effected; (4) the transmittance of the entrance optics (Υ_E) and spectrometer (Υ_S) are constant with respect to wavelength, and also the sensitivity (γ) of the photodetector is constant with respect to wavelength; (5) the wavelength interval $\lambda_u - \lambda_l$ is the spectrometer bandpass (in wavelength units) (λ_u = upper wavelength and λ_l = lower wavelength of bandpass); (6) the

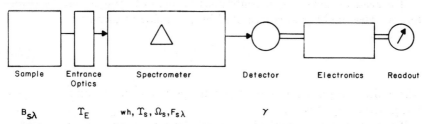

Sample Entrance Optics Spectrometer Detector Electronics Readout

$B_{s\lambda}$ Υ_E $wh, \Upsilon_s, \Omega_s, F_{s\lambda}$ γ

Fig. D.1 Arrangement of typical optical-electrical measurement system for spectroscopy.

spectrometer has identical entrance and exit slit widths and heights (apertures); and (7) the optical train of components is aligned.

If the spectrometer is set at λ_s, the measured photodetector signal S_{λ_s} (in A) is:

$$S_{\lambda_s} = \gamma \Upsilon_E \Upsilon_S h w \Omega \int_{\lambda_l}^{\lambda_u} B_{s_\lambda} F_{s_\lambda} d\lambda \qquad \text{(D-1)}$$

where all terms have been defined above, except F_{s_λ} which is the spectrometer function. If the slit apertures (entrance and exit) are identical and are greater than the diffraction-limited slit width, that is, $w > 2\lambda_s f/h$, where f is the collimator focal length, F_{λ_s} has approximately a triangular distribution:

$$F_{\lambda s} = 1 - \frac{|\Delta\lambda|}{\Delta\lambda_s} \qquad \text{(D-2)}$$

where $\Delta\lambda = \lambda - \lambda_s$, λ is any wavelength, and $\Delta\lambda_s$ is the spectrometer bandpass. If a spectral band with a half-width $\Delta\lambda \gg \Delta\lambda_s$ is measured,

$$\left(S_{\lambda_s}\right)_{\text{band}} = \gamma \Upsilon_E \Upsilon_S h w \Omega_s \Delta\lambda_s B_{s\lambda} \qquad \text{(D-3)}$$

where B_{λ_s} is the spectral radiance emitted, transmitted, or luminesced by the sample. If a spectral line with a half-width $\delta\lambda_1 \ll \Delta\lambda_s$ is measured,

$$\left(S_{\lambda_s}\right)_{\text{line}} = \gamma \Upsilon_E \Upsilon_S h w \Omega_s \delta\lambda_l B_{s\lambda} \qquad \text{(D-4)}$$

$$\left(S_{\lambda_s}\right)_{\text{line}} = \gamma \Upsilon_E \Upsilon_S h w \Omega_s B_L \qquad \text{(D-5)}$$

where B_L is the integrated spectral radiance of the line source. If the measured line or base has a half-width within 10-fold of the spectrometer bandpass, the integral (a convolution of $B_{s\lambda}$ and $F_{s\lambda}$) is not simple and must be solved mathematically. It should be pointed out that $B_{s\lambda}$ is also given by a convolution integral of a gaussian and a lorentzian broadening function.

If instead of measuring a continuous radiation impinging on the detector, an integrated signal is measured by an analog integrator,

$$Q_{\lambda_s} = \int_{t_1}^{t_2} S_{\lambda_s} dt \qquad \text{(D-6)}$$

where the integration is over the time period t_1 to t_2, and Q_{λ_s} is the integrated charge (in C).

If digital (counting) methods are used, the counting rate $R_{\lambda s}$ is related to

$S_{\lambda s}$ by:

$$R_{\lambda s} = \frac{S_{\lambda s}}{e\overline{G}\eta\varepsilon} \qquad (D\text{-}7)$$

and the number of counts $N_{\lambda s}$ is related to $Q_{\lambda s}$ by:

$$N_s = \frac{Q_{\lambda s}}{e\overline{G}\eta\varepsilon} \qquad (D\text{-}8)$$

where e is the charge of the electron (in C), $\overline{G}$ is the average gain of the photodetector (dimensionless), η is the cathode efficiency, (dimensionless), and ε is the efficiency of the dynode chain including the cathode t first dynode (dimensionless). If the signal $S_{\lambda s}$ is constant for a counting period t_c (in s),

$$N_{\lambda s} = \frac{S_\lambda t_c}{e\overline{G}\eta\varepsilon} \qquad (D\text{-}9)$$

SELECTED LIST OF SUPPLIERS OF SPECTROSCOPIC* INSTRUMENTATION

J. D. WINEFORDNER

Department of Chemistry
University of Florida
Gainesville, Florida

E.1 ATOMIC SPECTROSCOPY

Absorption	B2, C5, I1, J3, P2, P3, R1, S6, V2
Emission (flame)	B3, C3, I2, J2, J3, M1, P2, P5, R1, S3, S4, S6
Emission (arc, spark, Rf)	A4, B1, J2, J3, S3, S4, S5, S6, S9, S10

E.2 NUCLEAR SPECTROSCOPY

X-ray	C4, D1, E2, F3, K1, N1, N3, N5, P3, R1, S3, S5, T3
γ-ray	B1, C2, E2, G2, H3, I1, L1, L2, M3, N3, N4, N5, N6, O1, P1, P3, S3, S8, T3, E3

E.3 MOLECULAR SPECTROSCOPY

UV absorption	A1, A3, A7, B2, B4, C3, E5, F1, G4, H1, J5, M2, P2, P3, S7, T2, T4, V1, V2
Visible absorption	A1, A3, A7, B2, B4, C3, D1, D2, D3, E5, F1, F2, G1, G4, H1, J1, J3, J5, M1, M2, P2, P3, P4, P5, P6, R1, S1, S3, S4, S6, S7, T4, V1, V2
IR absorption	B2, B3, B4, C3, E4, E5, F1, G1, H1, J1, J2, J3, N2, P2, P3, P6, R1, S3, S4, S7, W1, W2
Polarimetry	J5, R4, V1
Circular dichroism, optical rotation	B3, D2, J2, S3, S4, V1

*Not all companies are listed; see list of companies for key to symbols used to denote suppliers.

Kinetics (stopped flow, temperature pump, continuous flow, and so on)	A3, B2, B3, B4, C1, C3, D3, G4, P2, P3, S8, T2
Fluorescence	A3, B1, B4, C3, F2, P2, P3, S5, T4
Phosphorescence	A3, B1, B4, F2, T1

INDEX OF COMPANIES

A1 Abott, Scientific Products Div., 820 Mission St., S. Pasadena, Calif. 91030. (213) 441-1171

A2 AEI Scientific Apparatus Inc., 500 Executive Blvd., Elmsford, N. Y. 10523. (914) 592-4620

A3 American Instrument Co., Div. of Travenol Labs., Inc., 8030 Georgia Ave., Silver Spring, Md. 20910. (301) 589-1727

A4 Angstrom, Inc., P. O. Box 248. Belleville, Mich. 48111. (313) 697-8058

A5 Applied Research Labs., 9545 Wentworth St. P. O. Box 129, Sunland, Calif. 91040 (213) 352-6011

A6 Associated Electrical Industries International Ltd., 33 Grosvenor Pl., London, S. W. 1, England BE-1234

A7 Avco Corp., 10700 E. Independence St., Tulsa, Okla. 74115. (918) 437-1776

B1 Baird-Atomic, Inc., 125 Middlesex Tnpk., Bedford, Mass. 01730. (617) 276-6000

B2 Bausch & Lomb, Inc., Analytical Systems Div., 820 Linden Ave., Rochester, N. Y. 14625. (716) 385-1000

B3 Beckman Instruments, Inc., 2500 Harbor Blvd., Fullerton, Calif. 92634. (714) 871-4848

B4 Biomed Instruments, Inc., 6 N. Michigan Ave., Suite 2000, Chicago, Ill. 60602. (312) 545-9060

B5 Block Engineering, Inc., 19 Blackstone St., Cambridge, Mass. 02139. (617) 868-6050

B6 Brinkmann Instruments, Inc., Cantiague Rd., Westbury, N. Y. 11590. (516) 334-7500

C1 Calibiochem, Clinical & Diagnostics Div., 10933 N. Torrey Pines Rd., LaJolla, Calif. 92037. (714) 453-7331

C2 Canberra Industries, Inc., 45 Gracey Ave., Meriden, Conn. 06450. (203) 238-2351

C3 Coleman Instruments, Div. of The Perkin-Elmer Corp., 42 Madison St., Maywood, Ill. 60153. (312) 345-7500

C4 Columbia Scientific Industries, 3625 Bluestein Blvd., Austin, Tex. 78762. (512) 926-7850

C5 Corning Glass Works, Houghton Pk., Corning, N. Y. 14830. (607) 974-9000

C6 CVC Products, Inc., 1775 Mt. Read Blvd., Rochester, N. Y. 14603. (716) 458-2550

D1 Diano Corp., Industrial X-Ray Div., 2 Lowell Ave., Winchester, Mass. 01890. (617) 729-5770

D2 Durrum Instrument Corp., 3950 Fabian Way, Palo Alto, Calif. 94303. (415) 321-6302

E1 EDAX International, Inc., 103 Schelter Rd., P. O. Box 135, Prairie View, Ill. 60069. (312) 634-3870

E2 Elscint Inc., P. O. Box 297. Palisades Park, N. J. 07650. (201) 461-5406

E3 E. I. du Pont de Nemours & Co., Inc., 1007 Market St., Wilmington, Del. 19898. (302) 774-2421

E4 The Ealing Corp., 2225 Massachusetts Ave., Cambridge, Mass. 02140. (617) 491-5870

E5 E.O.C.O.M., 19722 Jamboree Rd., Irvine, Calif. 92664. (714) 833-2781

E6 Ercona Corp., 2492 Merrick Rd., Bellmore, N. Y. 11710. (516) 781-2770

F1 F & J Scientific, 79 Far Horizon Dr., Monroe, Conn. 06468. (203) 268-3335

F2 Farrand Optical Co., Inc., Commercial Products Div., 117 Wall St., Valhalla, N. Y. 10595. (914) 428-6800

F3 Finnigan Corp., 595 N. Pastoria Ave., Sunnyvale, Calif. 94086. (408) 732-0940

G1 Gamma Scientific, Inc., 3777 Ruffin Rd., San Diego, Calif. 92123. (714) 279-8034

G2 General Electric Co., Analytical Measurements, 25 Federal St., West Lynn, Mass. 01905. (617) 594-7883

G3 General Electric Co., Space Technology Products, P. O. Box 8439, Philadelphia, Pa. 19101. (215) 962-8300

G4 Gilford Instrument Labs., Inc., 132 Artino St., Oberlin, Ohio 44074. (216) 774-1041

H1 Harrick Scientific Corp., Croton Dam Rd., P. O. Box 867, Ossining, N. Y. 10562. (914) 762-0020

H2 Hewlett-Packard, Avondale Div., Route 41, Avondale, Pa. 19314. (215) 268-2281

H3 Hewlett-Packard Co., (Calif.), 1501 Page Mill Rd., Palo Alto, Calif. 94304. (415) 493-1501

I1 INAX Instruments Ltd., 306 Moodie Dr., P. O. Box 6044, Stn. J, Ottawa, Canada (613) 829-5068

I2 Instrumentation Laboratory Inc., 113 Hartwell Ave., Lexington, Mass. 02173. (617) 861-0710

J1 Janos Optical Corp., P. O. Box 37, Brookline Rd., Newfane, Vt. 05345. (802) 365-7714

J2 Japan Spectroscopic Co., Ltd., 5-2967, Ishikawacho, Hachioji, Tokyo, Japan 192. (042) 642-9225

J3 Jarrell-Ash Div., Fisher Scientific Co., 590 Lincoln St., Waltham, Mass. 02154. (617) 890-4300

J4 JEOL Analytical Instruments, Inc., 235 Birchwood Ave., Cranford, N. J. 07016.

J5 J-Y Optical Systems, 20 Highland Ave., Metuchen, N. J. 08840. (201) 494-8660

K1 Kevex Corp., 898 Mahler Rd., Burlingame, Calif. 94010. (415) 697-6901

L1 LKB Instruments, Inc., 12221 Parklawn Dr., Rockville, Md. 20852. (301) 881-2510

L2 LND, Inc., 3230 Lawson Blvd., Oceanside, N. Y. 11572. (516) 678-6141

M1 McKee-Pedersen Instruments, P. O. Box 322, Danville, Calif. 94526. (415) 937-3630

M2 McPherson Instrument Corp., Sub. of GCA Corp., 530 Main St., Acton, Mass. 01720. (617) 263-7733

M3 Mech-Tronics Nuclear, 430A Kay Ave., Addison, Ill. 60101. (312) 543-9304

N1 New England Nuclear, 575 Albany St., Boston, Mass. 02118. (617) 426-7311

N2 Norcon Instruments Inc., 132 Water St., S. Norwalk, Conn. 06854. (203) 846-2224

N3 Nuclear Research Corp., Street Rd. & 2nd Street Pike, Southampton, Pa. 18966. (215) EL7-5015

N4 Nuclear Equipment Corp., 931 Terminal Way, San Carlos, Calif. 94070. (415) 591-8203

N5 Nuclear Measurements Corp., 2460 N. Arlington Ave., Indianapolis, Ind. 46218. (317) 546-2415

N6 Nuclear Semiconductor, 163 Constitution Dr., Menlo Park, Calif. 94025. (415) 325-4451

N7 Nuclide Corp., 642 E. College Ave., State College, Pa. 16801. (814) 238-0541

N8 The Nucleus, Inc., P. O. Box R, Oak Ridge, Tenn. 37830. (615) 483-0008

O1 ORTEC Incorporated, 100 Midland Rd., Oak Ridge, Tenn. 37830. (615) 482-4411

P1 Packard Instrument Co., Inc., 2200 Warrenville Rd., Downers Grove, Ill. 60515. (312) 969-6000

P2 Perkin-Elmer Corp., 702-G Main Ave., Norwalk, Conn. 06856. (203) 762-1000

P3 Philips Electronic Instruments, Inc., 750 S. Fulton Ave., Mount Vernon, N. Y. 10550. (914) 664-4500

P4 Process Analyzers, Inc. (Calif.), 4151 Middlefield Rd., Palo Alto, Calif. 94303. (415) 321-7801

P5 Process & Instruments Corp., 1943 Broadway, Brooklyn, N. Y. 11207. (212) 452-8380

P6 Pye Unicam Ltd., York St., Cambridge, England CA6-1631

R1 Rank Precision Industries, Inc., 411 E. Jarvis Ave., Des Plaines, Ill. 60018. (312) 297-7720

R2 Research Products International Corp., 2692 Delta La., Elk Grove Village, Ill.

60007. (312) 766-7330

R3 Rigakl/USA, Lakeside Office Park, Door 16, Wakefield, Mass. 01880. (617) 245-6014

R4 Rudolph Instruments, P. O. Box 161, Little Falls, N. J. 07424. (201) 256-1491

S1 Sargent-Welch Scientific Co., 7300 N. Linder Ave., Skokie, Ill. 60076. (312) 677-0600

S2 Scientific Research Instruments Corp., Sub. of G. D. Searle & Co., 6707 Whitestone Rd., Baltimore, Md. 21207. (301) 944-4020

S3 Scientific Resources, Inc., 3300 Commercial Ave., Northbrook, Ill. 60062. (312) 498-2920

S4 Shimadzu Seisakusho, Ltd., Foreign Trade Dept., 14-5, Uchikanda 1-chome Chiyoda-ku, Tokyo, Japan 296-2350

S5 Siemens Corp., 186 Wood Ave., S., Iselin, N. J. 08830. (201) 494-1000

S6 SMI, (SpectraMetrics, Inc.), 204 Andover St., Andover, Mass. 01810. (617) 475-7015

S7 Spectral Imaging, Inc., 572 Annursnac Hill Rd., Concord, Mass. 01742. (617) 259-8330

S8 Soltec Corp., 10747 Chandler Blvd. Hollywood, Calif. 91601. (213) 984-1100

S9 Spectro Products, Inc., 385 State St., N. Haven, Conn. 06473. (203) 281-0121

S10 Spex Industries, Inc., 3880 Park Ave., Metuchen, N. J. 08840. (201) 549-7144

T1 TCS, P. O. Box 141, Southampton, Pa. 18966. (215) 947-2275

T2 Teledyne Analytical Instruments, 333 W. Mission Dr., San Gabriel, Calif. 91776. (213) 576-1633

T3 Tracor Northern, Inc., 2551 W. Beltline Hwy., Middleton, Wis. 53562. (608) 836-6511

T4 G. K. Turner Associates, 2524 Pulgas Ave., Palo Alto, Calif. 94303. (415) 324-0077

V1 Varian Instrument Div., 611 Hansen Way, Palo Alto, Calif. 94303. (415) 493-8100

V2 Varian Techtron, 611 Hansen Way, Palo Alto, Calif. 94303. (415) 493-8100

V3 Veeco Instruments, Inc., Terminal Dr., Plainview, N. Y. 11803. (516) 681-8300

W1 Wiley Corp., 720 E. Fee Ave., Melbourne, Fla. 32901. (305) 727-2046

W2 Wilks Scientific Corp., P. O. Box 449, 140 Water St., S. Norwalk, Conn. 06856. (203) 838-4537

INDEX

Absorbance, 93
  definition of, 188
Absorbed fraction, 93
Absorption, induced, 86
  molecular, in atomic absorption, 161
  nonspecific, in atomic absorption, 160, 161
  x-ray continuum, 378
  x-ray line, 378, 379
Absorption coefficient, 16, 93
  effective, 96, 97
  linear, 96, 97
  mass, 96, 97, 364, 365, 378-380
  see also Extinction coefficient
Absorption edge, x-ray, 365-377
Absorption edge spectrometry, x-ray, 379, 380
Absorption transitions, $n \rightarrow \pi^*$, 191
Absorptivity, 188, 189
  molar, 188, 189
Acceleration potential, 391
Accelerator, 330, 331
  linear, 330
Accuracy, 1, 3, 149, 442
Activation, γ-rays, 331-336
Activity, induced, 280
  specific, 337
Allen correction, 36, 37
Alpha radiation, 279
Amalgam exchange, 340
Amplification reactions, 187, 188
Amplifier, AC, 51-56
  DC, 46, 50
  lock-in, 52-55
  in luminescence spectrometry, 244, 245
  tuned, 19, 51
Analyte addition method, 16, 41, 42
  activation analysis, 336
  x-ray fluorescence, 361, 362
Analytical curve, 39, 150, 151, 176
  calibration, 1, 15, 436
    least squares, 39
  evaluation, 436
  factor method, 40
Analytical function, calibration, 442, 443

evaluation, 442, 443
Analyzer, mass, 389-392
Arc, capillary, 139, 145
  free burning, 134-137
Arc discharge, DC, 134-137
Arc plasma, thermally pinched, 140
Ashing aid, 65
Ashing of sample, 156
Aspiration rate, in flames, 131
Atomic absorption, basis of, 152, 153
  wavelength selection, 159
Atomization, degree of, 151
  in arc, 135
  in flame, 128-130
  in spark, 139
  sputter, 157
  thermal, 157
Atomization efficiency, 128-131, 150, 151
Auger electrons, 381, 382
Auger spectroscopy, 103, 104
Average, statistical, 444
Azo compounds, 226, 227

Background, broad, 37, 38
  curved inflected, 37, 38
  flame, 132
  flat linear, 35, 36
  with maximum or minimum, 37, 38
  neutron activation, 281
  plasma, 140
  simple curved, 36
  sloped, linear, 36
  x-ray fluorescence, 358-360
Background correction, 35-38
  baseline method, 36
Background count, x-ray fluorescence, 358-360
Background emission, 152, 153
  flame, 168, 169
  furnaces, 156, 170
Bandwidth, 20-22
  diffraction, 111
  dispersion, 111
  electrical, 20-22, 48, 49
    effect on noise, 22